Atomic and Nuclear Methods in Fossil Energy Research

Atomic and Nuclear Methods in Fossil Energy Research

EDITED BY
ROYSTON H. FILBY
Nuclear Radiation Center
Washington State University
Pullman, Washington

ASSOCIATE EDITORS
B. STEPHEN CARPENTER
National Bureau of Standards
Washington, D.C.

AND
RICHARD C. RAGAINI
Lawrence Livermore National Laboratory
Livermore, California

PLENUM PRESS • NEW YORK AND LONDON

Library of Congress Cataloging in Publication Data

American Nuclear Society Conference on Atomic and Nuclear Methods in Fossil Fuel
  Energy Research (1980: Mayaguez, P.R.)
  Atomic and nuclear methods in fossil energy research.

  Includes bibliographies and index.
  1. Fossil fuels—Analysis—Congresses. 2. Radiochemical analysis—Congresses. I.
Filby, Royston H. II. Carpenter, B. Stephen. III. Ragaini, Richard C. IV. American
Nuclear Society. V. Title.
  TP321.A53   1980   621.402'3   81-21169
  ISBN 0-306-40899-6   AACR2

Proceedings of the American Nuclear Society conference on
Atomic and Nuclear Methods in Fossil Fuel Energy Research,
held December 1-4, 1980, in Mayaguez, Puerto Rico

© 1982 Plenum Press, New York
A Division of Plenum Publishing Corporation
233 Spring Street, New York, N.Y. 10013

All rights reserved

No part of this book may be reproduced, stored in a retrieval system, or transmitted,
in any form or by any means, electronic, mechanical, photocopying, microfilming,
recording, or otherwise, without written permission from the Publisher

Printed in the United States of America

PREFACE

The increased demand on fossil fuels for energy production has resulted in expanded research and development efforts on direct use of fossil fuels and conversion of fossil fuels into synthetic fuels. These efforts have focused on the efficiency of the energy production and/or conversion processes, and of the emission control technology, as well as delineation of the health and environmental impacts of those processes and their by-products. A key ingredient of these studies is the analytical capability necessary to identify and quantify those chemicals of interest in the process and by-produce streams from coal combustion, oil shale retorting, petroleum refining, coal liquifaction and gasification. These capabilities are needed to analyze a formidable range of materials including liquids, solids, gases and aerosols containing large numbers of criteria and pollutants including potentially hazardous polynuclear aromatic hydrocarbons, organo-sulfur and organo-nitrogen species, trace elements and heavy metals, among others.

Taking notice of these developments we sought to provide a forum to discuss the latest information on new and novel applications of a subset of those necessary analytical capabilities, namely atomic and nuclear techniques. Consequently, we organized the conference on Atomic and Nuclear Methods in Fossil Fuel Energy Research, which was held in Mayaguez, Puerto Rico from December 1 to December 4, 1980.

The wide range of atomic and nuclear techniques which were discussed in the invited and contributed papers included:

> neutron activation analysis
> plasma emission spectroscopy
> ion selection electrodes
> atomic absorption spectroscopy
> hydride generation
> fission track etching
> x-ray fluorescence analysis
> Rutherford backscattering
> gas chromatography

> mass spectroscopy
> photoelectron spectroscopy
> gamma-ray spectroscopy
> radioactive tracer studies
> electron microprobe analysis
> nuclear magnetic resonance spectroscopy

These papers were arranged into sessions according to the following major problem areas:

> methods of elemental analysis
> investigations of mechanisms and structures
> characterization of species

The problem areas were keynoted by excellent plenary state-of-the-art review papers given by J. W. Fruchter, W. D. Ehmann, R. A. Cahill, H. D. Buenafama, R. B. Finkelman, B. R. T. Simoneit, O. P. Strauz, H. L. Retcofsky, T. Novakov, R. P. Skowronski, J. R. Speight and C. B. Cecil, well-known authorities in their fields. The contributed papers followed by presenting recent experimental results in those areas reviewed in the plenary papers.

This conference provided an unusual opportunity for attendees with expertise and interests in the diverse areas of atomic analytical techniques and nuclear analytical techniques to communicate together and discuss topics of mutual interest. We feel these discussions were very fruitful and educational for all the attendees and will lead to beneficial collaborations in the future.

The Atomic and Nuclear Methods in Fossil Energy Research Topical Conference was co-sponsored by the American Nuclear Society, Isotopes and Radiation Division; the American Chemical Society, Divisions of Analytical Chemistry, Nuclear Chemistry and Technology, Petroleum Chemistry, and Fuel Chemistry; U.S. Department of Energy, Center for Energy and Environment Research; and University of Puerto Rico - Mayaguez Campus. The organizing committee for this conference was composed of R. H. Filby, General Chairman; B. S. Carpenter and R. C. Ragaini, Technical Co-chairmen; J. W. Morgan, K. B. Pedersen, E. Ricci, R. E. Sinclair, J. R. Vogt, and N. A. Wogman, members of the Technical Committee.

The editors would like to acknowledge the unselfish giving of time and talent of those whose help made this conference a success. We wish to express our added appreciation to the session chairmen for their overall contributions: J. R. Vogt, B. S. Carpenter, H. D. Buenafama, R. C. Ragaini, J. S. Fruchter, R. H. Filby, and R. P. Skowronski. We thank the following for their various support roles so important to the success of this conference: J. R. Hiegel and D. Price of the National Bureau of Standards; R. Farmakes and S. H. Krapp of the American Nuclear Society; and S. Shives and E. Macpher-

PREFACE

son of Washington State University. A special thanks is extended to S. Hendrickson of Lawrence Livermore National Laboratory for the compilation and preparation of the abstracts.

Although proper prearrangements are critical to any meeting, implementation is equally critical. For the invaluable on-site assistance we received during the various phases of the conference, we are indebted to many, but especially to K. B. Pedersen for arranging so much of the local assistance. We are indebeted specifically to the staff of the Center for Energy and Environment Research; S. E. Alemany, Chancellor, University of Puerto Rico - Mayaguez Campus; and M. Vargas-Cesani, Recinto Universitario de Mayaguez. We wish to express our appreciation to D. Bennett, C. Filby and N. Wogman for coordination of all the last minute crises, both before and during the meeting. Finally, we would like to thank R. E. Sinclair, Applied Physics Technology, for organizing the coffee breaks which were sponsored by Canberra, ORTEC, Nicolet, and Nuclear Data.

The location of the conference, Mayaguez, Puerto Rico, not only provided an excellent secluded atmosphere for informal working groups to be formed, but also was an equally enjoyable and superb setting for those who accompanied the attendees. The location was surpassed only by the cordiality of the hosts, J. Guzman, H. J. Santiago, J. A. Petrillo and staff, at the Mayaguez Hilton, the conference hotel. Competing with scenery and cordiality of the hotel was the exquisite cuisine prepared by D. Hannig, the hotel's chef.

The editors wish to thank the authors, who supplied manuscripts for all the papers included in these proceedings. We also extend our appreciation to J. Busis, Plenum Press, for the aid in publishing these proceedings. Finally, we would like to extend a "thank you" to each and every person connected with the conference for a job well done.

CONTENTS

METHODS OF ELEMENTAL ANALYSIS OF FOSSIL FUELS

Analysis of Oil Shale Products and Effluents
    Using a Multitechnique Approach 1
  J. S. Fruchter, C. L. Wilkerson, J. C. Evans,
  and R. W. Sanders

Nuclear Methods for Trace Elements in
    Petroleum and Petroleum Products 29
  Hector D. Buenafama

Trace Element Composition of Athabasca
    Tar Sands and Extracted Bitumens 49
  F. S. Jacobs and R. H. Filby

Uranium Content of Petroleum by Fission
    Track Technique 61
  A. S. Paschoa, O. Y. Mafra, C. A. N. Oliveira,
  and L. R. Pinto

Fast Neutron Activation Analysis of Fossil
    Fuels and Liquefaction Products 69
  W. D. Ehmann, D. W. Koppenaal and S. R. Khalil

Determination of Trace Element Forms in
    Solvent Refined Coal Products 83
  B. S. Carpenter and R. H. Filby

Comparison of Mainframe and Minicomputer Spectral
    Analysis Codes in the Activation Analysis
    of Geological Samples 97
  J. R. Vogt and C. Graham

Application and Comparison of Neutron Activation
    Analysis with Other Analytical Methods
    for the Analysis of Coal. 115
  R. A. Cahill, J. K. Frost, L. R. Camp, and R. R. Ruch

An Automated Multidetector System for Instrumental
    Neutron Activation Analysis of Geological
    and Environmental Materials 133
  S. R. Garcia, W. K. Hensley, M. M. Minor,
  M. M. Denton, and M. A. Fuka

Modes of Occurrence of Trace Elements and
    Minerals in Coal: An Analytical Approach 141
  R. B. Finkelman

Development and Calibration of Standards
    for PNAA Assay of Coal 151
  Y. Nir-El, B. Director, T. Gozani, H. Bernatowicz,
  E. Elias, D. Brown, and H. Bozorgmanesh

On-line Nuclear Analysis of Coal and its Uses 155
  D. R. Brown, H. Bozorgmanesh, and T. Gozani,
  and J. McQuaid

Analysis of Mineral Phases in Coal
    Utilizing Factor Analysis 163
  B. A. Roscoe and P. K. Hopke

The Analysis of Inorganic Constituents in
    the Groundwater at an Underground
    Coal Gasification Site 175
  R. R. Ireland, W. A. McConachie, D. H. Stuermer,
  F. T. Wang, and R. F. Koszykowski

Application of PIXE, RBS and High Energy
    Proton Microbeams to the Elemental
    Analysis of Coal and Coal Waste 189
  H. W. Kraner, A. L. Hanson, K. W. Jones,
  S. A. Oakley, I. W. Duedall, and P. M. J. Woodhead

INVESTIGATIONS OF MECHANISMS AND STRUCTURES IN FOSSIL FUELS

Investigation of Coal Hydrogenation Using
    Deuterium as an Isotopic Tracer · · · · · · · · · · · 207
  R. P. Skowronski, J. J. Ratto, and L. A. Heredy

The Reaction of Carbon-14-Labeled Reagents with Coal. . . . 229
  C. J. Collins, V. F. Raaen, C. Hilborn,
  W. H. Roark, and P. H. Maupin

CONTENTS

Natural Permeability Reduction in Porous Media
 Due to the Presence of Kaolinite 241
  W. Kubacki

Research Methodology in Used Oil Recycling 257
  D. A. Becker

The Application of Photon Induced X-Ray
 Fluorescence for the Simultaneous
 Determinations of Cobalt, Nickel and
 Molybdenum in Hydrodesulfurization
 Catalysts 271
  J. J. LaBrecque, C. A. Peña, E. Marcano,
  P. Rosales, and W. C. Parker

Quantitative Electron Probe Microanalysis
 of Fly Ash Particles 285
  R. L. Myklebust, J. A. Small, and D. E. Newbury

Spectroscopy and Asphaltene Structure 295
  J. G. Speight and R. B. Long

Geologic Factors that Control Mineral
 Matter in Coal 323
  C. B. Cecil, R. W. Stanton, F. T. Dulong,
  and J. J. Renton

Structure Analysis of Coals by Resolution
 Enhanced Solid State ^{13}C NMR Spectroscopy 337
  E. W. Hagaman and M. C. Woody

Applications of Nuclear Magnetic Resonance to
 Oil Shale Evaluation and Processing 349
  F. P. Miknis and G. E. Maciel

NMR Results Raise Questions on Coal
 Liquefaction Model Compound Studies 365
  D. C. Cronauer, R. I. McNeil, D. C. Young,
  and R. C. Ruberto

CHARACTERIZATION OF SPECIES IN FOSSIL FUELS

The Identification of Organic Compounds in
 Oil Shale Retort Water by GC and GC-MS 383
  D. H. Stuermer, D. J. Ng, C. J. Morris,
  and R. R. Ireland

Pyrolysis/High Resolution Mass Spectrometry
    with Metastable Peak Monitoring Applied
    to the Analysis of Green River Shale Kerogen 399
  R. Infante and G. G. Meisels

Advances in the Chemistry of Alberta Tar Sands:
    Field Ionization Gas Chromatographic/Mass
    Spectrometric Studies 409
  O. P. Strausz, I. Rubinstein, A. M. Hogg,
  and J. D. Payzant

Sulfur Forms Determination in Coal Using
    Microwave Discharge Activated Oxygen
    and Mössbauer Spectroscopy 443
  J. L. Giulianelli and D. L. Williamson

Analysis of Radionuclides in Airborne Effluents
    from Coal-Fired Power Plants 459
  G. Rosner, B. Chatterjee, H. Hotzl,
  and R. Winkler

Secular Equilibrium of Radium in Western Coal 473
  V. R. Casella, J. G. Fleissner, and C. E. Styron

Electron Beam Ionization for Coal Fly
    Ash Precipitators 481
  R. H. Davis, W. C. Finney and L. C. Thanh

Index 495

METHODS OF ELEMENTAL ANALYSIS
OF FOSSIL FUELS

ANALYSIS OF OIL SHALE PRODUCTS AND EFFLUENTS

USING A MULTITECHNIQUE APPROACH

> J. S. Fruchter, C. L. Wilkerson, J. C. Evans, and
> R. W. Sanders
>
> Pacific Northwest Laboratory operated by Battelle
> Memorial Institute
> Richland, Washington 99352

INTRODUCTION

Oil shale reserves of the Green River formation in Colorado, Utah and Wyoming have the potential to provide anywhere from 300 million to 2 trillion barrels of oil, depending on the various economic, technical and environmental assumptions made in the estimate. Even the lower very conservative figure amounts to more than five times the United States' proved petroleum reserves, and moderate estimates of 600-800 billion barrels compare favorably with both the United States' coal reserves and the entire world's proved reserves of petroleum. Thus, if the technological, economic and political barriers to the production of shale oil can be overcome, oil shale could become an important source of precious liquid hydrocarbons.

The study reported here was carried out at the Paraho Semiworks Retort, which has been described elsewhere (Jones, 1976). The objectives of the initial characterization study were to 1) calculate a representative mass balance for a number of trace and major elements of potential environmental significance; 2) obtain information on the physical and chemical forms of certain elements emitted from the retort; and 3) determine elemental distributions of particles emitted from the crushing and retorting operations.

INORGANIC ANALYTICAL METHODS

The Multitechnique Approach

Many of the samples obtained from oil shale retorts are chemically and physically complex, creating the potential for matrix effects as well as other types of interferences in many of the commonly used methods for chemical analysis. Therefore, the techniques employed for inorganic analysis of the Paraho samples were chosen where possible for their relative freedom from matrix effects as well as their sensitivity and precision. Because no one method can now meet all these requirements for all elements of interest, a multitechnique approach was adopted. This multitechnique approach to inorganic analysis also provided an opportunity to assess the strengths and weaknesses of the various methods for different samples. The major techniques used included instrumental neutron activation analysis (INAA), energy dispersive x-ray fluorescence analysis (XRF), dc arc plasma emission spectroscopy, flame atomic absorption spectroscopy, and graphite furnace atomic absorption spectroscopy. Other techniques including radiochemical neutron activation analysis, cold vapor atomic absorption spectroscopy, ion-selective electrodes, hydride generation, and various gas monitoring devices were used to supplement these techniques for specific elemental and speciation analyses. These methods are described in detail below.

INSTRUMENTAL NEUTRON ACTIVATION ANALYSIS

The solid samples were crushed, mixed, and split as described above. Sample aliquots of about 0.1 to 0.5 grams were weighed into 2/5-dram polyethylene vials, and the vials were heat sealed. Each 2/5-dram vial was then placed in a 2-dram polyethylene vial which was also heat sealed. Depending on the irradiation facility, liquid samples were either sealed in the 2/5-dram vials, placed directly into the 2-dram vials and then sealed in plastic bags, or placed in 10-ml screw-cap bottles.

Sequential Irradiations

The neutron activation procedures used in this laboratory have been described elsewhere (Laul 1979). The basic parameters controlling the determination of elements by sequential INAA are: 1) neutron flux in a reactor; 2) irradiation times, 3) delay interval prior to counting; 4) half-life and γ-ray energy of a radionuclide; and 5) the resolution and efficiency of the detector. Table 1 outlines the irradiation parameters, radionuclide (isotope) produced for each element, half-life, γ-ray energy selected, and optimum decay interval used in the analysis of geological samples. Also listed in Table 1 are the possible γ-ray interferences from other radionuclides and interfering reactions from other elements.

The sequential INAA procedure consists of two irradiation periods and four counting intervals as delineated in Table 1.

Table 1. Instrumental Neutron Activation Analysis (n, γ) Process

Isotope	Half-Life	γ-Energy Selected (keV)	Decay Interval	Possible Interference
I. ^{51}Ti	5.79 m	320	15 m	
^{27}Mg	9.46 m	1014		^{27}Al(n,p)
^{52}V	3.75 m	1434		
^{28}Al	2.32 m	1779		^{28}Si(n,p)
^{49}Ca	8.80 m	3084		
^{165}Dy	2.32 h	95	3-5 h	
^{56}Mn	2.58 h	847		^{56}Fe(n,p)
^{24}Na	15.0 h	1369		^{24}Mg(n,p), ^{27}Al(n,a)
II. ^{153}Sm	46.8 h	103	5-7 d	
^{152}Eu	12.7 y	122		
^{177}Lu	6.74 d	208		^{177m}Lu(208)
^{239}Nb(U)	56.3 h	228		^{177m}Lu(229)
^{175}Yb	4.21 d	396		
^{131}Ba	12.1 d	496		
^{147}Nd	11.1 d	531		
^{140}La	40.2 h	816, 1597		
^{24}Na	15.0 h	1369, 1733		^{24}Mg(n,p), ^{27}Al(n,a)
^{42}K	12.4 h	1524		
^{182}Ta	115 d	68, 1221	40-50 d	
^{152}Eu	12.7 y	122, 1408		
^{141}Ce	32.5 d	146		^{59}Fe(143)
^{160}Tb	72.1 d	299		^{233}Pa(300)
^{233}Pa(Th)	27.0 d	312		
^{51}Cr	27.8 d	320		
^{181}Hf	42.5 d	482		
^{85}Sr	64.0 d	514		
^{95}Zr	65.5 d	757		^{154}Eu(757)
^{134}Cs	2.05 y	796		
^{58}Co(Ni)	71.3 d	811		^{152}Eu(811)
^{46}Sc	83.9 d	889, 1121		
^{86}Rb	18.7 d	1078		
^{59}Fe	45.6 d	1099, 1292		
^{65}Zn	243.0 d	1116		^{152}Eu(1112), ^{46}Sc(1121)
^{60}Co	5.26 y	1173, 1332		
^{124}Sb	60.3 d	1691		

Parameters Neutron Flux - 6 x 10^{12} n/cm^2/sec
Irradiation I - 5 Minutes
Irradiation II - 6 Hours

This scheme is routinely used in this laboratory for geological or other types of samples and is similar to the schemes used in other laboratories. The irradiated samples and standards are counted at an appropriate geometry (shelf position) to limit the counting dead time to less than 15% on a 130 cc Ge(Li) detector (FWHM 1.8 keV for 1332 keV γ) coupled to a 2048-channel analyzer. A pileup rejector amplifier or a pulser is used in the system to minimize or account for the pulse losses (in high dead time cases). The computer code used to process the gamma spectra was CANGAS, which has been described elsewhere (Laul et al., 1979).

Short Irradiations

Samples and standards (with Cu foil flux monitors) were individually irradiated at a flux of 1×10^{13} n/(cm^2 · s) for 5 min. in the pneumatic facility ("rabbit" system) of the Washington State University reactor. After irradiation, the primary activity produced is that of ^{28}Al ($t_{1/2}$ = 2.3 min.), which decays rapidly compared with most of the other radionuclides. If the sample is counted immediately after irradiation, the ^{28}Al with its high activity and high γ-energy (1779 keV) would produce a very high Compton continuum, and thus decrease considerably the sensitivities of the other radionuclides. On the other hand, if 15 min. or so decay is allowed, then the other radionuclides are more readily measured. Immediately after irradiation, the γ-activity ratio of raw shale ^{28}Al/^{51}Ti is about 2400; after 15 min., the ratio is reduced to ∼160. Thus, decay time is necessary to optimize the detection of other radionuclides.

After the irradiation, the 2/5-dram polyethylene vial containing the sample was transferred into a clean 2-dram polyethylene vial, and, after 15 min., the samples and standards were counted on the Ge(Li) system for 5 min. each. A timer was installed in the counting system to yield the actual elapsed time during the count, which was later used to estimate the true decay correction factor for short-lived radionuclides. The elements Ti, Mg, V, Al, and Ca are measured in this counting interval. The major interfering reactions are ^{28}Si(n,p)^{28}Al and ^{27}Al(n,p)^{27}Mg. These interferences were evaluated and were found to amount to 20% Si = 0.14% Al, and 10% Al = 2.8% Mg. The oil shale samples had Si/Al ratios similar to the geological rock standards so the Al correction from Si was insignificant. The Mg correction from Al was significant in most of the solid samples, and therefore Mg concentrations by INAA were not reported.

Following the first counting interval, the irradiated samples were allowed to decay for 3 to 5 hours to let the short-lived

radionuclides decay completely, and then the samples were again
counted on the Ge(Li) system for 10 to 15 min. each. The elements
Dy, Mn, and Na are measured in the second counting. The major
interferences are ^{56}Fe(n,p)^{56}Mn, ^{24}Mg(n,p)^{24}Na, and ^{27}Al(n,α)^{24}Na,
respectively. The contribution to Mn from Fe was negligible. The
contributions for Na are 7.0 ppm from 1% Al in the rabbit system.
Thus, with 5 min. of irradiation and two optimum decay intervals,
it is possible to measure seven elements (Ti, V, Al, Ca, Dy, Mn,
and Na) in the soil oil shale samples.

Long Irradiations

To determine the elements characterized by longer half-life
isotopes, appropriate standards were irradiated at a thermal neutron flux of 1×10^{13} n/(cm^2 · s) for 5 to 7 hours in the Washington State University or Oregon State University reactors. The
samples were allowed to decay for 3 to 5 days to let the major
^{24}Ba activity decay considerably. The irradiated samples were
transferred into clean 2/5-dram polyethylene vials and each
2/5-dram vial was then placed in a 2-dram polyethylene vial. The
samples and standards were then counted for 30 to 100 min. each
on a Ge(Li) detector system. Only short- and medium-lived radionuclides with selected γ-energies (Table 1) for elements Sm, Eu,
Lu, U via ^{239}Np, Yb, Ba, Nd, La, Na, and K are measured in this
counting interval. The element Na is also determined in the short
irradiation and this provides an additional check on the Na value
obtained in this counting from the long irradiation.

After completion of the first count, the samples were allowed
to decay for 40 to 50 days to let the short- and medium-lived
activities decay completely or considerably. The samples and
standards were then counted for 100 to 1000 min. Fifteen
elements (Ta, Eu, Ce, Tb, ^{233}Pa(Th), Cr, Hf, Ba, Sr, Cs, Rb,
Fe, Zn, Co, and Sb) corresponding to the radionuclides (isotopes)
and selected γ-ray energies listed in Table 1 are measured in
this count.

After the completion of Ge(Li) counting, the samples and standards were later counted for 30 to 1000 min. on a Ge(Li)-Na(Tl)
coincidence and non-coincidence counting system. Figure 1 displays spectra of coincidence and non-coincidence γ-rays of the
BCR-1 standard counted for 30 min. after a 50-day decay interval.
The peak/Comptom ratios and detection sensitivities for many elements such as Ce, Cr, Ba, Sr, Nd, Rb, Zn, and Sb are greatly
increased in the non-coincidence spectrum. The 1116-keV peak of
Zn is better resolved by the reduced peak tail of the 1121-keV
peak of ^{46}Sc in the non-coincidence spectrum. The 146-keV peak of
^{141}Ce, unlike in normal Ge(Li) counting, contained only a small

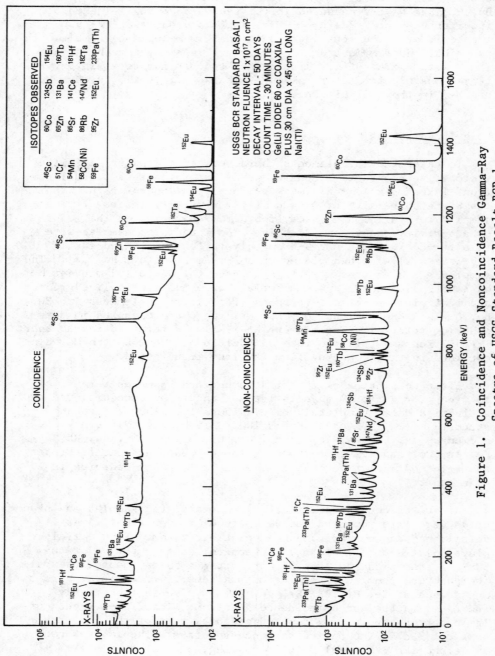

Figure 1. Coincidence and Noncoincidence Gamma-Ray Spectra of USGS Standard Basalt BCR-1

contribution from the 143-keV peak of ^{59}Fe. The element Ni, previously not measured in the normal Ge(Li) counting, was easily measured in the non-coincidence counting. Overall, the accuracy and relative precision for many elements was improved by a factor of about 3 by non-coincidence counting relative to the normal Ge(Li) counting. Sensitivity was improved by a factor of about ten.

X-RAY FLUORESCENCE ANALYSIS

Our group uses two XRF systems for rapid simultaneous elemental evaluation. The primary system is a KEVEX-910 multiple secondary unit; the other system is a PNL-built isotopic unit. These systems have been recently improved by the addition of several isotopic sources including 100-mCi isotopic sources of ^{55}Fe, ^{109}Cd, and ^{241}Am, as well as Zr and Ag secondary sources. The reason for using multiple sources is that the most efficient excitation of an element can be achieved by having the excitation energy or source energy slightly above that of the electron binding energy of the element being excited. These differences are illustrated in Figures 2 and 3. Data reduction of the multichannel analyzer output is accomplished by using FORTRAN computer programs SAP AND SAP 3 (Nielsen, 1977).

Sample Preparation for XRF Analysis

Solids. Samples of from 0.2 to 1.5 g of powder (<140 mesh) are pelletized in a 3.2-cm-dia., 27,000-kg laboratory press. They are either pelletized directly or are pelletized after first mixing with an equal weight of cellulose (TLC reagent grade, J. T. Baker).

Oils. A one-ml oil sample is pipetted onto a prepared slide with a 1/8-mil polypropylene seal on the flush side. The mounted sample is placed in a laminar flow hood and allowed to attain a highly viscous state. Normally about a day is required with a resultant 5 to 15% weight loss for most crude oils. The sample so prepared generates very little outgassing when exposed to vacuum in the x-ray system.

The analysis requires a 6-min. x-ray exposure using a tungsten x-ray primary beam with a filtered Zr or Ag secondary source.

If there is a concern for loss of volatile inorganics, the sample can be processed in air, eliminating the vaporative step. This procedure, however, requires the use of a prepared multi-element standard to compensate for the back-scattering and

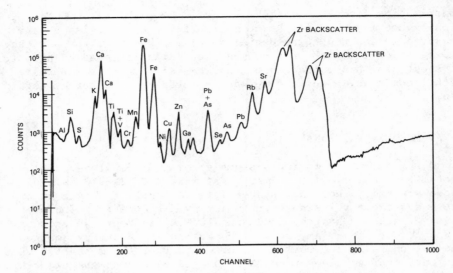

Figure 2. X-Ray Fluorescence Spectrum of a Spent Oil Shale Using a Zr Secondary Excitation Source

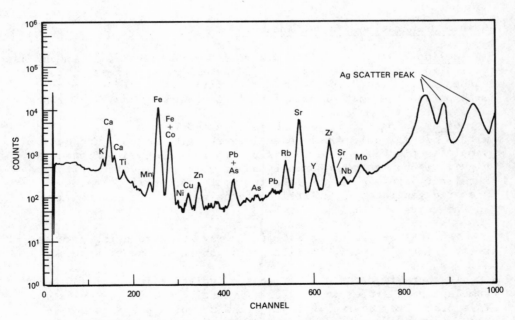

Figure 3. X-Ray Fluorescence Spectrum of a Spent Oil Shale Using a Ag Secondary Excitation Source

absorption caused by air. Sensitivities for the low Z elements are diminished by a factor of 5 or 6 for air compared to vacuum. Volatilization losses do not seem to be significant with most crude and shale oil samples.

Table 2 presents a comparison of neutron activation analysis (NAA) and XRF analysis of a crude oil from Wilmington, California, and a Paraho shale oil.

Comparison of these analyses shows that, except for the case of arsenic, the results are in good agreement, which provides an indication of the accuracy of the XRF method. It is possible that some volatile forms of arsenic are lost during the XRF sample preparation. The XRF determination also provides data on a number of elements not easily obtainable by other methods.

<u>Process Waters.</u> We have developed and validated three x-ray fluorescence procedures for use in the analysis of oil shale waters, including process waters and impinger solutions. One method involves evaporation of a small volume (100 µl) of solution into a 1-cm dia. area and placing the sample into the constant flux area of the x-ray beam. This method is rapid but requires a homogeneous sample. It also presents difficulties with low Z elements when there is mass absorption due to organics.

Another method involves freezing the solution into a thin wafer and exposing it to x-rays under vacuum for a short period. This application has the advantage of no volatile loss. The primary disadvantage is the necessity of relatively high scattering resulting from the mass of water.

The most useful and versatile procedure is addition of a known weight of solution (10 to 50 g) to 50 mg of cellulose. The sample is evaporated into a laminar flow hood at ambient temperature, homogenized, and pressed into a weighed, 3.18-cm-dia. wafer, and its weight is determined. Counting time is 60 min. at 40 kV and 13.5 mA using a Zr or Ag secondary source. The resultant spectra are reduced using a computer program that corrects for peak overlap, mass absorption, and elemental enhancement. Where comparison with determinations from other methods considered reliable were available, the agreement in the reported values was generally good with the difference between methods usually less than 10%.

DC ARC PLASMA EMISSION SPECTROSCOPY

The system in use in a Spectrospan III system employing a dc argon plasma excitation system and an ultrahigh resolution

Table 2. Comparison of Neutron Activation Analysis and X-Ray Fluorescence Analysis of a Crude Petroleum and a Crude Shale Oil

Elements ppm	Wilmington Crude Petroleum Method			Paraho Oil Collected 8-25-78 Method		
	NAA	XRF-Air	XRF-Vac	NAA	XRF-Air	XRF-Vac
	10% Error					
As	0.25	0.08 ± 0.04	0.14 ± 0.07	31 ± 2	23 ± 2	19 ± 2
Br	<1	0.11 ± 0.06	0.27 ± 0.09	<0.2		<0.14
Ca			<10			245 ± 29
Cl	6		<126	3.0 ± 0.9		<140
Co	1.21	1.31 ± 0.2	1.79 ± 0.36	1.39 ± 0.08		1.44 ± 0.42
Cr	0.15	<1.0	<1.47			1.55 ± 0.66
Cu	<10	1.50 ± 0.2	0.48 ± 0.27	1.1 ± 0.04	0.83 ± 0.12	0.97 ± 0.23
Fe	32	37 ± 3	40 ± 4		59 ± 5	74 ± 8
Ga		<0.08	<0.14			<0.11
Hg		<0.3	<0.51			
K			<20			41 ± 15
Mn	<0.2	<0.38	<0.67	0.4 ± 0.05	0.45 ± 0.25	<0.51
Mo		<0.19		<3		
Nb		<0.17				
Ni	54	58 ± 4	62 ± 6		2.72 ± 0.20	2.76 ± 0.36
Pb		<0.4	<0.60		<0.40	<0.40
Rb		<0.13	<0.17		<0.19	<0.14
S,%	1.58		1.63	0.99 ± 0.08		0.89 ± 0.10
Se	0.37	0.37 ± 0.07	0.38 ± 0.10		0.92 ± 0.07	0.72 ± 0.12
Sr		<1.2	<1.6		<1.3	<1.2
Ti			<2.5	2.6		4.1 ± 1.9
V	47	46 ± 3	56 ± 6			<1.29
Y		<0.14				
Zn	1.07	1.4 ± 0.1	1.7 ± 0.2	0.29 ± 0.01	0.57 ± 0.09	0.55 ± 0.13
Zr		<0.16		<0.70		

Eschelle grating spectrometer. The greatest value of the system has been for boron analysis; however, procedures have been tested and satisfactory results obtained in aqueous samples for the 32 elements shown in Table 3. Because of the very high temperature of the argon arc (6000 K), chemical interferences are minimal. Ionization interferences, on the other hand, are a problem and must be corrected for in many cases by addition of Li buffer, matrix-matched standards, or the method of standard additions.

Sensitivity drift in the excitation source was a major problem with the instrument in its original form. An improved three-electrode excitation source has recently been retrofitted on our instrument. This source has been operated for up to 5 hr. continuously without significant deterioration in the electrodes. Some increased sensitivity has been observed in the new source, but the main improvement has been its stability. Multielement operation now appears to be a very desirable option with this technique. With suitable modification, the system can become capable of simultaneous determination of a group of 20 elements.

Sample Preparation

Aqueous samples that were homogeneous, including impinger solutions, were analyzed directly with no pretreatment. The powdered solid samples were fused with a small amount of Ultrex sodium carbonate, and dissolved in Ultrex nitric acid. Approximately 3 g of Na_2CO_3 were used to fuse 500-mg samples of raw and retorted shale. Each fused sample was then dissolved in acid and diluted to 100 ml of solution.

ATOMIC ABSORPTION SPECTROSCOPY

Atomic absorption (AA) spectroscopy is a versatile technique that can be used to determine a number of major and minor elements in oil shale samples, including Li, Na, Mg, Al, Si, K, Ca, Fe, Cd, and Hg. Conventional flame atomic absorption was used to determine all of these elements except Hg, which was determined by flameless atomic absorption. We also experimented with the use of a graphite furnace AA technique for determinations of Cd directly in solid samples.

Conventional Flame Atomic Absorption

The instruments used were a Perkin Elmer 403 and an Instrumentation Laboratory Associates 257. Aqueous samples that were

Table 3. DC Arc Plasma Emission Spectroscopy – Approximate Detection Limits for Elements in Aqueous Samples

Elements with Detection Limits 1 to 5 ppb	Elements with Detection Limits 5 to 10 ppb	Elements with Detection Limits 10 to 100 ppb	Elements with Detection Limits 100 to 500 ppb
Be	Li	K	Si
Na	Ti	Fe	P
Mg	Co	In	As
Ca	Ni	Au	Se
Sr	B	Pb	Sn
Ba	Ag	Tl	
V	Cd		
Cr			
Mn			
Cu			
Zn			
Mo			
Ag			
Al			

homogeneous, including impinger solutions, were aspirated directly. Powdered solid samples were fused with lithium metaborate, with a ratio of $LiBO_2$/sample of 6:1.

Commercially prepared aqueous atomic absorption standards were used for most elements. A sample of U.S. Geological Survey standard rock BCR-1 prepared in a similar fashion to the solid raw and spent shale samples was also analyzed as a control.

Graphite Furnace Atomic Absorption

A new state-of-the-art graphite furnace atomic absorption system has recently been set up at our laboratory. The system is an Instrumentation Laboratory Associates Model 257 microprocessor-based spectrometer with D_2 arc background correction. The instrument is equipped with an IL 555B graphite furnace system and FASTAC auto-sampler. The system is capable of measuring a wide range of elements in both solids and liquids with sensitivities in the picogram range. One of the unique features of the instrument is the use of a tungsten resistance thermometer in contact with a graphite cuvette for direct temperature measurement from room temperature up to 3000°C. The temperature measurement is also used in a feedback circuit to control the heating rate.

Technique development is still under way with this new system; however, it has already been used successfully for the determination of cadmium in raw and retorted oil shale and NBS Standard (coal fly ash) 1633. For solid samples, a rectangular cuvette of pyrolytic graphite is used. One to two mg of sample is weighed out on pyrolytic graphite microboats and inserted into the cuvette. It was found that the samples could be charred up to 400°C without loss of Cd. Ramp atomization up to 1400°C removed all Cd from the Paraho samples. It was found that more highly sintered samples such as NBS fly ash required an atomization temperature of 2000°C for best sensitivity. Standardization was achieved simply by using solution standards dried on the microboats and analyzed under the same conditions using the peak area mode of the microprocessor. The value obtained for NBS fly ash was 1.44 $\pm$ 0.15 ppm Cd compared to the certified value of 1.45 $\pm$ 0.06.

By careful temperature programming, it was possible to determine the minimum temperature at which cadmium was released from the sample. About 500°C appears to be the threshold for Cd release from small, finely-ground samples in an inert atmosphere. Under actual retorting conditions, Cd release may be more retarded because of larger feedstock size (1/2 to 3 in.) and the presence of oxygen.

Flameless Atomic Absorption for Mercury Analysis

The procedure quantitatively separates Hg from 50 to 150-mg solid samples by a process of sample combustion and thermal decomposition. The separation train (Figure 4) channels Hg-free air to samples, which are subsequently heated to 950°C in an enveloping quartz tube. The high temperature and oxidizing conditions within the tube volatilize Hg from the sample matrix, and underlying 3 to 4-cm layer of gold-coated quartz beads convert the evolved Hg to elemental mercury vapor (Hg^o). To separate Hg^o from interfering organic vapors, the effluent sample gases are directed into an alumina column which selectively retains most of the organic vapors but passes Hg^o. The Hg^o is finally collected by amalgamation with a column of gold-coated glass beads maintained at room temperatures.

Following the amalgamation step, the column containing the trapped Hg is detached from the separation train and assembled into a flameless atomic absorption system. Here the amalgamated Hg is released by heating the gold-coated glass beads to 500°C in a N_2 gas stream. The Hg^o is swept into a long-pathlength gas absorption cell where it is measured by atomic absorption at 254 nm.

The effectiveness of the procedure has been tested by determining Hg in NBS coal (S.R.M. #1632) and NBS orchard leaves (S.R.M. #1571). These standard reference materials provided known Hg concentrations in representative environmental matrices. The resulting analyses together with Hg analyses for selected coal and oil shale materials are given in Table 4. The data for the two standard reference materials are in reasonable agreement with NBS certified values.

ANALYSIS OF AUGUST 24, 1977 SAMPLES

Special attention was given to the analysis of samples collected on August 24, 1977, particularly the raw and retorted shale. There were several purposes behind this additional effort. First, a multitechnique and intercomparison was carried out to enable objective assessment of the accuracy of the analytical data. The August 24 raw shale in particular has been rigorously characterized for some 53 elements. A large quantity (15 to 20 kg) of this material has been archived and will eventually be made available to other workers in the field as an analytical reference material. The data given in this report combined with additional analyses provided by colleagues at Lawrence Berkeley Laboratory will form the basis for standardization of the

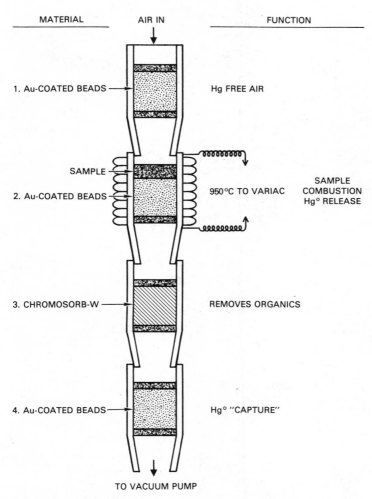

Figure 4. Separation Train for Volatilizing Mercury from Sample Materials

Table 4. Determination of Hg in NBS Standards and Selected Coal and Oil Shale Materials

Material	Total Hg Found, ppb	Mean Estimated Analytical Error, ppb	NBS Certified Value, ppb
NBS Coal	93,92,93,95,90,97	93 ± 10	120 ± 20
NBS Orchard Leaves	154,168,164,174,147	161 ± 15	155 ± 15
Feed Coal	172,179,155,165,148	164 ± 20	
Solvent Refined Coal	19,16,20	18 ± 3	
Coal Mineral Residue	<5,<5,<5	<5	
Oil Shale (1)	268,274,263	268 ± 20	
Oil Shale (2)	107,101,80,100,98,97	97 ± 10	
Spent Shale	19,12,13,14	15 ± 4	
Shale Oil	153,202,172,138,203	175 ± 30	

material. In addition, a more limited analysis has been carried out on two lots to verify the homogeneity of the material. Finally, the August 24, 1977 samples have been chosen as representative samples for a material balance study for several selected elements of interest.

Data Analysis

Tables 5, 6, and 7 give the complete analytical results for analysis of August 24, 1977 raw and retorted shale for 53 elements. In many cases several techniques were used and it is possible to intercompare results. In general, six replicate samples were analyzed by each analytical technique. The analytical errors shown in Tables 5 and 6 are thus derived from the precision of six analyses. In a few cases only a single determination was made. These include the radiochemical measurement for Cd, Se, Zn, and U, and the graphite furnace AA determination of Cd. The element analyses by the LECO Corporation shown in Table 7 are the result of triplicate determinations. To reduce the data to a form suitable for graphical representation, a simple computer code was used to perform the following operations:

1) A minimum error of 2% was assigned to all data. Any data with a reported error less than 2% were set equal to 2%.

2) An error-weighted average of the data was computed for each element.

3) If more than two analyses were reported, Chauvenet's criterion was then applied and, if necessary, the worst outlier rejected. A new error-weighted average was then computed. This procedure was only followed once per element for each sample because of the small size of each data set. Very few points were actually rejected in this manner. Those which were rejected are noted with an asterisk in Tables 5 and 6.

4) Once an appropriate average value was determined a percentage deviation of each individual value from the mean for that element was computed. These values together with the appropriate percentage standard deviation for each point are plotted in Figures 5 and 6. This is a convenient format for viewing all of the data at once. Only elements which were analyzed by more than one reliable technique are plotted; individual determinations rejected by Chauvenet's criterion (step 3) are omitted.

Table 5. Multielement Analysis of Paraho-Feedstock Shale Collected 8/24/77 (in ppm except as noted)

Element	INAA	RCAA	XRF	PES	FAA	GFAA	CVAA	Error Weighted Average
Al(%)	3.89 ± 0.14			3.76 ± 0.08	3.69 ± 0.11			3.77 ± 0.06
As	48.0 ± 0.7		41.6 ± 5					44.3 ± 0.6
B				96 ± 2				94 ± 2
Ba	483 ± 36			515 ± 8				512 ± 10
Br	0.57 ± 0.13							0.57 ± 0.13
Ca(%)	10.4 ± 0.5		10.7 ± 0.5	9.9 ± 0.1	11.0 ± 0.4*			10.1 ± 0.2
Cd		0.64 ± 0.03						0.64 ± 0.03
Ce	43.1 ± 0.9							43.1 ± 0.9
Co	9.0 ± 0.1							9.0 ± 0.1
Cr	36.7 ± 1.8		39.7 ± 9.3	33.8 ± 0.6				34.2 ± 0.6
Cs	3.84 ± 0.22							3.84 ± 0.22
Cu			40.3 ± 2.3	40.0 ± 5.5				40.3 ± 2.1
Dy	2.4 ± 0.4							2.4 ± 0.4
Eu	0.60 ± 0.02							0.60 ± 0.02
Fe(%)	2.08 ± 0.04		2.02 ± 0.10	2.01 ± 0.04	2.14 ± 0.04			2.07 ± 0.02
Ga			8.4 ± 0.8					8.4 ± 0.8
Hf	1.75 ± 0.05							1.75 ± 0.05
Hg							0.089 ± 0.005	0.089 ± 0.005
Ho	0.67 ± 0.11							0.67 ± 0.11
K (%)	1.69 ± 0.11		1.66 ± 0.02	1.79 ± 0.03	1.55 ± 0.03			1.61 ± 0.02
La	20.6 ± 0.7							20.6 ± 0.7
Lu	0.28 ± 0.03							0.28 ± 0.03
Mg(%)				3.42 ± 0.05	3.59 ± 0.03	0.61 ± 0.08		3.46 ± 0.06
Mn	312 ± 20		319 ± 20	314 ± 22				315 ± 12
Mo			24 ± 2.5	20.0 ± 1.9				22.0 ± 1.5

OIL SHALE PRODUCTS AND EFFLUENTS

Element				
Na(%)	1.68 ± 0.01		1.73 ± 0.07	1.69 ± 0.03
Nb		8.0 ± 0.7		8.0 ± 0.7
Nd	20.4 ± 2.1			20.4 ± 2.1
Ni	23.0 ± 5.3	24.2 ± 1.2*	27.6 ± 0.6	27.5 ± 0.6
Pb		26.5 ± 2.1		26.5 ± 2.1
Rb	74.9 ± 2.3	74.0 ± 2.7		74.5 ± 1.8
S		5730 ± 500		5730 ± 500
Sb	2.09 ± 0.08			2.09 ± 0.08
Sc	5.77 ± 0.16			5.77 ± 0.16
Se	2.1 ± 0.2	2.0 ± 0.1		2.03 ± 0.09
Si(%)		14.1 ± 0.7	15.2 ± 0.1	15.0 ± 0.3
Sm	3.10 ± 0.3			3.10 ± 0.3
Sr	674 ± 24	678 ± 21	712 ± 14	696 ± 11
Ta	0.55 ± 0.02			0.55 ± 0.02
Tb	0.37 ± 0.03			0.37 ± 0.03
Th	6.33 ± 0.13			6.33 ± 0.13
Ti(%)	0.18 ± 0.02	0.17 ± 0.02	0.18 ± 0.01	0.18 ± 0.01
U	4.2 ± 0.3	4.6 ± 0.2		4.5 ± 0.2
V	86 ± 6	95 ± 6	96 ± 3	94.2 ± 2.4
Y		14 ± 1		14 ± 1
Yb	1.26 ± 0.11			1.26 ± 0.11
Zn	67.2 ± 3.7	62.6 ± 2.3		63.6 ± 1.6
Zr		63 ± 3	73.2 ± 4.0* 36.2 ± 1.3	36.2 ± 1.3

* Deleted from error weighted average.

INAA - Instrumental Neutron Activation Analysis
RCAA - Radiochemical Neutron Activation Analysis
XRF - X-Ray Fluorescence Analysis
PES - Plasma Emission Spectroscopy (Sodium Carbonate Fusion)
FAA - Conventional Flame Atomic Absorption (Lithium Metaborate Fusion)
GFAA - Graphite Furnace Atomic Absorption
CVAA - Cold Vapor Atomic Absorption

Table 6. Multielement Analysis of Paraho Retorted Shale Collected 8/24/77 (in ppm except as noted)

Element	INAA	RCAA	XRF	PES	FAA	GFAA	CVAA	Error Weighted Average
Al(%)	4.83 ± 0.05*			4.46 ± 0.05	4.56 ± 0.17			4.48 ± 0.08
As	59.2 ± 0.9		59.8 ± 1.9					59.4 ± 1.0
B				107 ± 2				107 ± 2
Ba	593 ± 13			613 ± 12				604 ± 9
Br	0.80 ± 0.18							0.80 ± 0.18
Ca(%)	13.1 ± 0.5		13.9 ± 0.7	11.1 ± 0.2	13.2 ± 0.2			13.3 ± 0.2
Cd		0.90 ± 0.04				0.99 ± 0.13		0.91 ± 0.04
Co	51.5 ± 1.5							51.5 ± 1.5
Cr	11.1 ± 0.2		49.6 ± 8.9	41.0 ± 3.8				11.1 ± 0.2
	44.3 ± 0.9							44.2 ± 0.9
Cs	4.68 ± 0.21							4.68 ± 0.21
Cu			56.3 ± 1.0	46.9 ± 5.3				55.9 ± 1.1
Dy	3.5 ± 0.2							3.5 ± 0.2
Eu	0.73 ± 0.02							0.73 ± 0.02
Fe(%)	2.42 ± 0.04		2.56 ± 0.13	2.35 ± 0.03	2.47 ± 0.08			2.40 ± 0.030
Ga			11.6 ± 1.2					11.6 ± 1.2
Hf	2.11 ± 0.03							2.11 ± 0.03
Hg							0.035 ± 0.003	0.035 ± 0.003
Ho	0.88 ± 0.04							0.88 ± 0.04
K (%)	1.98 ± 0.20		1.94 ± 0.05	2.11 ± 0.05*	1.81 ± 0.03			1.86 ± 0.03
La	24.7 ± 0.3							24.7 ± 0.3
Lu	0.35 ± 0.03							0.35 ± 0.03
Mg(%)				3.88 ± 0.03	4.32 ± 0.09			4.07 ± 0.06
Mn	388 ± 23		420 ± 24	374 ± 28				396 ± 14
Mo			32.7 ± 1.4	41.3 ± 3.9				33.7 ± 1.3

Element				
Na(%)	2.15 ± 0.03		2.24 ± 0.03	2.19 ± 0.03
Nb				9.2 ± 1.5
Nd	22.3 ± 1.1	9.2 ± 1.5		22.3 ± 1.1
Ni	29.7 ± 4.6	32.1 ± 3.9	32.4 ± 1.8	32.1 ± 1.5
Pb		36.2 ± 2.0		36.2 ± 2.0
Rb	89.9 ± 4.1			88.4 ± 1.8
S		88 ± 2		6720 ± 620
Sb	2.63 ± 0.15	6780 ± 620		2.63 ± 1.5
Sc	6.84 ± 0.15			6.84 ± 0.15
Se		2.3 ± 0.1		2.3 ± 0.1
Si(%)		17.8 ± 1.8	18.2 ± 0.3	18.2 ± 0.4
Sm	3.68 ± 0.07			3.68 ± 0.07
Sr	820 ± 26*	866 ± 7	892 ± 14	879 ± 12
Ta	0.65 ± 0.02			0.65 ± 0.02
Tb	0.42 ± 0.04			0.42 ± 0.04
Th	7.55 ± 0.10			7.55 ± 0.10
Ti(%)	0.21 ± 0.04	0.24 ± 0.01	0.20 ± 0.03	0.24 ± 0.01
U	5.3 ± 0.2			5.10 ± 0.14
V	111 ± 13	139 ± 19	133 ± 7	129 ± 6
Y		16.4 ± 0.9		16.4 ± 0.9
Yb	1.61 ± 0.13			1.61 ± 0.13
Zn	89.2 ± 3.0	86.2 ± 5.5	93.6 ± 5.3*	82.3 ± 2.2
Zr		77 ± 4	36.2 ± 1.3	36.2 ± 1.3

* Deleted from error weighted average.

INAA – Instrumental Neutron Activation Analysis
RCAA – Radiochemical Neutron Activation Analysis
XRF – X-Ray Fluorescence Analysis
PES – Plasma Emission Spectroscopy (Sodium Carbonate Fusion)
FAA – Conventional Flame Atomic Absorption (Lithium Metaborate Fusion)
GFAA – Graphite Furnace Atomic Absorption
CVAA – Cold Vapor Atomic Absorption

Table 7. Major Elemental Concentration in Paraho Raw and Retorted Shale Collected August 24, 1977(a)

Element	Raw	Retorted	Retorted/Raw
Hydrogen (%)	1.50 ± 0.03	0.167 ± 0.004	0.111 ± 0.002
Carbon (%)	15.5 ± 0.16	6.11 ± 0.04	0.394 ± 0.005
Nitrogen (%)	0.66 ± 0.10	0.298 ± 0.004	0.45 ± 0.017
Oxygen (%)	39.6 ± 1.7	50.7 ± 1.7	1.18 ± 0.09
Sulfur (ppm)	6000 ± 240	5830 ± 250	0.97 ± 0.06

(a) Analysis by LECO Corporation.

5) Additionally, a percent root mean squared deviation is computed for each group of data (shown at top of Figures 5 and 6) and a chi-square test was applied to test the validity of the error analysis.

A number of observations can be drawn from Figures 5 and 6. Agreement between analytical methods is in general quite good, and in almost all cases error bars overlap. One notable exception is the determination of As in the raw shale by instrumental neutron activation and XRF. This is only noticeable since both methods have good precision. Even in this case the disagreement is only 10-15%. A wide range of precision is evident illustrating the advantages of using several different analytical techniques for multielement analysis.

ELEMENTAL MASS BALANCES FOR THE PARAHO SEMIWORKS RETORT

Mass balances for 31 elements have been completed for the Paraho Semiworks Retort as operated during direct mode retorting. The relative distribution of elements among the products and effluents was determined from: 1) accurate, high-precision measurements of each element in the raw shale, retorted shale, product oil, product water, and product gas and 2) the calculated equilibrium material balance between the mass of feedstock and product materials. The results are shown in Table 8.

The detailed chemical analysis obtained from samples collected on August 24, 1977 were used to compute the elemental mass balances described here. In the few cases where chemical characterization data were not available for this collection date, data from other days were substituted. It is assumed that such substitutions are reasonable and that the computed mass balance of each affected element is valid for the date of interest. This assumption is supported by reported feedstock variation studies (Wildeman and Heistand, 1979). The overall uniform comparison between the data of different days suggests that the input and redistribution of elements were rather constant during the time of this study.

SUMMARY AND CONCLUSIONS

Inorganic analysis of solid, liquid, particulate, and gaseous samples from the Paraho Semiworks Retort was completed using a multitechnique approach. Most of the techniques used instrumental methods, so that interferences from chemically

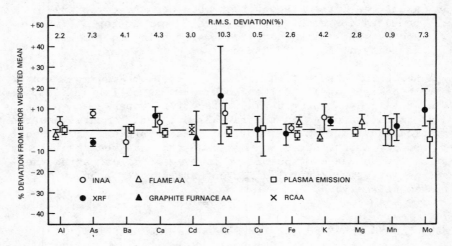

Figure 5a. Relative Performance of Analytical Techniques Used in the Multielement Analysis of Paraho Raw Oil Shale (8-24-77)

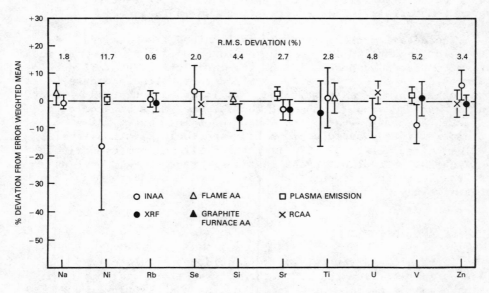

Figure 5b. Relative Performance of Analytical Techniques Used in the Multielement Analysis of Paraho Raw Oil Shale (8-24-77)

OIL SHALE PRODUCTS AND EFFLUENTS

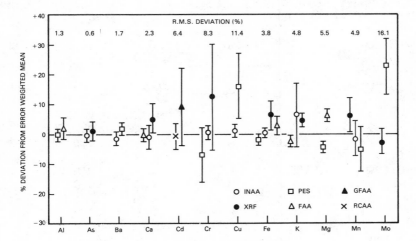

Figure 6a. Relative Performance of Analytical Techniques Used in the Multielement Analysis of Paraho Retorted Shale (8-24-77)

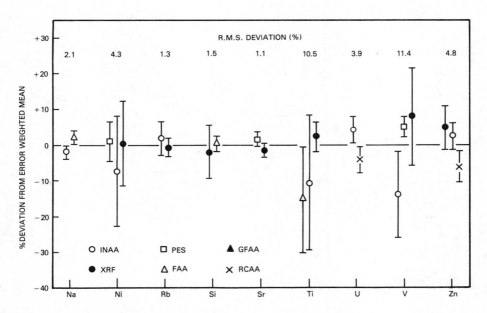

Figure 6b. Relative Performance of Analytical Techniques Used in the Multielement Analysis of Paraho Retorted Shale (8-24-77)

Table 8. Elemental Mass Balances for the Paraho Semiworks Retort During Direct Mode Retorting

	Elemental Abundance In Raw Shale, ppm	% Element Redistributed into Products and Effluents (August 24, 1977)				
		Retorted Shale	Product Oil	Product Water	Product Gas	Imbalance (%)
Al	37,700	106.7	<0.01	<0.01	<0.01	+6.7
As	44	111.7	5.90	0.05	0.08	+17.7
B	94	94.8	N.D.(1)	0.18	<0.01	-5.0
Ba	510	98.0	<0.03	<0.01	<0.03	-2.0
C	155,000	32.8	46.5	0.06	21.1	+0.5
Ca	101,000	109.7	<0.01	<0.01	<0.01	+9.7
Co	9.0	102.7	0.87	<0.01	<0.02	+3.6
Cr	34	107.8	<0.01	<0.01	<0.01	+7.8
Cu	40	116.6	0.09	<0.01	<0.01	+16.7
Fe	20,700	96.6	0.02	<0.01	<0.01	-3.4
H	15,000	9.4	65.4	2.7	40.0	+17.5
Hg	0.089	32.8	28.2	0.01	23.0	-16.0
K	16,100	96.2	<0.05	<0.01	<0.01	-3.8
Mg	34,600	104.0	<0.01	<0.01	<0.01	+4.0
Mn	315	104.7	<0.01	<0.01	<0.01	+4.7
Mo	22	117.4	<0.8	<0.01	<0.04	+17.4
N	6,600	37.9	30.3	1.3	17.9	-12.6
Na	16,900	107.9	<0.01	<0.01	<0.01	+7.9
Ni	28	95.2	0.82	<0.01	<0.01	-4.0
Pb	27	108.0	0.09	<0.01	<0.06	+8.0
Rb	75	97.7	<0.01	<0.01	<0.01	-2.3
S	6,000	80.9	10.4	2.5	12.7	+6.5
Sb	2.1	103.1	0.11	<0.01	<0.03	+3.2
Se	2.0	95.8	3.81	1.88	<0.8	+1.5
Si	150,000	101.1	<0.01	<0.01	<0.01	+1.1
Sr	700	104.7	<0.01	<0.01	<0.01	+4.7
Th	6.3	100.5	0.01	<0.01	<0.01	+0.5
Ti	1,800	111.1	0.01	<0.01	<0.01	+11.1
U	4.5	94.4	<0.06	<0.01	<0.02	-5.5
V	94	101.9	0.02	<0.01	<0.01	+1.9
Zn	64	106.7	0.05	<0.01	<0.01	+6.7

(1) Not determined.

complex matrices could be minimized. In many cases, analytical techniques were altered or improved in order to make them applicable to oil shale samples.

The techniques employed for most of the analyses were a combination of instrumental neutron activation, energy dispersive x-ray fluorescence, flame atomic absorption spectroscopy and dc plasma emission spectroscopy. Additional analyses were performed by graphite furnace atomic absorption, cold vapor atomic absorption, and radiochemical activation. The data were statistically analyzed to determine the precision of each method and to see how closely the various techniques compared. Better than 10% comparisons between two or more of the techniques were obtained for analyses of Al, Ba, Ca, Cu, Fe, K, Mg, Mn, Na, Rb, Si, U, V and Zn. Better than 20% comparisons were obtained for Cd, Cr, Mo and Ti. Systematic differences of about 15-20% between x-ray fluorescence and neutron activation determinations of arsenic were noted, with the neutron activation being generally higher. The environmentally interesting elements B, Hg, and Se could be reliably determined by only one method, so intercomparisons could not be made. Fluorine could not be reliably determined by any technique (selective ion electrode, colorimetry) that we tried.

The data were also used to construct mass balances for 31 trace and major elements in the various effluents, including the offgas for the Paraho retort operating in the direct mode. The computed mass balances show that approximately 1% or greater fractions of the As, Co, Hg, Ni, S and Se are released during retorting and are redistributed to the product shale oil, retort water, or product offgas. The fraction for these seven elements ranged from 1% for Co and Ni to 50 to 60% for Hg and N.

Approximately 20% of the S and 5% of the As and Se are released. The mass balance redistribution during retorting for Al, Fe, Mg, V and Zn was observed to be no greater than 0.05%. These mass balance figures are generally in agreement with previous mass balance studies made for a limited number of elements on laboratory-scale or smaller pilot retorts (Fox et al., 1977; Schendrikar and Faudel, 1978; Wildeman, 1977; Fruchter et al., 1978). Thus, the mass balances reported here for the Paraho retort may have some general validity for other types of oil shale retorting technologies.

REFERENCES

Fox, J. P., McLaughlin, R. D. Thomas, J. F., and Poulson, R. E., 1977, The partitioning of As, Cd, Cu, Hg, Pb, and Zn during simulated in situ oil shale retorting, in: "Proceedings of the 10th Oil Shale Symposium," Colorado School of Mines, Golden, Colorado.

Fruchter, J. S., Laul, J. C., Petersen, M. R., Ryan, P. W., and Turner, M. E., 1978, High precision trace element and organic constituent analysis of oil shale and solvent-refined coal materials, in: "Advances in Chemistry Series," 170:225-289, American Chemical Society.

Jones, J. B., Jr., 1976, The Paraho oil shale retort, in: "Quart. Colorado School of Mines," 71, 4:39-48.

Laul, J. C., Wilkerson, C. L., and Crow, V. L., 1979, Computer methodology and its applications to geological and environmental matrices, in: "Proceedings of the American Nuclear Society Topical Conference on Computers in Activation Analysis and Gamma-Ray Spectroscopy.

Nielsen, K. K., 1977, "Application of Direct Peak Analysis to Energy Dispersion X-Ray Fluorescence Spectra," BNWL-2277, Pacific Northwest Laboratory, Richland, Washington.

Schendrikar, A. D., and Faudel, G. B., 1978, Distribution of trace metals during oil shale retorting, Environ. Sci. and Tech., 12(3):332-334.

Wildeman, T., 1977, Mass balance studies in oil shale retorting, in: "Trace Elements in Oil Shale, Progress Report, June 1, 1975-May 31, 1976," Environmental Trace Substances Research Program, University of Colorado, Colorado State University, and Colorado School of Mines.

NUCLEAR METHODS FOR TRACE ELEMENTS IN PETROLEUM

AND PETROLEUM PRODUCTS

        Héctor D. Buenafama

        Instituto Tecnológico Venezolano del Petróleo
        INTEVEP S.A.
        Apartado 76343, Caracas 107, VENEZUELA

INTRODUCTION

    Environmental pollution, petroleum geochemistry and characterization, design and operation of refinery plants, and quality control of end products are some of the most important problems associated with trace metal content of petroleum (1-5).

    In addition to hydrocarbons, petroleum also contains the following types of components:

1. Heterocyclic molecules containing heteroatoms like S, N, and O.

2. Minerals and trace metals present as different types of molecules.

    Upgrading of crudes is to a great extent the elimination, by physical or chemical methods, of these non-hydrocarbon components as well as the removal of the high molecular weight fraction called asphaltenes. As a general rule, trace metal content in crude oil increases with its asphaltene yield. It is a well known fact that as the crude becomes lighter (higher API º), its metal content becomes lower. The API gravity increases as the oil becomes more mature. This can be explained by an extensive degradation of the older oil as a result of its subjection to additional physical and chemical changes. Lower trace metal content is related to the migrational history of the oil. During migration through rocks and clays, the high molecular weight and polar constituents with which trace metals are associated, has the tendency of being selectively adsorbed to the rocks and thus leaving the lighter, non-polar constituents to pass without essential hindrance. Accordingly, a lower trace metal content would be expected for aged crude oils. This fact explains why the Cretaceous oils, in general, contain lower amounts of trace metals than the younger Miocene oils.

It has been pointed out that for a large number of crude oils, the high sulfur content, correlates with high vanadyl porphyrin content, high vanadium content and high ratios V/Ni.

From all metals that have been identified in crude oil, V and Ni are the most important. The V/Ni ratio has been used as a parameter of age, but it still does not appear to be a well defined and rigorous correlation. The V/Ni ratio may range from 0.05 to 15 (6).

In 1934 the existence of V and Fe petroporphyrins was first established (7). The now well known chemical association between chlorophyll in marine plants and the process of petroleum formation was also established (8).. It was not until 1948 that the second major metallic porphyrin was isolated from crudes (9).

The first quantitative data on trace elements in crude oils were presented in 1931 (10). Since then, up to about 40 elements have been identified in different crude oils as presented in TABLE I.

With the exception of V and Ni, very little is known about their presence and only their concentration values have been reported for particular crudes. The nature and role of other elements in crudes may provide valuable clues to the origin and the geochemistry of petroleum.

For a valid geochemical interpretation of trace elements data in crude oils it is also important to know the metal speciation, a matter which is still far from being completely elucidated. On the other hand, it is only necessary to take into account those metals closely related to petroleum genesis, and not those contained as minerals or contaminants incorporated by drilling, piping and transportation. It is now recognized that V and Ni may occur as porphyrin and non-porphyrin forms, and the ratio of these two forms may vary over a wide range. In TABLE 2 are briefly summarized the different types of species with which trace metals in crudes may be associated. The vanadyl macrocycle has been of both a practical and theoretical interest because it could serve as a geochemical tracer that would link geochemistry with biology. A sample with a vanadyl porphyrin skeleton subject to either an interruption of the conjugation in the outer ring (hydroporphyrins, such as chlorins and mesochlorins) or additional aromaticity resulting from extra aryl substitution or fusion (arylporphyrins) can be classified as non-porphyrins. Other non-porphyrin species

Table 1. Trace Elements Identified in Crude Oils

Group	Elements (concentration range)
Row 1	Na 2.0–5; Mg 0.4–3.0
Row 2	K 0.2–5; Ca 10–100
Row 3	Cs* 0.8–6; Ba 0.2–1.4

Sc*	0.3–1.0
Ti	0.7–2.2
V	0.5–1200
Cr	0.02–1.0
Mn	0.05–2.5
Fe	0.5–80
Co	0.003–9
Ni	0.7–110
Cu	0.1–4.5
Zn	0.7–40
Ga	0.01–0.3
As	0.02–0.5
Se	0.01–1.2
Br	0.04–1.5
La	0.1–2.2
Mo	0.02–0.8
Ag	—
Cd*	0.5–150.0
Sn	0.07–0.7
Sb	0.01–1.2
I	0.8–2
Al	0.1–1.1
Au*	0.3
Hg	0.02
Pb	0.07–1.6
Cl	1.2–25
Ce	—
Eu*	0.4–3
Th*	1–10
U*	0.7–70

NUMBERS REFER TO APROXIMATE CONCENTRATION RANGE
p.p.m. or p.p.b. (*)

Table 2.- The Nature of Trace Metals in Petroleum

I.- Metalloporphyrin type chelates
Etioporphyrin III (Etio III)
Deoxophylloerythroetioporphyrin (DPEP)

II.- Complexes of tetradentate mixed ligands
β - Ketimines
β - Diketones
Monothio β - Diketones

III.- Organometallic compounds

IV.- Minerals and other inorganic species.
Silica, Alumina, etc.

include degraded porphyrin products as well as those tetradentates made of mixed ligands of 2N2O, 2N2S, 3N1O etc. One of the properties of these vanadyl complexes is the ease with which they undergo demetallation and, unlike true porphyrins which exhibit the well known Soret band ($a_{1u} \rightarrow e_g^*$ transition), they only exhibit other different transitions ($d_{xy} \rightarrow d_{x^2y^2}$, $d_{xy} \rightarrow d_{z^2}$).

It has been proposed that the V/Ni ratio decreases with increasing maturity (11); other workers have proposed the inverse correlation (12, 13) and still others have found no correlation between age and V/Ni ratios (14, 15). The fact that both metals may be present in crudes as porphyrin and non-porphyrin species may partially explain some of the controversy concerning the correlation of V/Ni ratios with age of reservoir rock. Recently, some new trace metal indices have been proposed for crude oil characterization (16, 17). The vanadium content was compared to that of antimony, nickel and magnesium in several crudes. The data obtained suggest that there is a very close correlation between different metal indices and age, allowing an approximate classification of Miocene and Cretaceous oils on the basis of such indices. On the other hand, trace metal indices, in heavy ends including Sb, Ni and V, appear to be useful in correlating their occurence with the genesis of the source oil, from which those heavy ends were obtained.

Since it has been proposed that as the age of the crude increases the petroporphyrins proceeds from deoxophylloerythroetioporphyrin (DPEP) to etioporphyrin (ETIO) (18), the ratio DPEP/ETIO would also be useful as a marker for age of crudes.

Porphyrins isolated from different rock extracts and determined by spectrophotometry, also confirm that a considerable part of V and Ni (30% - 50%) is bound in the non-porphyrin compounds (19).

The spectrophotometric method of determining metalloporphyrins, which is based on the integration of the 410 nm absorption peak (20) may be inaccurate since at least four types of porphyrin complexes may be present in petroleum as shown in Figure 1. A novel chromatographic separation allows the porphyrinic metals to be eluted while the non-porphyrinic type species are retained in the column (21). The Ni porphyrins are concentrated in the aromatic fraction while the vanadyl porphyrins remain in the resin fraction. Assuming that only porphyrinic metals are eluted, a value of 45% of Ni porphyrin and 46% of vanadyl porphyrin was obtained for a Kuwait residue. These values are much higher than the figure of 14% measured by absorption spectrophotometry.

These facts clearly reveal that much work has still to be done and special efforts need to be concentrated in better separation methods with concomitant reliable analytical techniques for quantitation.

On the other hand, the variance of V/Ni porphyrins in crudes may also suggest that the two metals were incorporated independently during the diagenesis and maturation processes. The finding that nickel porphyrins are contained in early sediments but not the vanadium porphyrins may indicate an early introduction of nickel. A proposed route to the late entry of vanadium has been suggested, assuming that all vanadium, sulfur compounds and asphaltene arise through the action of sulfur bacteria in the reservoir. Profile analysis including all trace metals for fingerprinting may also be very useful as a complementary technique (22).

The object of this presentation will be threefold: to review nuclear methods of trace metal analysis in petroleum and petroleum products, to present some methodological aspects of trace analysis and elemental determination in crudes, and finally to propose some future trends in trace metal determination via neutron activation and other alternative techniques.

DETERMINATION OF TRACE METALS

Sampling and Storage

Only recently has the need of accurate and reliable analytical methodology for determining ppb levels of trace elements in petroleum been recognized. Sampling procedures designed to fulfill such requirements have not yet been precisely defined and studied.

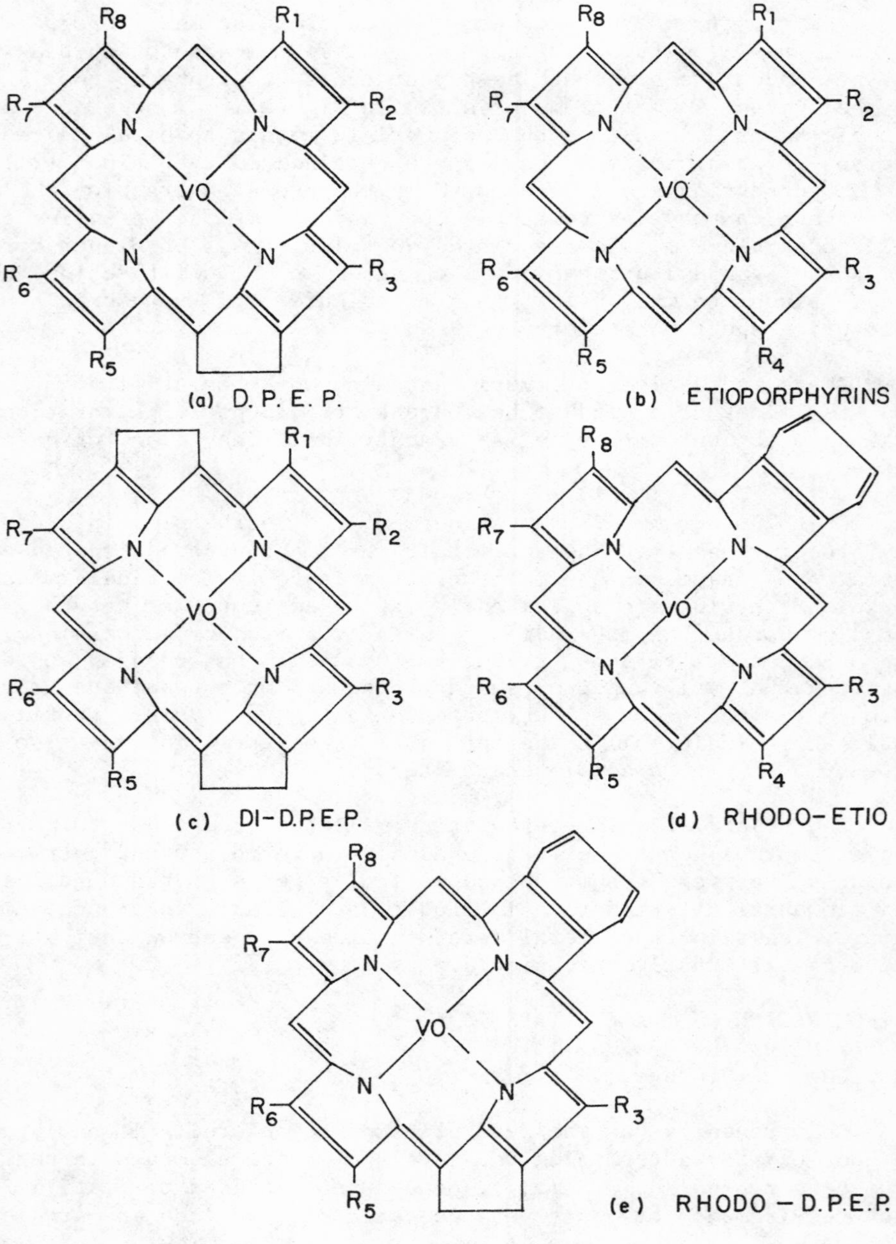
(a) D. P. E. P.
(b) ETIOPORPHYRINS
(c) DI-D.P.E.P.
(d) RHODO-ETIO
(e) RHODO – D.P.E.P.
VO
N
R1
R2
R3
R4
R5
R6
R7
R8

Figure 1. Structure of Vanadylporphyrins Present in Petroleum

Personal judgment, as well as skill and sampling experience, have been recognized to play an important role in the process of obtaining a representative sample (23). However the contamination introduced by piping, transportation, storage, sampling and aging or weathering of petroleum and petroleum products has not been studied in detail yet. Inaccurate data may be obtained if high quality sampling, storage and homogenization procedures are not carefully observed.

Data obtained for some representative elements from a typical crude oil as determined by N.A.A. are presented in TABLE 3. This table partially shows the results of a study of stability and homogeneity of crudes concerning its trace element determination (24). Data are normalized to those obtained in a sample completely homogenized prior to sampling for analysis and compared to those obtained in the same type of crude, as received, and stored for eight months. The disagreement in some results may suggest some type of mechanism leading to an inhomogeneous sample after storage or the loss of the representative character of the sample within different aliquots.

Research was undertaken to find the best container for use in shipping samples for the purpose of cross-checking analytical methods. Postirradiation losses of Hg were observed but not of As in the irradiated polyethylene sample containers. A mechanism of mercury reduction to the elemental state in which the solvent played an important role was proposed (25).

Modern analytical methods have extended the determination limits for certain elements to a point that the sensitivity of an analytical technique is greatly limited by the analytical blank value. Although often ignored at higher levels of concentration, blank values may deteriorate the final results at sub-ppm levels.

Standards

Standards are used both for recovery or yield corrections and for quantitation. Standard reference samples for multiple trace metals in petroleum are rather scarce, and thus any attempt for cross-checking or technique evaluation may be limited. A standard sample may also be prepared by adding a known amount of an organometallic standard to a petroleum matrix. Since the nature of the organometallic, the petroleum matrix and possibly the container may affect the stability of the prepared standard, the influence of each of those factors must be clearly established before its use.

Table 3.- Trace Elements Determination in a Heavy Crude Relative Concentrations

	V (ppm level)	Au (ppb level)	Ba (ppm level)	Ni (ppm level)
Original Sample	1.00 ± 2.0	1.00 ± 35.3	1.00 ± 15.2	1.00 ± 8.5
Stored Sample	0.99 ± 4.3	1.35 ± 39.7	0.76 ± 19.6	0.88 ± 10.9

Std. Dev. as % of the mean.

Stabilizers are usually advised to be incorporated in the standard solutions of organometallics, but even with the use of such stabilizers and at 500-800 ppm level, vanadium standards prepared by using gas oil or base oil diluent deteriorate with time. Finally when tracers or carriers are added to the irradiated sample prior to digestion it may be necessary to take into account that as they may differ greatly in nature respect to the irradiated matrix, their use can lead to a false recovery factor. The different aspects of the stability of organometallic standards have been widely studied (26).

Sample Preparation

One of the main advantages of N.A.A. is that it can be performed in a purely nondestructive form and thus minimize contamination. However, even with the use of solid state detectors it is sometimes mandatory to perform post-irradiation radiochemical separations to eliminate interfering isotopes and thus improve sensitivity. Dry ashing or burning, reagent-aided ashing and wet digestion are the most widely used procedures for sample mineralization. A judicious selection of a mineralization procedure cannot be done without consideration of the elements to be determined, their concentration level and the nature of the matrix. Some results comparing different mineralization techniques concerning the most important elements are presented in TABLES 5 and 6. Different amounts of organometallic nickel standards and a well standardized crude were mineralized in a Bethge apparatus using a H_2SO_4 - $HClO_4$ - HNO_3 mixture. In all cases quantitative recoveries

Table 4.- Recovery of Nickel After Wet Digestion

Sample Amount (gr)	Ni concentration (ppm) theoretical	Ni concentration (ppm) experimental
2	6.2	5.8
	116.7	112.1
8	6.2	6.4

Table 5.- Recovery of Vanadium After Different Mineralization Procedures

Sample type	V Concentration (ppm) Procedure # 1	V Concentration (ppm) Procedure # 2
Gas oil 1	0.46	0.63
Gas oil 2	0.07	0.11
Cut Crude S.N.	0.64	0.85
Cut Crude L.V. Plus 1µg v/g of crude	1.04	1.44
Cut Crude L.V.	0.26	0.39

Procedure # 1: Standard Method DIN 51790
Procedure # 2: H_2SO_4 Treatment and muffle ashing at 550ºC

were obtained, indicating that for nickel, this mineralization procedure may be considered to be free of losses. Regarding vanadium, some type of loss may appear to take place when using the DIN 51790 German standard method, applied to low concentration samples as compared to the standard method based on H_2SO_4 treatment and muffle ashing.

When using INAA, a common practice of sample preparation prior to irradiation is to filter the sample as received or previously diluted with a proper solvent. This is especially recommended when dealing with process products, which may be in some cases contaminated with catalyst particles (in case of fluidized bed reactors) or products of process equipment corrosion. Some products of hydrodesulfurization, on the other hand, may become so unstable and heterogeneous in nature that a stratification of the samples may take place as evidenced by its vanadium content. These samples should not be stored prior to sampling for analysis. The data corresponding to this conclusion are shown in TABLE 6 for a product of catalytic hydrodesulfurization sampled at 400°C.

In some instances it is necessary to transfer the irradiated sample to a clean vial prior to counting and extreme care should be exercised to assure that no trace metal or any product of sample radiolysis associated with metals remains absorbed on the container walls.

EXPERIMENTAL ASPECTS

Determination of Cadmium

Cadmium levels in crudes, as determined by N.A.A. (27), may vary over a wide range; from 0.5 to 1500 ppb. Due to the very low concentration of Cadmium in crudes, it cannot be determined instrumentally. A novel radiochemical separation has been developed at INTEVEP to determine cadmium at sub-ppm levels which has proven to be quantitive and precise. Samples and standards were submitted to irradiation for 40 hr. at a neutron flux of 1×10^{12} $n.cm^2.sec^{-1}$. After irradiation, each sample was quantitatively transfered to a Bethge digestion apparatus and cadmium carrier was added. Mineralization was performed using a mixture of H_2SO_4 - HNO_3 - $HClO_4$. The final solution is extracted with n-hexane and partially neutralized to pH of 6. The dissolved oxygen was expelled by stripping the solution with nitrogen and 600 mg of cadmium amalgam were added. After stirring the solution, it was discarded, the amalgam rinsed and extracted with a 0.5 M solution of Tl (I). Finally a known fraction of the solution was used to measure the 336 KeV γ transition of 115 m In after equilibration has been established. Standards were treated in identical fashion. Radiochemical yields are assumed to be quantitative and constant, as previously determined using either spiked solutions and organometallic standards. In TABLE 7 some results of cadmium determinations in crude oils are presented.

Computer Controlled Determination of Vanadium for the Analysis of

Numerous Samples

The discovery of one of the world's largest heavy crude oil deposits (10° API gravity) named the "Orinoco Tar Belt" has required the design of a system capable of analyzing up to 25,000 samples per year in order to support prospecting and pilot plant

Table 6.- Distribution of Vanadium in a Hydrodesulfurizated Product.

Sample	Vanadium concentration (ppm)
As received	222.4
After centrifugation	
Top	141.0
Bottom	582.6

Table 7.- Cadmium Contents of Some Crude Oils

Sample type	Cd Concentration (ppb)
Boscan II	17.9 ± 2.0
Cerro Negro	8.5 ± 1.1
Melones	6.2 ± 0.7

studies for the processing of these crudes. A computer controlled system has been designed for the automatic N.A.A. of vanadium in petroleum using a 252 Cf source and with the concept of complete unattended operation (28).

Irradiation Assembly

The irradiation assembly consists of a 1.6 m^3 stainless steel tank which is placed in a concrete lined hole. The tank has two concentric and centrally located stainless steel tubes. The inner tube houses the 252 Cf source. The outer tube serves as a guide for two movable stainless steel crosses which hold the aluminum irradiation tubes.

Counting Assembly

A 8180 Canberra 4096 MCA is used to count the irradiated samples. The MCA is coupled to a Digital PDP 11/04 computer provided with 28 K of core memory, real time clock, ROM bootstrap, TTY interface and a DL 11D communication interface. Programs and accumulated data are stored in a DEC RX-11 Dual Drive Floppy

Disk. A horizontally mounted Ge (Li) detector 1.9 KeV FWHM, 15.4% efficiency and 37:1 peak/compton ratio is used. Canberra Laboratory Automation Soft-ware System (CLASS) is used as the programing language.

Calibration and Data Analysis

A linear regression analysis is used for the 50 points calibration curve. Mandell's method of determining the uncertainities in the calibration curve is applied in the mathematical interpolation analysis (29). Peak smoothing, integration and variance estimation are performed by using the SPECTRAN III programs (30). Two different calibration curves are used for crudes or aqueous solutions which are linear up to about 40 mg total vanadium. In TABLE 8 the results of nine aliquots of deasphalted crude oil, performed two weeks apart, are presented. Different standards were used for each calibration curve. The relative standard deviation of 2.92% and 2.82% obtained is good. The fact that it may be necessary to increase sample weight as to increase sensitivity required to study the effect of sample size upon the analyzed vanadium content. The results shown in TABLE 9 indicate that there is no difference in using about twice the sample weight. In TABLE 10 are presented the results obtained by analyzing prepared standards and the concentration calculated from the weights of vanadium. In the range 19-5000 ppm the largest difference obtained is 4%.

Other different approaches in the use of 252 Cf sources are worthy of mentioning. An on-stream N.A.A. system for vanadium determination has also been proposed (31). The detection limit is 10 ppm, but very long counting times are necessary even for medium-level vanadium containing samples.

A lower detection limit of 0.1 ppm has been obtained by optimizing the geometrical conditions for activation and counting, using a 24 µg source (32). However, this system does not appear to be suitable for large scale analytical projects.

An on-line system, using a 660 µg source has also been proposed to monitor the vanadium level in different process products (33).

MODERN TRENDS AND APPLICATIONS

In 1972 an intercompany project was initiated between five U.S. petroleum companies to develop techniques for measuring trace concentration of metals in crudes (26). In this project, N.A.A. was used to check the data where it was applicable. The elements which were checked by N.A.A. were Sb, As, Co, Mn, Hg, Mo and Se. It is interesting to point out that the majority of cases led to results up to 20% higher than those obtained by any

TRACE ELEMENTS IN PETROLEUM

Table 8.- Reproducibility of the Analysis of Deasphalted Crude Oils

Sample Code	V Concentration (ppm)	
1	637.0^a	641.8^b
3	656.2	660.6
4	657.9	661.0
5	690.1	694.2
6	644.7	678.5
7	634.3	638.7
8	672.1	676.0
9	681.0	684.2
10	661.6	665.1
Average	659.1 ± 19.2	666.3 ± 18.7
R.S.D.	2.92	2.82

a.- Analysis based on calibration using standard Conostan I.

b.- Analysis based on calibration using standard Conostan II.

Table 9.- Irradiation of Crude Oil Samples of Different Weights

Sample code	Weight (gm)	V Concentration (ppm)
40	5.674	18.6 ± 3.7
41	11.072	18.1 ± 4.2
1181 P	5.899	23.4 ± 5.4
1181 PP	12.197	21.2 ± 4.9
43	5.697	205.6 ± 10.4
43 S	11.412	209.3 ± 11.8

Table 10.- Accuracy and Detection Limit Studies of Prepared Standards

Sample Code	Sample Weight (gm)	V Concentration (ppm) Analyzed	V Concentration (ppm) Calc. from Weight
41	11.072	19.2 ± 3.2	19.80
42	10.840	107.5 ± 6.1	103.76
43	11.412	218.7 ± 9.8	209.85
44	11.553	300.0 ± 17.0	300.08
45	10.985	4827.9 ± 201.6	5000.00
41 S	10.726	2574.4 ± 116.2	2499.90
45 S	11.156	404.2 ± 19.6	409.41
NBS STD		299.0 ± 5.7	320 ± 15 [a]
1 [b]		21.20 ± 1.89	21.34
2 [b]		50.20 ± 1.91	50.20
3 [b]		11.84 ± 1.31	10.97

a.- Value reported by National Bureau of Standards
b.- Analyzed in triplicate

other method selected by the different laboratories of the companies.

The Environmental Protection Agency, on the other hand, has initiated a program to monitor the elements in fuels and related emissions into the atmosphere. As part of this program, nine laboratories, using different analytical methods were asked to determine the concentration of 28 elements in the same fuel. Among the elements investigated were Hg, Be, Pb, Cd, As, V, Mn, Cr and Fe. The analytical methods were N.A.A., A.A., X.R.F. and O.E.S. The results from the interlaboratory comparisons were evaluated to assess the capability of various methods applied to those matrices (34). The wide range in reported concentrations, even for those laboratories that had used I.N.A.A., indicates that different sample preparation and analytical techniques lead to erroneous results and they are a solid argument for the need to develope standard reference materials certified in trace elements which can be used for evaluating analytical methods and quality control.

TRACE ELEMENTS IN PETROLEUM

Since residues of crude oils found in beaches are possibly the result of oil spillage at sea, and although these residues could also arise from natural oil spillage, trace metal data can help to identify oil spills. According to this, it was determined for asphalt the precision of the two most sensitive methods available: flameless atomic absorption and N.A.A. (35). The elements chosen were iron and nickel. Results of this clearly show that there are great difficulties in obtaining reliable results and that they are still greatly dependent upon the technique used.

Using two neutron irradiations and four gamma spectrometric measurements, 23 elements have been determined in different crudes by I.N.A.A. (36). In this work, blank values of Conostan base oil, commonly used as solvent for organometallic standards solutions, have also been determined. On the other hand our results for Fe and Zn in that base oil are presented in Table 11. The concentration of these elements, show that great care has to be taken in using that solvent for standards preparation at ppb levels.

Using different irradiations and decay times, 49 commercial petroleum products for 22 elements were analyzed (37). The trace metals concentration in the different distillation fractions allowed to establish four groups of elements, according to their enrichment factor in each fraction, which was found to range from 10 to more than 400.

As an effort for using multielement trace metal data for fingerprinting and identifying crude oil residues, Zn, Co, Sc, Br, Cr and Sb were determined in beach asphalts (38). Choosing one sample as an arbitrary standard and ratioing the elements in other samples to that of the standard, it can be concluded that no two samples have all the six elements in the same ratio as the standard.

Pattern recognition techniques were applied as an effort to determine the source of an oil spill when the field sample has undergone weathering (39). Vanadium was by far the most significant element for oil source identification of the 22 elements considered followed by nickel and sulfur. It is rather difficult to expect accuracy in that technique since the analysis of the spill sample as received from the field will differ from the analysis of the true source for at least three reasons: weathering, contamination and analytical errors.

I.N.A.A. has been applied for the determination of several elements in oil shale (40). Results, when possible, were compared to those obtained by X.R.F. Speciation for As and Hg was proposed for the off-gas from the oil shale retort.

Several elements were also determined in a petroleum distil-

Table 11.- Blank Values of Conostan Base Oil

Element	Concentration (ppm)
K	0.6
Fe	< 1.26
Zn	0.19

lation unit (41). The concentration of 10 elements were determined in different fractions from light naphta to the residues.

A prompt gamma ray spectrometry system for the determination of sulfur has also been proposed (42). A detection limit of 0.48% has been obtained with a 0.24 µg 252 Cf source. An increase in source activity would surely allow a lower detection limit.

X-RAY FLUORESCENCE AND OTHER COMPLEMENTARY TECHNIQUES

Systematic studies for the multielemental analysis of crude oils by X.R.F. have not appeared frequently in the literature. Energy dispersive X.R.F. determination of V, Ni, Fe and Mo in NBS fuel oil were performed with a minimum sample preparation and standardization via the use of spiked solutions (43).

A simultaneous determination of V and S was performed by energy dispersive X.R.F. using radioisotope excitation (44).

Using solid or aqueous standards, Zn, Fe and Pb were determined in petroleum products by X.R.F. (45). Normalization involving corrections for scattering from the sample cell was used.

A review on the application of E.D.X.R.F. on wear metals determination in lubricating oils has been published (46).

Atomic absorption spectrometry, both in flame and flameless mode, appear to be a powerful technique, as it can exhibit very good detection limits. Direct flame analysis may give erratic and inaccurate results as well as direct flameless analysis (47). The complexity of the matrix as well as the great differences between crudes make the signals dependent on the type of matrix, even with the use of electrothermal atomization. Problems associated with the selection of the proper organometallic standards were also reported (48, 49).

FUTURE TRENDS

Several years ago, the petroleum industry was confronted

with an exceptionally challenging problem in trace metals analysis, brought about by the need of meeting pollution control requirements. It was necessary to monitor a wide variety of materials for mass balance studies of trace metals. Because of the large quantities of material involved, even concentrations below ppm could be significant. Therefore, unusually accurate and reliable data were required in concentration ranges of 1-1000 ppb.

The need of improving all existing techniques for trace metal analysis in crude oils in order to attain still lower detection limits with better accuracy and precision appears to be a challenge which will demand the combination of all our future efforts. The necessity of standardizing the analytical methodology to achieve that goal, in different laboratories and using different techniques, is to a great extent dependent of the universal availability of different types of standard crudes. Round-robin studies and multitechnique analysis as those performed with NBS Coal and Fly Ash (50, 51) are still missing in connection with crude oils of different nature, as a means of obtaining large amounts of Standard Reference Crudes in the near future.

A comprehensive study of sampling, contamination and stability of crudes and standards, specially at sub-ppm levels has to be done, in addition to the efforts already made, to elucidate and quantify an important problem, not always taken into consideration at the time of performing an analysis.

Metal speciation is a very important field in which much research needs to be performed. Very recently, gallium porphyrins have been isolated from bituminous coal (52). This unexpected fact should make us think, also, about the ultimate nature of trace metals in oils, from which very little is known aside from V and Ni. Separation techniques have also to be improved and work in close connection with trace metal determination in order to obtain more realistic data about metallic species fractionation. More realistic than data concerning trace metal distribution in the different flows of a distillation plant, are those data obtained from metal distribution in different components of crudes such as parafins, aromatics, acids, bases, etc.

Metal speciation and improved separation methods will give us more complete information about the role, nature and origin of trace metal in crude oils. This is still much more important when dealing with heavy and extra-heavy crudes, which in many aspects require a very different approach for analysis.

REFERENCES

1. T.F. Yen, in "The Role of Trace Metals in Petroleum", T.F. Yen, ed., Ann Arbor Science, Ann Arbor, Michigan (1973).-

2. G.W. Hodgson, B.L. Baker, E. Peake, "Fundamental Aspects of Petroleum Geochemistry", B. Nagy, U. Colombo, eds. p177, Elsevier, Amsterdam (1967).-

3. J.S. Ball, W.J. Wenger, H.J. Hyden, C.A. Horr, A.T. Myers, J. Chem. Eng. Data. 5:5533 (1967)

4. D.A. Skinner, Ind. Chem. Eng. 44:1159 (1952)

5. W.H. Zoller, G.E. Gordon, E.S. Gladney, A.G. Jones, "Trace Elements in the Environment", E.L. Kothny, ed., Amer. Chem. Soc., Washington (1973).-

6. B.P. Tissot, D.H. Welte, "Petroleum Formation and Occurence" Chap 1, Part IV, p.364, Springer - Verlag, Berlin (1978).-

7. A. Treibs, Ann. der Chim 509:103 (1934), 510:42 (1934)

8. G. Bertrand, R. Paulais, Comp. Rend., 203:683 (1936).-

9. E. A. Gleborskaya, M. V. Vol'kenshtein, J. Gen. Chem., (U.R.S.S.) 18:440 (1948)

10. W. B. Shirey, Ind. Eng. Chem., 23:1151, (1931).-

11. G. W. Hodgson, Bull. Amer. Assoc. Petrol. Geol. 38:2537 (1954).

12. H.J. Hyden, U.S. Geol. Survey Bull, 1100.B (1961)

13. H. Al-Shahristani, M.J. Al-Atiya, Geochim. et Cosmochim. Acta., 36:929 (1972).-

14. P.Y. Demenkova, L.N. Zakharenkova, A.P. Kurstalskaya, Tr. Vses. Neft. Nank. Issled. Geol. Inst., 123: 59 (1958).-

15. A.F. Gilmashin, M.G. Gazinov, E.N. Berturina, Tr. Tater. Neft. Nank. Issled. Inst. 18:113 (1971).-

16. M. Abu-Elgheit, S.O. Kalil, A.O. Barakat, Amer. Chem. Soc., Div. Petr. Chem., Preprints, 24: 793 (1979).

17. M. Abu - Elgheit, M.J. Ijam, Amer. Chem. Soc., Div. Petr. Chem., Preprints 24:1014 (1979).-

18. T.F. Yen, S.R. Silverman, Amer. Chem. Soc., Div. Petr. Chem. Preprints 14: (3) E-32 (1969).-

19. M. Popl, V. Dolansky, G. Sebor, M. Stejskal, Fuel, 57:565 (1978).-

20. J.M. Sugihara, R.M. Bean, Chem. Eng. Data., 7:269 (1962).-

21. A.J.G. Barvise, E.V. Whitehead, Amer. Chem. Soc., Div. Petr. Chem. Preprints 25:268 (1980).

22. R.H. Filby, K.R. Shah, Proc. Amer. Nucl. Soc. Conf. on Nucl. Meth. in Environ. Res., p68, Columbia Univ. of Mo. Press, Missouri (1971).

23. American Chemical Society for Testing Materials, Manual on Measurement and Sampling of Petroleum and Petroleum Products, D 270-65, D 1265-55, D1145-53 (1970).-

24. H.D. Buenafama, J.A. Lubkowitz, J. Radioanal. Chem. 39:293

25. J.O. Larson, Anal. Chem. 47:1159 (1975)

26. R.A. Hofstader, O.I. Milner, J.H. Runnels eds., "Analysis of Petroleum for Trace Metals" Chap. 2, p.6, Amer. Chem. Soc., Washington (1976).-

27. R.H. Filby, K.R. Shah, Amer. Chem. Soc., Div. Petr. Chem., Preprints 18: 615 (1973).-

28. J.A. Lubkowitz, H.D. Buenafama, V. Ferrari, Anal. Chem.52: 233 (1980).-

29. J. Mandell, F. Linning, Anal. Chem. 29:743 (1957).-

30. Canberra Industries Inc., Spectran III, Users Manual, Meriden, Conn. (1977).

31. S.M. Al-Jobori, S. Szeged: J. Csikai, Radiochem. Radioanal. Lett. 25:87 (1976).-

32. H. Meier, G. Zeitler, P. Menge, J. Radioanal. Chem. 38:267 (1977)

33. H. Braun. Chemic. Ing. Techn., 47:903 (1975).-

34. D.J. von Lehmden, R.H. Jungers, R.E. Lee, Anal. Chem.46:239 (1974).-

35. L.A. May, J.B. Preley, Anal. Chem. Acta 83:385 (1976).

36. C. Block, R. Dams, Anal. Lett, 10:1233 (1977)

37. C. Block, R. Dams, J. Radioanal. Chem. 46:137 (1978).-

38. L.A. May, B.J. Pesley, J. Radioanal. Chem. 27:439 (1975).

39. D.L. Duewer, B.R. Kowalski, T.F. Schatzki, Anal. Chem. 47: 1573 (1975)

40. J.S. Fruchter, J.C. Laul, M.R. Petersen, P.W. Ryan, Amer. Chem. Soc., Div. Petr. Chem., Preprints 22: 793 (1977).

41. C. Bergerioux, J.L. Galinier, L. Zikovsky, J. Radioanal. Chem. 54:255 (1979).

42. A.R. Pouraghabagher, A.E. Porfio, Anal. Chem. 46:1223 (1974)

43. H. Kubo, R. Bernthal, T.R. Wildeman, Anal. Chem. 50:899 (1978)

44. A.N. Yousif, H. Al-Shahristani, Int. J. Appl. Rad. Isotop. 28:759 (1977).

45. T.V. Krishnan, Indian J. Pure Appl. Phys., 15:345 (1977).

46. H. Meier, E. Unger, J. Radioanal. Chem. 32:413 (1976).

47. P.L. Grizzle, C.A. Wilson, E.P. Ferrero, H.J. Coleman, Report 1977 BERC/RI - 77/8 (1977).

48. G. Sebor, I. Lang, Anal. Chem. Acta, 89:221 (1977).

49. I. Lang, G. Sebor, V. Sychra, D. Kolikova, O. Weisser, Anal. Chem. Acta, 84:299 (1976).

50. R.A. Nadkarni, Anal. Chem. 52:929 (1980).

51. J.M. Ondov, W.H. Zoller, I. Olmez, N.K. Aras, G.E. Gordon, L.A. Ranchitelli, K.H. Abal, R.H. Filby, K.R. Shah, R.C. Ragaini, Anal. Chem. 43:1102 (1975).

52. R. Bonnet, F. Czechowski, Nature, 283:465 (1980).

TRACE ELEMENT COMPOSITION OF ATHABASCA

TAR SANDS AND EXTRACTED BITUMENS

F. S. Jacobs and R. H. Filby

Nuclear Radiation Center
Washington State University
Pullman, Washington 99164

ABSTRACT

The Athabasca Tar Sands represent a potential petroleum resource in excess of 10^{11} bbl oil and thus form one of the world's major deposits of petroleum. The deposits consist of sandstones impregnated with a viscous heavy bitumen similar to a very heavy petroleum. Although the origin of the tar sand bitumen has not been established in detail, most theories involve the water washing and/or bacterial alteration of a conventional crude oil and that the Athabasca deposits are genetically related to other Cretaceous W. Canada Basin heavy oils.

Except for Ni, V, and Fe there is little information on the trace element contents of the bitumen extracted from the tar sand. Both Ni and VO porphyrins have been identified in Athabasca bitumen, but other metal complexes, if present, have not been reported. The preliminary work reported here is part of a research project to determine the trace element composition of the bitumen, to determine possible metal species in the bitumen, and to relate these to modes of origin.

Neutron activation analysis has been used to determine V, Ni, Fe, Co, Mn, Cr, Zn, Na, Rb, Cs, Se, Hg, Br, As, Sb, Cl, Se, Eu, and Ga in samples of McMurray Formation bitumen extracted with toluene in a Soxhlet extractor (syncrude product). Analyses of the bitumen show that it contains higher concentrations of V, Ni, Fe, and other trace elements than most conventional crude oils.

INTRODUCTION

Oil sands (also known as tar sands) are deposits of heavy oils that impregnate predominantly sandstone strata. The oil sands represent an important global petroleum resource and in-place reserves of known oil sand deposits of between 1.7 and 4.0 trillion barrels (bbl) of oil (1). The Venezuelan Oficina-Temblador and the W. Canada oil sands together represent more than 90% of the world total reserves of oil in oil sands. In-place reserves of the Venezuelan deposits are estimated to be 0.7 to 3.0×10^{11} bbl and those of W. Canada are in excess of 1×10^{11} bbl of oil. Although U. S. reserves total approximately 2×10^9 bbl the major oil sand deposits in the United States are the Asphalt Ridge, Utah (9×10^8 bbl), Sunnyside, Utah (5×10^8 bbl), Whiterocks, Utah (2.5×10^8 bbl), and Edna, California (1.7×10^8 bbl) deposits (1). Of the major North American oil sand deposits, only the Athabasca oil sands in Alberta, Canada have been exploited commercially by surface mining techniques and at present are producing between 150,000 to 200,000 bbl of synthetic crude oil per day.

The oil sands differ significantly from conventional crude oil in that oil cannot be pumped by conventional primary recovery methods but the petroleum component is a bitumen similar to the heavy ends of conventional crude oils. Table 1 lists some of the chemical and physical characteristics of Athabasca oil sands. The Athabasca oil sands consist of loosely compacted sandstones of the Lower Cretaceous McMurray-Wabiskaw Formation with the sand grains coated with bitumen but with a water film between the mineral phase and the bitumen. Thus the sands are water wet in contrast to the Utah oil sands which are oil wet. In addition to the sand, water and bitumen there are variable amounts of clays and other accessory minerals which affect the bitumen content of the formation (see Table 1).

There are compositional similarities and differences between oil sand bitumens and crude oils and two major theories of origin of the oil sand bitumen have been proposed. Montgomery et al., (2) have proposed that the oils are young immature oils similar to other Lower Cretaceous oils and that alteration processes such as biodegradation and water washing have played only a minor role. On the other hand there is considerable evidence (3, 4, 5) that biodegradation and water washing have converted a conventional Lower Cretaceous oil to heavy oil by selective removal of light alkanes and aromatic hydrocarbons, thus increasing the relative contents of resins and asphaltenes. This process also accounts for the high sulfur content of the bitumen (see Table 1).

There have been very few studies of the trace element content of the Alberta oil sands and other associated bitumens. Scott et al., (6), measured Ni, V, and Fe in the raw McMurray oil sand, in the extracted bitumen and in other Cretaceous crude oils. They found little

TABLE 1: Chemical and Physical Properties of Athabasca Oil Sand

Property	Value
Light Mineral Components (%)	99
Major Occurrence	
Quartz, Chert, Chalcedony, Clays (Kaolinite, Illite)	
Minor Occurrence	
Feldspar, Dolomite, Calcite, Micaceous minerals	
Heavy Mineral Components (%)	1
Major Occurrence	
Tourmaline, Iron Minerals, Titaniferous, Staurolite, Epidote	
Minor Occurrence	
Corundum, Garnet, Kyanite, Apatite, Magnesite, Andalusite, Zoisite, Sillimanite, Pyroxenes, Amphiboles, Spinel, Rhodochrosite, Chalcopyrite, Bornite, Cassiterite	
Water Content (%)	0 - 10
Density (g cm^{-3})	1.86 - 2.36
Bitumen content (%)	5.0 - 20
Bitumen	
Specific Gravity (°API)	6.5 - 8.6 $\simeq$1.02 g cm^{3}
Elemental Analysis (wt %)	
Carbon	82.9 - 83.6
Hydrogen	10.3 - 10.6
Oxygen	0.2 - 1.3
Nitrogen	0.29 - 0.6
Sulfur	4.2 - 5.5
Hydrocarbon Types (%)	
Asphaltenes	18 - 25
Resins	29 - 35
Oils	45 - 49

variation in the V, Ni, and Fe contents of extracted bitumens from widely separated localities and from different depths provided surface exposures were not included. The authors also found that there was an almost linear relationship between the Ni and V contents for all Cretaceous oils, including the McMurray bitumens, thus providing evidence for a similar mode of origin for these oils. Boyd et al., (7), and Bowman (8), have analyzed oil sand solids for several elements and the latter author has also reported Ni and V contents in the bitumens. Hitchon, Filby and Shah (9) have reported data on 22 elements in 88 Alberta crude oils, including two McMurray Formation bitumens from the SYNCRUDE mining operation. Factor analysis of the W. Canada trace element data indicated that the concentrations of Ni, V, S, Na, Zn, Co, Mn, Se and As in the conventional W. Canada crudes are controlled by maturation processes rather than migration processes (9).

The presence of Ni, V, and other metals in crude oils is of geochemical interest (10, 11) and also is important in refining operations. Because heavy oils must be upgraded to produce gasoline, diesel fuel, etc., the nature of metal species in the bitumen may affect demetallation processes (12). In addition, the coking residues have been proposed as potential sources of some important elements, e.g., V, Ni which may be extracted by leaching processes (13).

The study reported here is a preliminary study of the trace element composition of a McMurray Formation oil sand and the bitumen extracted from the oil sand.

EXPERIMENTAL PROCEDURE

The raw oil sand was obtained from the Syncrude Canada Ltd. mining site (McMurray Formation). The bitumen was removed from the oil sand by exhaustive Soxhlet extraction of 20 - 40 g oil sand in double-walled Whatman ultrafine extraction thimbles with analytical grade toluene. The extraction was continued until the remaining sand in the extraction thimble was clean and free flowing. The bitumen was recovered from the toluene extract by evaporation of the solvent in a rotary evaporator.

The methanol-acetone (12:88) extract was obtained by Soxhlet extraction of 20 - 40 g oil sand with the solvent mixture. The solvent was removed by rotary evaporation.

Neutron activation analysis of the oil sand, the extracted bitumen, and the methanol-acetone extract was performed using a modification of the methods described by Filby et al., (14).

Samples of oil sand (0.2 - 0.5 g), bitumen (0.5 - 1.0 g) and methanol-acetone extract were sealed in 2/5 dram polyethylene vials. Standards used were National Bureau of Standards Coal (SRM 1632, 1635), Fly Ash (SRM 1633) and Orchard Leaves (SRM 1571) and U.S.G.S.

Standard Rock GSP-1. Samples and standards were irradiated in the Washington State University TRIGA-III fueled research reactor at an approximate thermal neutron flux of 6 x 10^{12} neutrons cm^{-2} sec^{-1}. Irradiation periods, decay times and count periods are shown in Table 2.

Gamma-ray spectroscopy was carried out with Ge(Li) detectors interfaced to a Nuclear Data ND-6620 analyzer and data were stored on 9-track magnetic tape. Gamma-ray spectra were processed using the FOURIER and SPANAL data reduction program on an Amdahl 470 computer. Corrections for decay during counting (for $T_{1/2}$ <10 min), overlapping γ-ray peaks (e.g., $^{203}Hg/^{75}Se$ at 280 keV), and flux variations between irradiation levels were made in the FOURIER and SPANAL programs.

TABLE 2: Irradiation and Measurement Conditions

Nuclide Measured	Irradiation Time	Decay Time	Count Time (seconds)
^{51}Ti, ^{28}Al, ^{27}Mg, ^{52}V	3 minutes	1 - 6 minutes	180
^{56}Mn, ^{38}Cl, ^{24}Na, ^{42}K	3 minutes	5 - 45 minutes	1000
^{24}Na, ^{42}K, ^{99}Mo, ^{198}Au, ^{152m}Eu, ^{64}Cu, ^{122}Sb, ^{72}Ga, ^{82}Br, ^{76}As, ^{140}La	8 hours	30 - 60 hours	4000
^{141}Ce, ^{60}Co, ^{51}Cr, ^{134}Cs, ^{154}Eu, ^{179m}Hf, ^{177}Lu, ^{50}Cu (for Ni), ^{86}Rb, ^{124}Sb, ^{46}Sc, ^{75}Se, ^{85}Sr, ^{182}Ta, ^{160}Tb, Th (daughter ^{233}Pa), ^{140}La (for U), ^{169}Yb, ^{65}Zn, ^{95}Zr	8 hours	21 - 30 days	80000

RESULTS AND DISCUSSION

Table 3 shows the trace element composition of McMurray Formation oil sand as determined by INAA. It should be noted that the composition reflects a typical sandstone containing minor amounts of clay minerals and other accessory minerals. Table 4 shows elemental data for 24 elements in the raw tar sand, the toluene extracted bitumen and the methanol-acetone extracts. The yields of these extracts were found to be 12.3 ± 0.3% (mean of 3 determinations) and 10.0 ± 0.7% (mean of 24 determinations respectively). Because the weight yields of the extraction procedures were measured by difference (wt of raw oil sand - wt extracted sand), the yields have not been corrected for the water content of the tar sand (in this particular sample, <5% by weight). In the case of the methanol-acetone extract the water phase was removed with the solvent and hence the yield includes this component. It should also be noted that organic acids and organic acid salts have been reported in the aqueous phase of the oil sand (6). Such compounds would be expected to be soluble in the methanol-acetone solvent but not appreciably soluble in the toluene.

TABLE 3: Trace Element Composition of Raw Tar Sand (Ft. McMurray-Syncrude)

Element	Concentration (ppm)	Element	Concentration (ppm)
Ti	2500 ± 385	V	40.3 ± 1.8
Ca	482 ± 150	Ni	14.4 ± 1.0
Mg	1380 ± 340	Fe	4190 ± 15
Al	15600 ± 300	Co	2.95 ± 0.01
Na	319 ± 2	Cr	15.1 ± 0.09
K	2230 ± 19	Mn	69.4 ± 1.5
Rb	20.3 ± 0.4	As	2.77 ± 0.20
Cs	0.651 ± 0.01	Se	0.676 ± 0.03
Sc	2.22 ± 0.003	Sb	0.168 ± 0.004
Tb	0.303 ± 0.009	Hg	<0.0005
Eu	0.453 ± 0.007	Br	0.418 ± 0.06
Ce	26.0 ± 0.05	Cl	23.1 ± 2.0
La	17.6 ± 0.21	Ga	2.93 ± 0.46
Sr	37.1 ± 2.0	Cu	0.350 ± 0.031
Ba	153 ± 3.0	Zn	11.6 ± 0.20

TABLE 4: Comparison of Extracted Bitumen and Porphyrin Extract with Tar Sand

Element	Concentration (ppm)		
	Raw Tar Sand	Toluene Extract (Bitumen)	MeOH/Acetone Extract
Weight Fraction	100%	12.3%	10.0%
V	40.3 ± 1.8	140 ± 5.9	75.9 ± 3.2
Ni	14.4 ± 1.0	30.1 ± 1.7	19.7 ± 1.1
Fe	4190 ± 15	127 ± 0.9	3.89 ± 0.19
Co	2.95 ± 0.014	0.481 ± 0.002	0.070 ± 0.001
Cr	15.1 ± 0.09	0.719 ± 0.007	0.042 ± 0.003
Mn	69.4 ± 1.5	1.40 ± 0.02	0.091 ± 0.01
As	2.77 ± 0.20	0.286 ± 0.067	<0.1
Se	0.676 ± 0.027	0.167 ± 0.005	0.186 ± 0.006
Sb	0.168 ± 0.004	0.013 ± 0.001	0.002 ± 0.001
Hg	<1	0.686 ± 0.04	0.938 ± 0.05
Br	0.418 ± 0.064	0.194 ± 0.026	0.478 ± 0.032
Cl	23.1 ± 2.0	7.53 ± 0.64	173 ± 0.64
Cu	0.350 ± 0.001	0.726 ± 0.247	0.029 ± 0.002
Zn	11.6 ± 0.2	0.615 ± 0.02	0.619 ± 0.01
Ti	2500 ± 385	50.5 ± 21	<50
Al	15600 ± 300	352 ± 6	11.3 ± 0.5
Na	319 ± 2	32.7 ± 0.3	150 ± 0.9
K	2230 ± 19	42.7 ± 1.4	<10
Ga	2.93 ± 0.46	0.286 ± 0.06	0.12 ± 0.03
La	17.6 ± 0.21	0.546 ± 0.02	<0.1
Sc	2.22 ± 0.003	0.066 ± 0.001	<0.001
Rb	20.3 ± 0.37	0.476 ± 0.02	<0.04
Cs	0.651 ± 0.01	0.022 ± 0.001	0.001 ± 0.0002
Eu	0.453 ± 0.007	0.011 ± 0.001	<0.001

*Data for Hg are in ppb

Table 5 shows the percentage of the element present in the raw oil sand that is extracted by the solvent extraction procedures. The toluene extraction removes 42.7% V and 25.7% Ni indicating that the extracted elements are present in the oil sand as porphyrin and non-porphyrin complexes as has been described previously (15). It is interesting to note that Cu is extracted to a similar degree as Ni. Inspection of the data for the methanol-acetone extract shows that Ni and V are appreciably extracted, presumably as porphyrin complexes plus some resin type non-porphyrin Ni and V species. Less than 1% of the Cu in the oil sand is extracted by the methanol-acetone indicating that any Cu species present in the bitumen are probably complexed into the asphaltene structure as has been observed by Filby (16) for conventional crude oils. The extraction of 74.9% of Cl, 11.4% Br, and 4.71% Na by methanol-acetone indicates that these fractions of the elements are present as simple salts in the aqueous phase of the oil sand.

The content of VO porphyrin and Ni porphyrins in Athabasca Oil sand bitumen has been reported as 500 ppm and 45 ppm respectively by Hodgson et al., (15) and as 411 ppm and 20.5 ppm respectively by HajIbrahim (17). Assuming a skeletal molecular weight of 476 (most abundant mass for many porphyrins), the porphyrin concentrations measured by HajIbrahim (17) correspond to 38.6 ppm V (as porphyrin) and 2.25 ppm Ni (as Ni porphyrin). Corresponding values for the concentrations given by Hodgson et al., (15) are 47.0 ppm V and 4.94 ppm Ni. It thus appears that the methanol-acetone extraction removes some non-porphyrin Ni and V, either as Ni and V present in the aqueous phase or as Ni and V bound in other organic complexes soluble in methanol-acetone.

Table 6 compares the toluene extracted bitumen with two laboratory prepared SYNCRUDE bitumens, with the Ni, V, and Fe values quoted for McMurray Formation bitumens by Scott et al., (6) and with the mean values obtained for W. Canada crude oils published by Hitchon et al., (9). In general the values are in agreement with other data, except perhaps for Ni and Fe. The Ni value in the bitumen extracted in this study is significantly lower than the other values quoted and the reason for this is not known. More recent work in this laboratory on toluene extracted bitumens has given Ni contents ranging between 68.1 to 79.4 ppm Ni, which is similar to previously published data. The value for Fe of 127 ppm is higher than that found by Scott et al., (6) but these authors found that Fe contents of the bitumen were very variable if the oil sand had been exposed to surface oxidation or percolating ground waters. Iron contents of up to 1500 ppm were found.

Further work is in progress on the trace element characterization of the McMurray oil sand bitumen to determine the abundances of Ni and V porphyrins and the distributions of metals among asphaltenes and maltenes and chromatographic fractions.

TABLE 5: Percentages of Elements Extracted from Raw Oil Sand by Solvent Extraction

Element	Percent Removed by Toluene	Percent Removed by Methanol-Acetone
V	42.7	18.8
Ni	25.7	13.7
Fe	0.370	0.009
Co	2.01	0.24
Cr	0.59	0.03
Mn	0.25	0.01
As	1.27	<0.5
Se	3.04	2.75
Sb	0.95	0.12
Br	5.71	11.4
Cl	4.01	74.9
Na	1.26	4.71
Cu	25.5	0.83
Zn	0.65	0.53
Ti	0.25	<0.2
Al	0.28	0.01
K	0.24	<0.05
Ga	1.20	0.41
La	0.38	<0.06
Sc	0.37	<0.01
Rb	0.29	<0.02
Cs	0.42	0.025
Eu	0.30	<0.02

TABLE 6: Comparison of Athabasca Bitumen Trace Element Concentrations with W. Canada Crude Oil Averages

Element	Concentration (ppm)				
	SYNCRUDE Bitumen (9)	SYNCRUDE Bitumen (9)	Mean Value (6)	Toluene Lab Extract	W. Canada Oils Mean (9)
V	n.d.	177	181	140	13.6
Ni	74.1	71.9	71.2	30.1	9.38
Fe	142	254	63.6	127	10.8
Co	1.35	2.00	n.d.	0.481	0.0537
Cr	1.01	1.68	n.d.	0.719	0.0933
Mn	n.d.	3.85	n.d.	1.40	0.100
Zn	n.d.	n.d.	n.d.	0.615	0.459
Cu	n.d.	n.d.	n.d.	0.725	n.d.
As	0.400	0.321	n.d.	0.286	0.111
Se	0.286	0.517	n.d.	0.167	0.052
Sb	0.031	0.028	n.d.	0.013	0.0062
Hg*	n.d.	n.d.	n.d.	0.686	50.9
Br	0.155	0.104	n.d.	0.194	0.491
Cl	n.d.	8.00	n.d.	7.53	39.3
Ti	n.d.	n.d.	n.d.	50.5	n.d.
Na	40.3	20.9	n.d.	32.7	3.62

n.d. = not determined

*Values for Hg are in ppb

REFERENCES

1. P. H. Phizackerley and L. O. Scott, "Major Tar-Sand Deposits of the World", in *Bitumens, Asphalts and Tar Sands*, G. V. Chilingarian and T. F. Yen, Editors. Elsevier Sci. Publ. Co. New York, pp. 57-92 (1978).
2. D. S. Montgomery, D. M. Clagston, A. E. George, G. T. Smiley and H. Sawatzky. Can. Soc. Petroleum Geol., Mem. 3, 168 (1974).
3. C. R. Evans, M. A. Rogers and N. J. L. Bailey, Evolution and Alteraction of Petroleum in Western Canada. Chem. Geol. $\underline{8}$, 147-170 (1971).
4. N. J. L. Bailey, A. M. Jobson and M. A. Rogers, Bacterial Degradation of Crude Oil: Comparison of Field and Experimental Data, Chem. Geol. $\underline{11}$, 203-221 (1973).
5. Rubinstein and O. P. Strausz, "The Biodegradation of Crude Oils: The Origin of the Alberta Oil Sands" in *The Alberta Oil Sands: Energy for the Future*, Strausz, ed. Verlag Chemie Int. Inc., 177-189 (1978).
6. J. Scott, G. A. Collins and Hodgson, G. W., Trace Metals in the McMurray and Other Cretaceous Reservoirs of Alberta, Oil in Canada $\underline{6}$, 34-50 (1954).
7. M. L. Boyd, and D. S. Montgomery, Mines Branch Research Report R. 78, Ohawa, Dept. 8, Mines and Technical Surveys (1961).
8. C. W. Bowman, Molecular and Interfacial Properties of Athabasca Tar Sands, No. 13 in Proceedings of the 7th World Petroleum Congress, Vol. 3, Elsevier Pub. Co., pp. 583-603 (1963).
9. B. Hitchon, R. H. Filby and K. R. Shah. Preprints ACS Div. Petro. Chem. $\underline{18}$, 623 (1973).
10. G. W. Hodgson and B. L. Baker, Vanadium, Nickel Porphyrin in the Thermal Geochemistry of Petroleum, Bull. Am. Assoc. Petroleum Geol. $\underline{41}$, 2413-26 (1957).
11. G. W. Hodgson, B. Baker and E. Land Peake, Geochemistry of Porphyrins in *Fundamental Aspects of Petroleum Geochemistry*, B. Nagy and U. Colombo, Editors. Elsevier Publ., New York, 117-259 (1967).
12. A. J. G. Barwise and E. V. Whitehead, Characterization of Vanadium Porphyrins in Petroleum Residues, Preprints ACS Div. Petro. Chem. 25(2), 268-279 (1980).
13. T. R. Jack, E. A. Sullivan and J. E. Zajic, Leaching of Vanadium and Other Metals from Athabasca Oil Sands Coke and Coke Ash, Fuel 58, 589-94 (1979).
14. R. H. Filby, W. A. Haller and K. R. Shah. J. Radioanal. Chem. $\underline{5}$, 277-290 (1970).
15. G. W. Hodgson, E. Peake and B. L. Baker, The Origin of Petroleum Porphyrins: The Position of Athabasca Oil Sands, M. A. Carrigy (ed), Research Council of Alberta, Edmonton, 75-100 (1963).
16. R. H. Filby. Preprints ACS Div. Petroleum Chem. $\underline{18}$, 630 (1973).
17. S. K. HajIbrahim, Development of HPLC for Fractionation and Fingerprinting of Petroporphyrins. Proc. 13th Int. Symposium Chromatography, Cannes, France, June 30-July 4 (1980).

URANIUM CONTENT OF PETROLEUM BY FISSION TRACK TECHNIQUE

A.S. Paschoa*, O.Y. Mafra**, C.A.N. Oliveira** and
L.R. Pinto**

*Pontifícia Universidade Católica do Rio de Janeiro
Depto. de Física, C.P. 38071, Z.C. 19
Rio de Janeiro, RJ, 22453, Brasil

**Instituto Militar de Engenharia
Seção de Energia Nuclear
Praia Vermelha, Rio de Janeiro, RJ, Brasil

ABSTRACT

The feasibility of the fission track registration technique to investigate the natural uranium concentration in petroleum is examined. The application of this technique to petroleum is briefly described and discussed critically. The results obtained thus far indicate uranium concentrations in samples of brazilian petroleum which are over the detection limit of the fission track technique.

INTRODUCTION

The main objective of this work is to apply the fission track registration technique to investigate the natural uranium concentration in samples of brazilian petroleum. Thus it is possible to provide the basic technical information to establish the feasibility of using the fission track registration technique for a potential future uranium impact assessment in the exploration, refining, combustion and other uses of petroleum and its derivatives.

The fission track registration technique has already been used to determine the uranium concentration in several media, as for example, rocks[1], biological materials[2], sediments[3], sea water[4], and even petroleum[5]. The uranium content of some oil field brines was reported to be about 200 ppm[6] well over the detection limits of the fission track technique (∿2ppb).

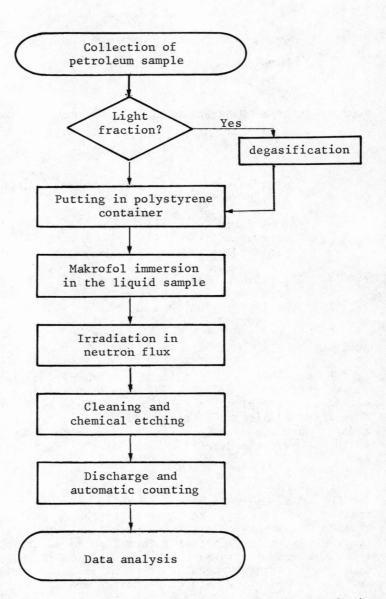

Fig. 1. Flowchart of the steps to be undertaken to apply the fission track registration technique to determine uranium concentration of petroleum.

More recent reports[7-9] indicate uranium concentrations in crude oils and petroleum associated materials, like retort water[9], within levels which are easily detectable by fission track registration technique.

In the present work petroleum samples in natural and solution forms were irradiated by thermal neutrons and subsequently the fission tracks were registered and counted.

THE FISSION TRACK REGISTRATION TECHNIQUE

The flowchart of the main steps to determine the uranium concentration in a petroleum sample by applying the fission track technique is shown in Fig. 1. The flowchart of Fig. 1 should be regarded as a simplification of the following subsequent procedures: (i) sample collection; (ii) sample preparation; (iii) sample irradiation; (iv) etching the detectors; and (v) track counting.

Sample Collection

The petroleum samples used in the present work were provided by Petrobrás S.A. from stored petroleum extracted from oil fields located in Ceará and Campos. The samples were not acidified upon collection, so it is likely that ion exchange with the internal surfaces of the storage container might have caused a reduction in the uranium concentration in the petroleum samples along the storage time. Quantitative results of uranium concentrations in petroleum by fission track registration technique may be more accurate when it is possible to acidify the petroleum samples immediately after collection, and to reduce the time between collection and analysis.

Sample Preparation

The petroleum samples can be irradiated by a neutron flux without any requirements other than to immerse the Makrofol polycarbonate KG, 10μm thickness detector film in the liquid sample. However, when a light fraction is present in the sample, degasification becomes necessary before irradiation. An alternative way to treat the petroleum samples is to prepare a solution for neutron irradiation, according to the following steps:
1. an aliquot of 45g is taken from the petroleum sample and dried in a sand bath;
2. the dried sample is put in an oven for about 10 hours with the temperature being slowly raised to 800°C;
3. once constant weight is attained, a mixture of nitric acid (HNO_3) plus hydrochloric acid (HCl)1:3 is added to dissolve the ashes;
4. the solution is then transferred to a teflon beaker and attacked chemically with fluoridric acid (HF);

5. after drying again, the sample is dissolved in HNO_3 (1:10), and the solution obtained is ready for irradiation together with the Makrofol detector mounted in a lucite frame.

Sample Irradiation

The neutron sources used to cause fission in uranium atoms present in the irradiated petroleum samples were the Argonauta Research Reactor at the Instituto de Engenharia Nuclear (IEN), Rio de Janeiro, with a thermal flux of 1.2×10^9 n/cm^2 sec, and the IEA-R1 reactor at the Instituto de Pesquisas Energéticas e Nucleares (IPEN), São Paulo, with a neutron flux of 7.2×10^{12} n/cm^2 sec. The IEN Research Reactor is used for samples containing uranium concentrations higher than 1 ppm, while the IPEN Reactor has to be used for samples containing uranium concentrations at the ppb level.

Polystyrene cylindrical containers with 50 ml volume capacity were used to hold the petroleum samples and the Makrofol detectors inside the reactor during irradiation. This arrangement was used in order to have a large contact area between the detector and the sample, and to achieve a geometry of easy reproduction[10].

When the Makrofol detector is exposed to a solution containing uranium and subsequently to a neutron flux, fission of uranium atoms will occur. The number of fissions can then be registered in the detector as fission tracks.

The fission track density can be related to the uranium concentration in the solution through the following equation:

$$DT = K \frac{N_0 I_A}{A} C \sigma_f^{(A)} \phi t \qquad (1)$$

where: DT is the fission track density (cm^{-2}); N_0 is the Avogrado's number; A is the atomic mass of ^{235}U(g); I_A is the percent natural isotopic abundance of ^{235}U (0.72%); $\sigma_f^{(A)}$ is the microscopic cross-section for fission of ^{235}U for thermal neutrons (cm^2); ϕ is the flux of thermal neutrons (n.cm^{-2}.sec^{-1}); t is the irradiation time (sec); C is the uranium concentration in the solution (g/cm^3); and K is a coefficient which expresses the total efficiency for detection (cm).

The irradiations were performed by using fluxes with predominance of thermal neutrons, which have a fission cross-section for ^{235}U equal to 579 barns. However, since the neutron fluxes were not composed exclusively of thermal neutrons, fissions from fast neutrons should also be taken into account for ^{238}U and ^{232}Th.

The ^{238}U contribution to the total number of fissions is small because ^{238}U is fissionable only for neutrons with energy higher than 1.4 MeV and even so with a fission cross-section of about 0.56 barns. The ^{232}Th microscopic fission cross-section for neutrons with energy higher than 1.2 MeV is 0.078 barns. Therefore, unless great amounts of thorium are present in the sample, as in the case for some thorium rich minerals, the ^{232}Th contribution to the number of fission tracks produced can be neglected for convenient irradiation conditions.

Each petroleum sample was irradiated simultaneously with a known standard uranium solution made of uranyl nitrate, $UO_2(NO_3)_2 6H_2O$. These standard solutions were used for calibration. Fig. 2 shows a calibration curve for an irradiation made with a thermal neutron flux of 1.2×10^9 n.cm^{-2}.sec^{-1} by using petroleum contaminated with known amounts of uranium in the order of ppm. This was done to check the experimental procedures against the possibility of the existence of neutron poisons, like boron and vanadium, in petroleum at levels high enough to interfere with the uranium determination by fission track technique.

Etching the detectors

Petroleum remains attached to the Makrofol film detector. As a consequence, it becomes necessary to clean the detector with a "petroleum-ether" solution, which is a mixture of n-hexane - $CH_3(CH_2)_4CH_3$ - plus n-heptane - $CH_3(CH_2)_5CH_3$ - before attacking the detectors with chemical reagents. The detectors were etched by a 24% solution of potassium hydroxide (KOH) at 60°C for 12 minutes[10]. The main parameters to be controlled when etching the detectors are time, temperature and the chemical properties of the processing reagents[10].

Track counting

An automatic discharge chamber was used for counting the fission tracks according to the technique developed by Cross and Tommasino[11]. The working voltage was 600 volts applied across an individual counting area of 2.54 cm^2. Each detector provided six exposed areas for counting in the automatic discharge chamber.

RESULTS AND DISCUSSION

Table 1 presents the minimum uranium concentrations of the petroleum samples analysed so far.

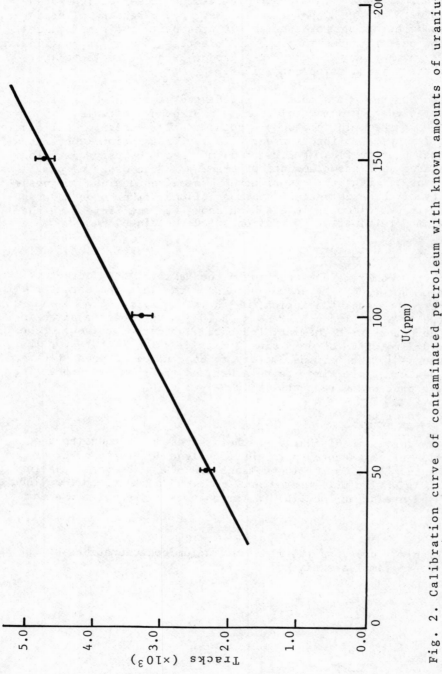

Fig. 2. Calibration curve of contaminated petroleum with known amounts of uranium.

Table 1. Minimum uranium concentrations in two samples of brazilian petroleum.

Origin	Characteristics	U(ppb)
Campos	light	23±1*
Ceará	heavy	42±2*

* Statistical error only.

These rather limited results indicate that the fission track registration technique can be successfully applied to the determination of the uranium content of petroleum.

The petroleum sample from the Lagoa Feia formation in Campos had a light fraction and needed to undergo a lengthy degasification procedure, indicated in Fig. 1, before being irradiated. This was done to avoid significant gas expansion during irradiation which might cause the collapse of the polystyrene cylindrical containers inside the reactor.

A calibration curve obtained by irradiating uranium standards made of uranyl nitrate was obtained for each irradiation to eliminate the effects of any uncontrolled variations in the neutron flux, ϕ, and the irradiation time, t, used in equation (1) to calculate the fission track density, D_T. This procedure is advantageous in that it enables one to irradiate the samples without careful monitoring of the neutron flux and the irradiation time.

The results presented in Table 1 appear as minimum values because the petroleum samples were not acidified immediately after collection and the time elapsed between collection and analysis was not known. However, for the purposes of this work that was not very important, since the minimum value determined was about ten times higher than the lower limit of detection for the technique. Therefore, the fission track registration technique can be used to determine uranium concentrations in petroleum down to levels of about 10 ppb (i.e., > 4 times the lower detection limit).

Although the feasibility of using the fission track registration technique to determine uranium concentrations in petroleum samples has been demonstrated, further research is still needed to apply routinely the technique for this purpose.

ACKNOWLEDGEMENTS

The authors wish to express their gratitude to Petrobrás S.A. Centro de Pesquisas e Desenvolvimento (CENPES) for providing the petroleum samples, and to Comissão Nacional de Energia Nuclear (CNEN) and Instituto de Pesquisas Energéticas e Nucleares (IPEN) for making available the reactors IEN-Argonauta and IEA-R1, respectively.

REFERENCES

1. J. D. Kleeman and J.F. Lovering, Uranium distribution in rocks by fission track registration in lexan plastic, Science, 156:512 (1967).
2. B. S. Carpenter and C.H. Cheek, Trace determination of uranium in biological material by fission track counting, Anal. Chem., 42:121 (1970).
3. K. K. Bertine, L.H. Cham, and K.K. Turekian, Uranium determination in deep sea sediments and natural waters using fission tracks, Geochim. Cosmochim. Acta, 34:641 (1970).
4. T. Hashimoto, Determination of the uranium content in sea water by a fission track method with condensed aqueous solution, Anal. Chim. Acta, 56:347 (1971).
5. I. G. Berzina, D.P. Popenko, and Yu. S. Shimelevich, Determination of trace amounts of uranium in petroleums from fission fragment tracks, Geokhimiya, 8:1024 (1969).
6. A. P. Pierce, J.W. Mytton, and G.B. Gott, Radioactive elements and their daughter products in the Texas Panhandle and other oil and gas fields in the United States, in "Proc. 1st. U.N. Conf. Peaceful Uses Atomic Energy", Geneva, Pergamon Press, London, p. 527 (1955).
7. H. J. Hyden, Distribution of uranium and other metals in crude oils, U.S. Geol. Survey Bull., 1100-B:17 (1961).
8. U. S. Atomic Energy Commission, Radon emanometry of starks salt dome, Calcasieu Parish, Louisiana, Division of Production Materials Management, RME-4114:53 (1972).
9. P. T. Dickman, M. Purdy, J.E. Doerges, V.A. Ryan, and R.E. Poulson, Analysis of oil shale and oil shale products for certain minor elements, in "Nuclear Methods in Environmental and Energy Research", CONF. 771072, p. 394 (1977).
10. L. P. Geraldo, O.Y. Mafra, and E.M. Tanaka, Determination of uranium concentration in water samples by fission track registration technique, J. Radioanal. Chem., 49:115 (1979).
11. H. G. Cross and L. Tommasino, A rapid reading technique for nuclear particle damage tracks in thin foils, Rad. Effects, 5:86 (1970).

FAST NEUTRON ACTIVATION ANALYSIS OF FOSSIL

FUELS AND LIQUEFACTION PRODUCTS

William D. Ehmann, David W. Koppenaal and Samir R. Khalil

Department of Chemistry and Institute for Mining and
Minerals Research, University of Kentucky
Lexington, Kentucky 40506

ABSTRACT

Fast neutron activation analysis has been applied to the direct, instrumental determination of oxygen and nitrogen in solid fossil fuels, their liquefaction products and various liquid and solid by-products. The matrices and the contents of oxygen and nitrogen of these materials are highly variable. Hence, no single set of comparator standards is suitable for the analysis of all samples. We have considered problems associated with neutron absorption/thermalization, gamma-ray self-absorption, and variable irradiation and counting geometries associated with the compositions, densities and physical states of the samples and standards. Two sets of liquid organic reagent primary standards and several solid standards have been selected and evaluated for use in the determination of oxygen and nitrogen in coals, coal conversion liquids, and residual solids. In the selection process we have also considered factors such as availability of the standard in high purity, well-defined stoichiometry, non-hygroscopic nature and simple elemental composition. Analyses of a number of coals, conversion products and NBS reference standards are presented. Problems associated with selecting a reproducible pre-analysis drying procedure for oxygen determinations in coal are discussed. It is suggested that a brief freeze-drying procedure may result in minimal matrix alteration and yield reproducible values for bulk oxygen contents of coals.

INTRODUCTION

Fast neutron activation analysis is a method uniquely suited for the direct, sensitive, accurate and purely instrumental determination of oxygen and nitrogen in fossil fuels and their liquefaction products. In coal, organic oxygen-containing sites are believed to be active sites in the conversion process. In liquefaction products, low oxygen contents may gauge the efficiency of the conversion process. The presence of nitrogen in conversion feed-stocks or intermediates may affect catalytic processes and the release of nitrogen compounds in the production or utilization of liquefaction products may have adverse environmental effects. Oxygen in coal has conventionally been determined indirectly by difference or by a pyrolytic method employing radiofrequency heating.[1] Nitrogen is normally determined by the tedious and time consuming Kjeldahl or Dumas techniques.

Instrumental 14 MeV neutron activation analysis (INAA) has been used to determine oxygen and nitrogen in a wide variety of matrices.[2-13] 14 MeV INAA applications to fossil fuel conversion studies are complicated by the unavailability of suitable certified comparator standards which have matrices similar to the wide range of sample compositions encountered. Several of the variables encountered include neutron absorption and thermalization, gamma-ray self absorption, and problems associated with maintenance of constant sample and standard irradiation and counting geometries. No single set of standards for oxygen and nitrogen are suitable for the full range of solid feedstocks, conversion intermediates, fuel products and residual wastes encountered in the conversion process.

We have selected two sets of organic reagent liquids and evaluated them for use as comparator standards in the INAA determination of oxygen and nitrogen in coal conversion liquids, shale oils and petroleum-based fuel oils.[14-16] We have also selected solid standards suitable for the determination of these elements in coal and fly ash.

Using these selected comparator standards we have analyzed NBS coal and fly ash standards (SRM's 1632, 1633, 1632a, 1633a and 1635) for oxygen using INAA. These standards have been certified for a wide variety of minor and trace elements, but little data exist for major abundance elements, such as oxygen and nitrogen. Recent publications from the group at the University of Maryland[17,18] present data for many elements including carbon, hydrogen and nitrogen in SRM's 1632a and 1635, but no oxygen data are reported. Nadkarni[19] reports oxygen data for several of the NBS standards, but does not describe in detail the pre-analysis sample handling procedures used in his study.

We feel standardized pre-analysis handling procedures are especially critical for determination of oxygen in solid fossil fuel samples. This view has also been stated by Volborth.[13]

The problem is due to the instability of the coal matrix when exposed to atmospheric conditions (in handling or in storage) and to volatile loss and oxidation in oven-drying procedures which may precede the analyses. We have investigated the effect of various drying procedures on the experimentally determined oxygen content of coals. The results of these studies suggest that a complex interplay of effects can result in biased oxygen data when conventional oven-drying procedures in air, in vacuum or in a nitrogen atmosphere are used prior to analysis. We suggest that the freeze-drying technique may introduce the least matrix alteration, yields constant weight samples in a minimal drying time, and is a reproducible drying method on which to base oxygen contents of coal samples.

EXPERIMENTAL

The analysis of oxygen by instrumental 14 MeV neutron activation has been performed routinely in our laboratory for many years and is described in detail in previous publications.[9-12] The technique is based on the $^{16}O(n,p)^{16}N$ nuclear reaction induced by exposure of the samples to 14 MeV neutrons produced by the $^{3}H(d,n)^{4}He$ reaction in a Cockcroft-Walton generator. The product radionuclide ^{16}N has a half-life of 7.1 seconds and emits highly energetic 6.13 and 7.1 MeV gamma rays which are detected using paired 7.6 cm x 7.6 cm NaI(Tl) scintillation detectors mounted at 180° with respect to a uniaxially spinning sample holder. Each detector is connected through a linear amplifier to a sum/invert amplifier and the output is fed to a single channel analyzer gated to the energy region between 4.5 and 8.0 MeV. Neutron flux variations are continuously monitored every 0.8 seconds during the irradiation with a BF_3 neutron proportional counter. Both the output of the BF_3 counter and the single channel analyzer are gated to a ND 2200 multichannel analyzer operating in the multi-scaler mode. The recorded time spectrum from the analyzer is computer processed to yield the oxygen data. The only potentially serious interfering elements are boron and fluorine. Due to the low abundance levels of these elements in coals and conversion products, interferences are negligible.

Our procedure for the determination of nitrogen by the 14 MeV INAA technique is similar to that described by James et al.[10] The technique is based on the $^{14}N(n,2n)^{13}N$ nuclear reaction induced by exposing the sample for 5 minutes to 14 MeV neutrons. The product radionuclide ^{13}N has a half-life of 9.96 minutes and decays via positron emission. The nitrogen content is determined by coincidence counting the 0.511 MeV positron annihilation photons using two solid 7.6 cm x 7.6 cm NaI(Tl) detectors. To discriminate against the potential interference from the annihilation radiation of ^{30}P produced by $^{31}P(n,2n)^{30}P$ reaction, a 12 minute decay time was used before the 20 minute counting period. Other significant interferences due to "knock on" recoil proton reactions, $^{12}C(p,\gamma)^{13}N$, $^{13}C(p,n)^{13}N$ and $^{16}O(p,\alpha)^{13}N$, were evaluated experimentally for our system by using appropriate standards.

All samples and standards are packed in cleaned 2-dram virgin polyethylene vials under an inert atmosphere of helium immediately prior to irradiation.

RESULTS AND DISCUSSION

In our previous work, NBS SRM 136b potassium dichromate and USGS Standard Rock BCR-1 have been used as primary and/or reference standards for oxygen determinations in silicate rocks and other geological materials.[12] For nitrogen determinations, N-1 naphthylacetamide has been considered a satisfactory standard for nitrogen determinations in a coal matrix because it has a bulk elemental composition very similar to that of bituminous coals.[10] In our recent work, the above oxygen standards have been found to be less satisfactory for the analysis of low to moderate ash content coals[20] or coal-derived liquids.[14-16] Similarly, the solid nitrogen standard was unsatisfactory for analysis of liquid conversion products. For 14 MeV INAA, samples and standards should be very similar with respect to macroscopic density, physical state, mean atomic number and content of low mass neutron thermalizing nuclei in order to minimize problems associated with incident neutron absorption/thermalization, gamma-ray self-absorption and variable irradiation and counting geometries. When dealing with short lived radionuclides it is also desirable to match the standard concentrations to the expected sample concentrations for the elements determined in order to avoid gross differences in activity levels and the associated instrumental deadtimes. For these reasons, the proper selection of standard sets is very important in obtaining accurate values for oxygen or nitrogen in coals and coal conversion liquids.

Conversion Liquids

In our recent studies two sets of organic reagent primary standards were selected for oxygen and nitrogen determinations in the coal conversion liquid matrix.[14,15] The standard liquids were selected using criteria of high boiling point, well-defined stoichiometry, high purity, nonhygroscopic nature and simple C-H-O or C-H-N elemental composition. These two sets of standards are listed in Tables 1 and 2 along with their molecular formulae and stoichiometric O and N contents. Mixtures of standards or dilutions with n-hexane were used to fill gaps, or to extend the range to lower levels.

For each organic liquid selected, NBS Standard Reference Material 136b, $K_2Cr_2O_7$ and N-1 naphthylacetamide were first used

Table 1. Standards for INAA Oxygen Determinations in Coal Conversion Liquids.[14]

Samples	Molecular Formula	% Oxygen by Stoichiometry
Methyl alcohol	CH_4O	49.49
1,2,5-Trihydroxyhexane	$C_6H_{14}O_3$	35.77
1,5-Pentanediol	$C_5H_{12}O_2$	31.30
2-(2-n-Butoxyethoxy) ethanol	$C_8H_{18}O_3$	29.59
2-Propanol	C_3H_8O	26.68
1-Butanol	$C_4H_{10}O$	21.62
2-Propanol and 1-Octanol mixture (1:1)	--	19.87
4-Methyl-2-pentanol	$C_6H_{14}O$	16.39
1-Octanol	$C_8H_{18}O$	13.05
Benzyl ether	$C_{14}H_{14}O$	8.88
Benzyl ether and n-hexane mixture (2:5)	--	3.55

as preliminary comparator standards of oxygen and nitrogen, respectively. In case of nitrogen determinations, important interferences are due to the proton recoil reactions, $^{12}C(p,\gamma)$, $^{13}C(p,n)$ and $^{16}O(p,\alpha)$, which produce the same desired indicator radionuclide product, ^{13}N. The recoil protons are produced by 14 MeV neutron collisions with protons in the polyethylene vials and in the hydrogen-rich samples themselves. Correction factors have been determined for these interferences by irradiating and counting pure hydrocarbons (n-hexane and naphthalene), deionized water and high purity quartz samples under the same conditions used for the nitrogen analyses. The experimentally determined factors in our system are 1.48×10^{-4} apparent g N/g C for carbon and 1.30×10^{-3} apparent g N/g O for oxygen. It should be noted that these correction factors are unique to our packaging and irradiation configurations and must be experimentally evaluated in each facility. No correction is required in the oxygen analyses for the potential fluorine and boron interferences, due to their very low levels in coal or coal conversion liquids. Triple coincidence summing of lower energy gamma rays (e.g., 1.8 MeV gamma rays of ^{56}Mn) has not been found to be a problem in the oxygen analyses of coal or coal

Table 2. Standards for INAA Nitrogen Determinations in Coal Conversion Liquids.[15]

Samples	Molecular Formula	% Nitrogen by Stoichiometry
Dimethylcyanamide	$C_3H_6N_2$	39.96
1,2-propanediamine	$C_3H_{10}N_2$	36.66
Dimethylcyanamide and aniline mixture (3:2)	--	29.98
Dimethylcyanamide and quinoline mixture (2:3)	--	22.23
Pyridine	C_5H_5N	17.53
Aniline	C_6H_7N	15.02
N-heptylamine	$C_7H_{17}N$	12.04
Di-n-butylamine	$C_8H_{19}N$	10.81
Quinoline	C_9H_7N	10.41
Di-n-octylamine	$C_{16}H_{35}N$	5.63
Quinoline and n-hexane mixture (2:9)	--	1.89

products with our dual detector multiscaler system.

After the recoil proton interference and blank corrections were made, the apparent nitrogen and oxygen contents of the organic liquids were calculated, based on the two solid standards. Linear relationships between "apparent" and stoichiometric abundances were obtained. However, systematic deviations between the experimental values and the stoichiometric contents of the organic liquids were observed, due to the matrix dissimilarity of the standards and samples. A single geometry/self-absorption correction factor was computed for each set and applied to the raw data to permit the final least squares fitted line for the experimental content vs. stoichiometric content plot to pass through the origin. Excellent agreement between the adjusted experimental values and the known stoichiometric nitrogen and oxygen contents was then obtained. The correlation coefficient is 0.9995 for the ten different compounds and mixtures covering a range of oxygen content from 3.55 to 49.94%.[14] For eleven organic liquids covering a range of nitrogen contents from 1.89 to 39.96%, a correlation coefficient of 0.9999 was obtained.[15]

Examples of the application of our techniques for the determination of oxygen and nitrogen in typical coal conversion liquids, shale oils and various standards are presented in Table 3. A precision of approximately ±1.5%, relative, for the lower oxygen con-

Table 3. Oxygen and Nitrogen Concentrations in Liquid Fossil Fuels and Conversion Products.

Sample Description	Oxygen, % ± σ_m	Nitrogen, % ± σ_m
Atmospheric still overhead[a]	1.42 ± 0.04	0.699 ± 0.008
Atmospheric still bottom[a]	1.70 ± 0.04	0.626 ± 0.023
Vacuum still overhead[a]	1.50 ± 0.04	0.879 ± 0.016
Vacuum still bottom[a]	9.48 ± 0.13	2.26 ± 0.08
Shale oil retort liquid sample[b]	1.46 ± 0.02	2.00 ± 0.03
Coal liquid product[c]	5.16 ± 0.03	---
Filtered liquid product[c]	3.88 ± 0.03	---
Liquid filtrate[c]	28.89 ± 0.39	---
NBS SRM 912 urea	---	46.28 ± 0.67 (46.64)
NBS SRM 148 nicotinic acid	---	11.37 ± 0.08 (11.38)
JPI 75N9502, liquid[d]	---	0.039 ± 0.007 (0.052)
JPI 75N1002, liquid[d]	---	0.111 ± 0.005 (0.103)
JPI 75N3002, liquid[d]	---	0.286 ± 0.015 (0.303)
Petroleum residuum[22]	---	0.211 ± 0.012 (0.195)

( ) Certified or compilation values.

[a] Samples obtained from H-Coal process development unit, Hydrocarbon Research, Inc., Trenton, NJ.

[b] Samples obtained from Lawrence Livermore Laboratory Retort.

[c] Samples obtained from Solvent Refined Coal Pilot Plant, Wilsonville, Alabama.

[d] Japan Petroleum Institute residual fuel oil standards.[21]

tents was obtained. The precision varied in the range of ±1.1 to 3.7%, relative, for nitrogen determinations in the conversion liquids and shale oil. Except for the lowest nitrogen content standard oil sample, the differences between our INAA results and the compilation values by the Kjeldahl method for the Japanese oil[21] and Exxon petroleum residuum[22] standards do not exceed 7.8% and average ±5.5%, relative.

Coals and Solid By-Products

For oxygen determinations in coal, the proper selection of calibration standards is very important in order to minimize the matrix problems discussed previously. Semi-empirical macroscopic density, atomic number and thermalization factors for a number of materials are listed in Table 4. For the listed values of coal, ash contents of 6.33 to 20.7% were assumed and an oxygen value corresponding to the organic oxygen content plus 47% of the ash content was taken.[9] The values in Table 4 indicate that reagent grade benzoic acid is an acceptable standard material for oxygen determinations in coals with low to moderate ash contents. Benzoic acid was also one of the materials selected by Volborth[13] as an oxygen standard for coals. In the case of silicate-rich materials such as fly ash, NBS SRM 136b, $K_2Cr_2O_7$ and USGS Standard Rock BCR-1 are still considered the most suitable oxygen standards.[20] For extremely high ash content coals, interpolation between values based on both benzoic acid and SRM 136b may be the best approach.

For nitrogen determinations in coal, N-1-naphthylacetamide pressed firmly with a die to yield a macroscopic density factor of 0.92 or NBS SRM 148, nicotinic acid, are acceptable primary standards. We also use NBS SRM 912, urea, as a secondary reference standard. The precision of our replicate nitrogen analyses on solid NBS SRM's 912 urea and 148 nicotinic acid is better than ±1.5%, relative, using N-1-naphthylacetamide as a primary standard. Agreement with certified levels is excellent.

Sample Handling Procedures

Pre-analysis handling techniques are important in the determination of oxygen in coals which may contain varying amounts of moisture and differ in their hygroscopic characteristics. Thus, it is essential that any oxygen data for coal be reported on a standardized basis. A simple "moisture-free" basis is the obvious solution for reporting reproducible oxygen concentrations, but obtaining "moisture-free" samples of coal may be a very difficult task. Conventional oven-drying at 105° to 110°C in air, vacuum, or under an inert gas atmosphere are methods that have been applied to coal samples. Freeze-drying procedures have also been used. In our laboratory, we applied all of these drying techniques to a set of coals that had been selected from a set of samples obtained from

Table 4. Semi-Empirical Comparison Factors for Various Standard and Sample Matrices.[20]

Sample ID	Macroscopic Density Factor[a]	Atomic Number Factor[b]	Thermalization Factor[c]
BCR-1, USGS Standard	1.49	2.97	1.82
$K_2Cr_2O_7$, NBS SRM 136b	1.67	2.91	1.43
Plastic rod, methyl methacrylate	1.45	3.25	9.02
Urea, NBS SRM 912	0.865	3.21	8.02
N-1-naphthylacetamide	0.62-0.92	3.19	8.13
Nicotinic acid, SRM 148	0.844	3.13	6.85
Benzoic acid	0.957	3.14	7.37
Coal (low to moderate ash content)	0.983	3.11 to 3.18	6.27 to 7.40

[a] The amount, in grams, of the as received material, that can be packed firmly into our standard 2/5 dram irradiation rabbit. This factor is related to the neutron and gamma-ray self-absorption effects for the material.

[b] The number of atoms of each of the elements above $\sim 1\%$ in one gram of the material, each times the atomic number of the element, summed for all the elements and divided by 10^{23}. This factor relates to the gamma-ray self-absorption effect for the material.

[c] The number of atoms of each element having $A \leq 20$ in one gram of the material, summed for each of these elements and divided by 10^{24}. This factor reflects the 14 MeV neutron thermalization effect for the material.

the Pennsylvania State University Coal Data Bank. The selected coal set (Table 5) has a wide range of moisture and ash contents representing ranks from anthracite to lignite. Tables 6 and 7 present oxygen data for "as received" samples and dried samples under differing drying conditions for the six selected coals.

The data in Table 6 indicate that the three drying methods, conventional air-drying, vacuum-drying and nitrogen atmosphere-drying, may produce highly variable results. This is probably due to a complex interplay of chemical reactions induced by the heating, coupled with the loss of volatiles other than water. More evalua-

Table 5. Selected Coal Set

Sample No.	State	Seam	Rank	Reported Moisture %[a]	Ash%[a]
PSOC 217	Kentucky	#9	HVB	5.09	17.66
PSOC 230	Montana	Rosebud	Subbit. B	19.84	9.16
PSOC 247	North Dakota	Noonan	Lignite	30.4	11.3
PSOC 465	Wyoming	Deadman #2	Subbit. A	18.1	11.5
PSOC 627	Pennsylvania	#2	Semi-Anthr.	1.6	16.5
PSOC 640	Kentucky	#6	HVA	3.8	8.1

[a] Values from Pennsylvania State University Coal Data Bank.

Table 6. Oxygen Percentages in "As Received" and Oven-dried Coals[a]

Sample No.	As-received	Air-dried	Vacuum-dried	N_2-dried
PSOC 217	20.3	18.6	20.3	16.6
PSOC 230	32.9	29.2	29.9	26.8
PSOC 247	36.7	30.3	31.8	26.1
PSOC 465	32.6	25.4	26.4	21.1
PSOC 627	11.8	12.0	11.5	12.5
PSOC 640	17.7	14.8	14.8	13.4

[a] Standard used is benzoic acid.

tions of these techniques are currently underway in our laboratory in order to understand the processes involved.[24] Oxygen concentrations in separate aliquants of freeze-dried and "as-received" coal samples are presented in Table 7. In our freeze-drying procedures, constant weight was attained in less than one hour and no further loss was observed over a 48 hour period (Table 8). The freeze-dry values agree closest with the values obtained by oven-drying under a N_2 atmosphere. Current research suggests that the freeze-drying

Table 7. Oxygen Percentages in "as received" and Freeze-Dried Coals[a]

Sample No.	"As received"[b]	Freeze-dried
PSOC 217	21.1 ± 0.2	17.4 ± 0.1
PSOC 230	33.5 ± 0.2	27.9 ± 0.2
PSOC 247	36.6 ± 0.2	28.4 ± 0.2
PSOC 465	33.7 ± 0.5	21.1 ± 0.6
PSOC 627	13.5 ± 0.1	12.4 ± 0.1
PSOC 640	18.1 ± 0.1	13.7 ± 0.1

[a] Standard used is benzoic acid, 26.20% O.

[b] These values vary slightly from those in Table 6 due to the fact that separate aliquants were used and this experiment was run several months later. Apparently, handling and atmospheric exposure of the samples resulted in slightly elevated oxygen contents for several of the samples.

Table 8. Freeze-Drying of Standard Coals (%)

Freeze-Dry Time (Hours)	PSOC 217	PSOC 230	PSOC 247	PSOC 627
As Received	21.1 ± 0.2	33.5 ± 0.2	36.6 ± 0.2	13.5 ± 0.1
1	17.8 ± 0.1	27.8 ± 0.2	28.9 ± 0.2	12.4 ± 0.1
24	17.4 ± 0.1	27.9 ± 0.2	28.4 ± 0.2	12.4 ± 0.1
48	17.8 ± 0.1	27.8 ± 0.2	28.5 ± 0.3	12.5 ± 0.1

Note: 48 hour sample was for uninterrupted period.

technique may give the most reliable oxygen concentrations on a moisture free basis, although this conclusion may be somewhat dependent on coal rank.

National Bureau of Standards coal and fly ash Standard Reference Materials 1632, 1633, 1632a, 1633a, and 1635 were analyzed according to the above procedures. Two to four sample splits of the NBS SRM materials were freeze-dried for 24 hours prior to analysis. In addition, several commercially available coal standards were analyzed. Results are tabulated in Table 9 along with literature values and moisture contents of the samples, based on weight loss. Samples of "as-received" 1632a and 1635 coal standard from previously unopened bottles were also analyzed. Our data on freeze-dried SRM's 1632 and 1633 do not agree with those of Nadkarni.[19] His very low oxygen value for NBS 1633 fly ash (0.25%) is clearly

Table 9. Oxygen Contents of Coal and Fly Ash Standards[a]

Sample Description	% Oxygen Found	% Oxygen, Literature or Calculated Value (Reference)	% Moisture[b]
NBS SRM 1632, Coal	15.05 ± 0.11	8.03 (19)	2.60
NBS SRM 1632a, Coal	18.31 ± 0.23	18.4 ± 0.7 (19)	1.62
NBS SRM 1632a, Coal[c]	19.80 ± 0.32		
NBS SRM 1635, Coal	20.79 ± 0.19	33.4 ± 1.6 (19)	14.0
NBS SRM 1635, Coal[c]	34.99 ± 0.32		
NBS SRM 1633, Fly Ash	47.02 ± 0.08	0.25 (19)	0.17
NBS SRM 1633a, Fly Ash	47.66 ± 0.36	--	0.35
AR 216, Coal[23]	7.25 ± 0.08	--	0.84
AR 217, Coal[23]	15.53 ± 0.25	--	2.13
AR 219, Coal[23]	15.28 ± 0.06	--	1.75
AR 220, Coal[23]	13.73 ± 0.35	--	1.93
Benzoic acid[d]	26.36 ± 0.35	26.20 (calc.)	--
NBS SRM 136b, $K_2Cr_2O_7$[e]	38.42 ± 0.21	38.07 (calc.)	--

[a] Based on sample after freeze-drying for 24 hr., unless otherwise noted.

[b] As determined by weight loss during freeze-drying.

[c] Based on sample "as received" from fresh bottle.

[d] Additional two splits run against our pre-packed benzoic acid standards.

[e] Using USGS BCR-1 as a standard.

not correct, considering the alumino-silicate matrix of fly ash is ~47% oxygen. Our data on freeze-dried SRM 1635 also disagrees with Nadkarni's value, but our "as-received," result is in general agreement with his. It appears that Nadkarni's result may be based on an "as-received" sample of SRM 1635. No description of pre-analysis handling procedures was given by Nadkarni. Stoichiometric calculations on the oxygen differences between our freeze-dried and "as-received" results are consistent with our weight-loss values for moisture content on SRM's 1632a and 1635. Finally, it should be noted that the oxygen contents presented here are total oxygen values. Determination of organic oxygen contents based on a difference method (through analyses of both bulk samples and LTAs), or by the direct analysis of acid-treated coals are currently undergoing further study in our laboratory.[24]

REFERENCES

1. K. Kinson, C. B. Belcher, Determination of oxygen in coal and coke using a radio-frequency heating method, Fuel, 54:205 (1975).
2. C. Block, R. Dams, Determination of silica and oxygen in coal and coal ash by 14 MeV neutron activation, Anal. Chim. Acta, 71:53 (1974).
3. H. Sevinli, H. Ozyol, E. Barutcugil, S. Dinger, Turk A.E.C., Tech. J., 1:15 (1974).
4. D. M. Bibby, H. M. Champion, The determination of nitrogen and phosphorus in raw materials for animal feeds, using fast neutron activation analysis, Radiochem. Radioanal. Letters, 18:177 (1974).
5. D. E. Brayn, Oxygen analysis by fast-neutron activation, IRT Corporation Report, IRT4437-002, 1976.
6. W. Herzog, J. Fahland, A rapid calibration method for the determination of traces of oxygen by 14 MeV neutron activation analysis, Anal. Chim. Acta, 89:271 (1977).
7. D. J. Schlyer, T. J. Ruth, A. P. Wolf, Oxygen content of selected coals as determined by charged particle activation analysis, Fuel, 58:208 (1979).
8. W. D. James, W. D. Ehmann, G. H. Sun, C. E. Hamrin, 14 MeV neutron activation analysis of oxygen and nitrogen in coal, "Proc. 4th Conference of Scientific and Industrial Applications of Small Accelerators," Inst. for Electrical and Electronic Engineers, Inc., 1976, p. 281.
9. C. E. Hamrin, A. H. Johannes, W. D. James, G. H. Sun, W. D. Ehmann, Determination of oxygen and nitrogen in coal by instrumental neutron activation analysis, Fuel, 58:48 (1979).
10. W. D. James, W. D. Ehmann, C. E. Hamrin, L. L. Chyi, Oxygen and nitrogen in coal by instrumental neutron activation analysis, implications for conversion, J. Radioanal. Chem., 32:195 (1976).

11. W. D. Ehmann, Applications of INAA and PIXE to the analysis of coal, "Proc. 3rd Int. Conf. on Nuclear Methods in Environmental and Energy Research," Columbia, MO (October 1977) ERDA CONF-771072, 1978, p. 284.
12. J. W. Morgan, W. D. Ehmann, Precise determination of oxygen and silicon in chondritic meteorites by 14 MeV neutron activation with a single transfer system, Anal. Chim. Acta, 49:287 (1970).
13. A. Volborth, Fast neutron activation analysis for oxygen, nitrogen and silicon in coal, coal ash and related products, in: "Analytical Methods for Coal and Coal Products," C. Karr, ed., Academic Press, New York (1980).
14. S. R. Khalil, D. W. Koppenaal, W. D. Ehmann, 14 MeV INAA oxygen determinations in coal conversion liquids, J. Radioanal. Chem., 57:195 (1980).
15. W. D. Ehmann, S. R. Khalil, D. W. Koppenaal, 14 MeV INAA nitrogen determinations in coal conversion liquids, J. Radioanal. Chem., 59:403 (1980).
16. S. R. Khalil, D. W. Koppenaal, W. D. Ehmann, Studies of oxygen and nitrogen in coal conversion liquids by 14 MeV INAA, "Proc. 4th Int. Conf. Nuclear Method in Environmental and Energy Research," Columbia, MO, April 1980, in press.
17. J. M. Ondav, W. H. Zoller, L. Olmez, N. K. Aras, G. E. Gordon, L. A. Rancitelli, K. A. Abel, R. H. Filby, K. R. Shah, R. C. Ragaini, Elemental concentrations in the National Bureau of Standard's environmental coal and fly ash standard reference materials, Anal. Chem., 47:1102 (1975).
18. M. S. Germani, I. Gokmen, A. C. Sigleo, G. S. Kowalezyk, I. Olmez, A. Small, D. L. Anderson, M. P. Failey, M. C. Gulovali, C. E. Chaquette, E. A. Lepel, G. E. Gordon, W. H. Zoller, Concentrations of elements in the National Bureau Standard's bituminous and subbituminous coal standard reference materials, Anal. Chem., 52:241 (1980).
19. R. A. Nadkarni, Multitechnique multielemental analysis of coal and fly ash, Anal. Chem., 52:931 (1980).
20. S. R. Khalil, D. W. Koppenaal, W. D. Ehmann, Oxygen concentrations in coal and fly ash standards, Anal. Letters, 13 (A12): 1063 (1980).
21. Standard sample information is reported in the Journal of the Japan Petroleum Institute, Vol. 18, No. 12.
22. H. V. Drushel, Exxon Research and Development Laboratories, Baton Rouge, Louisiana, private communication (1980).
23. Coals certified for use as proximate and ultimate analysis standards from Alpha Resources, Inc., P. O. Box 577, Xenia, OH.
24. C. E. Hamrin, D. W. Koppenaal, W. D. Ehmann, S. R. Khalil, Unpublished data, University of Kentucky.

DETERMINATION OF TRACE ELEMENT FORMS IN SOLVENT REFINED COAL PRODUCTS

B. S. Carpenter

National Measurement Laboratory
National Bureau of Standards
Washington, D. C. 20234

and

R. H. Filby

Nuclear Radiation Center
Washington State University
Pullman, Washington 99164

ABSTRACT

The Solvent Refined Coal Processes SRC I and SRC II are designed to produce low ash, low sulfur solid (SRC I) and liquid fuels (SRC II) from coal. Both processes are currently undergoing scale-up to a 6000 tons per day demonstration plant stage. The fate and distribution of Ti, V, Ca, Mg, Al, Cl, Mn, As, Se, Sb, Hg, Br, Ni, Co, Cr, Fe, Na, Rb, Cs, K, Sc, Eu, Sm, Ce, La, Sr, Ba, Th, Hf, Ta, Zr and Cu in the SRC I and SRC II processes have been determined using neutron activation analysis. The nature of the chemical species of several elements has been investigated using fission track analysis for U and a combination of gel permeation chromatography, HPLC, activation analysis and atomic absorption spectroscopy for other elements. In solid SRC I, U was measured to be 0.266 ppm and it was also found that the U was distributed inhomogeneously. The concentration of U on mineral particulates was found to be approximately 76 ppm, thus ruling out U minerals. In SRC I it was established that Ti, V, As, Ga, Fe, Zn, and Se showed distinct organic affinity and that these elements were probably present as metal-organic complexes or complexed in the asphaltene or pre-asphaltene structure. The nature of these complexes could not be established, but for Ti and V there is a strong possibility of phenolic-type complexes.

INTRODUCTION

Coal conversion processes, principally gasification and hydro-liquefaction, are currently undergoing a period of concerted development in order to reduce dependence on imported oil and to exploit abundant U.S. coal reserves in an environmentally satisfactory manner. Coal liquefaction shows promise of providing utility boiler fuels, chemical feedstocks and synthetic crude oils (syncrudes) which have lower sulfur, trace element, and mineral matter contents than the coals from which they are derived.

The Solvent Refined Coal Processes (SRC I and SRC II) developed by the Pittsburg & Midway Coal Mining Company under contract with the U.S. Department of Energy have been designed to produce either a low sulfur, low ash solid fuel (SRC I process) for use in utility boilers or a liquid product (SRC II process) for use as petroleum based fuel oil substitutes. Both processes have been developed and tested in a 50-ton per day pilot plant at Ft. Lewis, Washington and are now undergoing scale-up to the 6,000-ton per day demonstration plant stage. Both processes have been designed primarily for use with high pyritic sulfur bituminous coals, although non-catalytic solvent refining may be applied to other coals, e.g. lignites.

The fate of trace elements present in coal during coal conversion processes is of concern because of potential environmental emissions or detrimental effects on product quality. The fate and distribution of trace elements in gasification processes has been determined by Forney et al., (1) for 65 trace elements in the SYNTHANE process and by Craun and Massey (2) for the CO_2 Acceptor process. Koppenaal et al., (3) have published additional data for the SYNTHANE process. Thermodynamic predictions of the behavior of volatile elements such as As, Se, Hg, B, and Pb in coal in pressurized Lurgi, HYGAS, and Koppers-Totzek gasifiers have been made by Anderson, Hill, and Fleming (4). Their studies have shown that there could be significant mobilization of volatile elements, particularly Hg, Se, and As from coal during gasification and that these elements require specific processing techniques for solids, waste waters and gas streams. Forney et al., (1) also computed material balances for trace elements in the SYNTHANE process. In general it has been shown that gasification processes mobilize the elements Hg, Se, As, Sb, Pb, B, and possibly other elements from coals.

Several studies of trace elements have been carried out on liquefaction processes. Shults (5) reported on the behavior of 47 trace elements determined by spark source mass spectrometry in the COED process and Hildebrand et al., (6) also reported data on COED process streams. Yavorsky and Akhtar (7), Schultz et al., (8), Lett et al., (9) and Schultz et al., (10) have analyzed SYNTHANE process streams and products. The most extensive studies of trace element behavior in liquefaction have been made for the SRC processes. Early work by

Jahnig (11) concerned analyses of two bench-scale SRC I process products and Filby, Shah, and Sautter (12) reported neutron activation analysis data for 22 elements on SRC products and streams from bench-scale studies and initial products from the 50-ton per day pilot plant. Coleman (13), Koppenaal and Manahan (14) and Fruchter et al., (15) have also published data on trace elements in SRC I products from both the Ft. Lewis pilot plant and the Wilsonville, Alabama SRC I pilot plant. Recently Ruch et al., (16) have published data on 71 trace elements in liquefaction process coals and residues, including SRC I and SRC II materials to determine the possible economic potential of liquefaction residues as sources of heavy metals.

Detailed knowledge of the behavior and fate of trace elements in the SRC processes (SRC I and SRC II) is important because of possible effects of trace elements on the liquefaction process (i.e., catalytic effects), effects on product properties, and potential environmental effects. Some of the important environmental aspects of trace element behavior in the SRC processes are
  i) the potential for release of volatile toxic element species from gas streams, e.g., H_2Se, Hg, AsH_3,
 ii) the possible formation of volatile carbonyls, e.g., $Ni(CO)_4$ and release to the plant environment,
iii) the formation of toxic metal organic complexes or true organometallic species and incorporation in products,
 iv) the mobilization of toxic elements from coal to process waters and incorporation into waste waters or other waste products, and
  v) the effects of trace metal forms in fuels on the incorporation into different fly ash particulate sizes during combustion.

The determination of trace elements and their fate in the SRC I and SRC II processes has been reported in several recent publications and presentations (12, 17-20).

The Solvent Refined Coal Processes

Both the SRC I and SRC II processes involve the non-catalytic hydrogenation of coal at elevated temperature and pressure in the presence of a donor solvent. In the SRC I process coal is pulverized, mixed with a recycle solvent (with donor hydrogen properties) and is heated at 450°C with H_2 at 1500 psig for an effective reaction time of approximately 0.5 hours. Although the liquefaction chemistry is complex and incompletely understood, in the initial pyrolysis step the coal matrix is partially depolymerized and free radicals formed are hydrogenated via donor solvent molecules (e.g., tetralin, decalin, etc.) to give a mixture of low to high molecular weight hydrocarbons, hydrocarbon derivatives, heterocyclic compounds, CO, CO_2, H_2O, H_2S and NH_3 plus mineral matter and unreacted coal. After gas removal by pressure let-down the slurry is filtered to remove mineral matter and unreacted coal, the solvent is distilled off the coal filtrate to recycle in the process and the resulting solid product is low-sulfur, low-ash, SRC I. In the SRC II mode, the process is similar except that a higher hydrogen partial pressure (up to 2000 psig) is used and part of the reacted coal solution is recycled to the reactor to give a longer effective residence time (0.8 - 1.0 hrs) of the coal in the reactor than in the SRC I. After reaction, light gases are removed by pressure reduction, the liquid (+ mineral matter) stream is fractionated into light distillate (naphtha), middle distillate, heavy distillate fractions and a residual material (vacuum bottoms) that contains most of the mineral matter plus high molecular weight organic material and unreacted coal. A comparison of the process conditions for typical SRC I and SRC II runs and a comparison of product yields for two processes is shown in Table 1.

In the SRC I and SRC II processes a significant reduction in sulfur content compared to the original coal is achieved. In the process pyrite, FeS_2 is converted to pyrrhotite, FeS, thus:

$$FeS_2 + H_2 \rightarrow FeS + H_2S$$

In addition to a reduction of mineral bound sulfur, some of the organically bound sulfur reacts to form H_2S. This conversion is probably due to reaction of H_2 with aromatic -SH groups and by reaction with -S-S bridging groups in the coal structure. However, some sulfur functionalities, particularly ring-S (e.g., thiophenic S), are retained in the products. What is not clear is what happens to minor elements present in the pyrite. Both As and Hg have been considered as associated with pyrite in some coals (21) and their fate during the conversion of pyrite to pyrrhotite is of interest. Possible reactions of pyrite bound As and Hg are:

$$FeS_2(As)_{(s)} + H_{2(g)} \rightleftharpoons H_2S_{(g)} + FeS_{(s)} + \tfrac{1}{4}As^\circ_{4(g)}$$

$$FeS_2(Hg)_{(s)} + H_{2(g)} \rightleftharpoons H_2S_{(g)} + FeS_{(s)} + Hg^\circ_{(g)}$$

The release of $As^\circ_{4(g)}$ and $Hg^\circ_{(g)}$ would be controlled by diffusion out of the pyrrhotite lattice and would be facilitated by the conversion of the pyrite crystal lattice to very finely divided pyrrhotite. Other mineral species, e.g., quartz (SiO_2), calcite ($CaCO_3$), sphalerite (ZnS), and clay minerals are probably unreactive under liquefaction conditions, However, a number of volatile species of certain elements may be formed during hydrogenation. In addition, organically bound metals in the coal may react to form, or may be converted to, simpler metal-organic species during the depolymerization and hydrogenation of the coal. Also some metal-organic compounds may be formed by reaction with mineral species. Some of these trace element species may be toxic, volatile, or potentially harmful to catalysts, furnace linings, etc., and hence undesirable.

TABLE 1: Comparison of Typical Conditions and Products of SRC I and SRC II Processes*

	SRC I	SRC II
Conditions		
Coal	Bituminous	Bituminous
Mode	no recycle	slurry recycle
Average Dissolver T°C	499	456
Dissolver Pressure (psig)	1500	1880
Nominal Residence Time (hrs)	0.5	0.87
Products		
Gases	C_1-C_4 hydrocarbon, CO, CO_2, H_2O, NH_3, H_2S	C_1-C_4 hydrocarbon, CO, CO_2, H_2O, NH_3, H_2S
Liquids	light oils (10%)	naphtha (7%) middle distillate (18%) heavy distillate (10%)
Solids	SRC I (70%)	vacuum bottoms residue (for gasification)

*Data taken from references (17, 20). Percentages in parentheses refer to approximate yields (coal = 100%)

Chemical speciation of trace elements in the SRC processes is important in evaluating effects of trace elements in waste streams, on product quality, on catalysts used in product upgrading, etc. At present very little is known about trace element species in coal liquefaction products and waste streams, except for those elements present in known mineral forms that do not react in the hydrogenation process. Several techniques have been used in studying trace element forms in environmental materials, coal gasification and oil shale retorting waste streams. Techniques include combined gas chromatography-atomic absorption spectroscopy (GC-AAS) for volatile

Hg (22) and As species (23), LC-AAS, EXAFS for V in coals and coal conversion products (24) and SEM-EDX for identification of mineral host for trace elements in coal liquefaction residues.

The data reported here are from preliminary studies demonstrating the utility of fission track analysis for identifying U species in coal liquefaction products and size exclusion chromatography (SEC) applied to determining trace element distributions in solvent refined coal products. A great variety of organic compounds have been identified in SRC I products (25), therefore a separation scheme based on molecular size fractionation was used for separation of SRC I into size fractions. This technique has been applied to the determination of trace element distributions in crude oils (26) for which similar characterization difficulties exist. Because SRC products exhibit much greater functionality than does petroleum it was decided to use organic polymer GPC materials instead of controlled pore glass (CPG) substrates. The GPC materials used were BioRad BioBeads SX-1, and the more functional rigid resin SM-2.

EXPERIMENTAL

Solid SRC I was obtained from the Pittsburg & Midway Coal Mining Company Pilot Plant at Ft. Lewis, Washington. This material had been previously analyzed for trace elements by neutron activation analysis (17). Neutron activation analysis was used for determining trace elements and the techniques have been described elsewhere (12, 17).

Size exclusion chromatography of SRC I in tetrahydrofuran-pyridine (4:1) was carried out using BioRad SX-1 resin that has a nominal exclusion limit of 14,000. The rigid macroporous gel SM-2 with a nominal exclusion limit of 14,000 was also used. Details of the chromatographic separation scheme are given by Filby et al., (27).

Fission track analysis was applied to SRC I materials to determine particulate U, U minerals and organically bound U. A solution of SRC I in pyridine was applied to a 2.5 cm diameter paper filter and the solution air-dried. The filter paper was then contacted with a Lexan (General Electric) track detector, packaged with an isotopically normal U standard (solution evaporated on filter paper) and irradiated in the NBS 10 MW research reactor for 15 min at 10^{13} neutrons/cm^2 sec. After irradiation the Lexan detectors were allowed to cool for several hours and were then etched in 6.5 M NaOH at 50°C for 55 min. After etching the detectors were mounted on microscope slides and the track densities measured by optical microscopy. Track densities in tracks/field of view were computed for SRC I and the U standard and the concentration of U in SRC I computed from

$$U\ (\mu g/g) = \frac{T_{sa} \cdot W_u}{T_{st} \cdot M_{sa}}$$

where T_{sa} = av. no. of tracks/field of view for SRC I

T_{st} = av. no. of tracks/field of view for U standard

W_u = μg U in standard deposited on filter paper

M_{sa} = weight (g) of SRC I on sample filter paper.

RESULTS AND DISCUSSION

Table 2 compares the trace element composition of solid SRC I to that of the feed coal from which it was derived (Kentucky No. 9/14). Also listed in Table 2 are the elemental concentration ratios (SRC I/coal) and the percentage reduction of the element from coal to SRC I. Most metals are greatly depleted in the SRC I material compared to coal and it may be assumed that this reduction is a result of filtration of the mineral matter from the coal during the SRC process. Thus elements that are found predominantly in mineral rather than maceral components of the coal are efficiently removed, e.g., Fe, Na, K, Rb, Cs, Al, Ca, Mg and the rare earth elements. The behavior of Ti, however, is quite different in that most of the Ti present in the coal remains with the SRC I. This small depletion of Ti in the SRC I process is consistent for a given type of coal (17, 20) and several explanations for this behavior are possible:

i) Ti occurs in coal as very finely divided TiO_2 (rutile) which passes through the filtration system unlike other minerals which are effectively removed.

ii) Organic Ti compounds are present in the coal and are soluble in the reacted coal solution.

iii) Organic Ti compounds are formed from inorganic TiO_2 during the hydrogenation process.

Miller and Given (2) have postulated that Ti may be present in coal bound to the organic matrix and Ti has been shown to form strong bonds with humic acids in aqueous solution (29). However, in bituminous coals used in the SRC I process, it appears likely that most of the Ti present in the coal is TiO_2. Thus explanations (i) and (iii) above may explain the behavior of Ti.

Size exclusion chromatography was carried out on SRC I using an SX-1 column (nominal exclusion limit of 14,000). Fractions collected were analyzed for Ti and V and the data are shown in Table 3. The results of fractionation on SX-1 show that most of the Ti and V is contained in the first 3 fractions representing the highest molecular weight material (excluded for the first 2). The fractionation data indicate that V is distributed more uniformly among the first 3 fractions than is Ti, which is concentrated in the first three fractions. This suggests that a larger proportion of V than Ti is bound to lower MW SRC I material. Although these data suggest that Ti and V are associated with the high MW organic material, it is possible that particulate V and Ti may be excluded with the high MW organic

TABLE 2: Trace Element Distributions in SRC I Relative to Coal

Element	Concentration in Coal (µg/g)	Concentration in SRC I (µg/g)	Concentration Ratio SRC I/Coal	Percentage Reduction Compared to Coal
Ti	530.1	465.0	0.88	12
V	30.1	4.63	0.15	85
Ca	330	72.8	0.22	78
Mg	1160	89.0	0.08	92
Al	11800	200	0.02	98
Cl	260.1	159.5	0.62	38
Mn	34.0	20.3	0.60	40
As	12.5	2.0	0.16	84
Sb	0.76	0.06	0.08	92
Se	2.00	0.12	0.06	94
Hg	0.113	0.0396	0.35	65
Br	4.56	7.74	1.70	+70
Ni	14.9	<3	-	-
Co	5.88	0.22	0.04	96
Cr	13.7	1.64	0.12	88
Fe	21100	300	0.01	99
Na	137	4.23	0.03	97
Cs	0.75	0.02	0.03	97
K	1550	4.72	0.003	100
Sc	2.59	0.57	0.22	78
Tb	0.39	0.045	0.12	88
Eu	0.26	0.055	0.21	79
Sm	2.62	0.29	0.11	89
Ce	20.9	0.45	0.02	98
La	7.55	0.13	0.02	98
Th	2.00	0.22	0.11	89
Hf	0.51	0.084	0.16	84
Ta	0.14	0.046	0.33	67
Ga	3.56	1.79	0.50	50
Zr	62.9	16.0	0.25	75

material. Consequently GPC was carried out with the resin SM-2 in which it was found that the behavior of SRC I was a combination of adsorption and size exclusion. Figure 1 shows the relative elemental concentrations plotted with UV absorbances (254 nm) versus elution volume (fraction number). These figures show that no UV-absorbing organic material appears in the eluate until fraction 5 was collected. The elemental data for Ti show two peaks corresponding to inorganic material eluting in fractions 2 and 3 (probably particulate TiO_2)and Ti associated with organic material in fractions 5-8. The data for Na

TABLE 3: Distribution of Ti and V in SRC I Separated on SX-1 in THF/Pyridine (4:1)

				Mass of Element (µg) in Fraction No.				
Element	F1	F2	F3	F4	F5	F6	F7	F8
Wt SRC I in Fraction (g)	0.0504	0.0531	0.0261	0.0269	0.0507	0.0480	0.0643	0.181
Ti	50.2	11.2	9.3	6.09	4.28	3.91	3.14	5
V	0.991	0.591	0.302	0.208	0.363	0.250	0.220	0.244

and Al show that the major part of each element is associated with an inorganic phase. Vanadium, however, appears to be associated only with the organic component of the SRC I.

The chromatographic data appear to confirm that Ti is present in SRC I as both TiO_2 (or some other inorganic material) and as Ti-organic compounds with the latter being of greater quantitative significance. Thus the behavior of Ti is intermediate between a purely inorganic association exhibited by Al and Na and a purely organic association as shown by V. Although the SM-2 data show an obvious inorganic component for Na, Al, and part of the Ti (or most unlikely, organic compounds with no 254 nm UV absorbance) the association of V and most of the Ti in SRC I with the organic component cannot be taken as proving the existence of organic complexes or compounds of Ti and V. Similar behavior would be shown by inorganic forms of these elements (e.g., finely divided oxides) that were strongly bonded to high MW organic material. Work in progress, however, confirms an organic association of Ti and V.

Fission track determination of U in SRC I provides a unique opportunity to study the distribution of U in SRC I by measuring the relative concentration of U in the bulk SRC I matrix or in particles with cross sectional areas of 10^{-5} cm^2 or smaller. The U data obtained on SRC I are shown in Table 4. Figures 2a and 2b show two particles observed in the SRC I that exhibit higher U concentrations than the surrounding background in which there are relatively few fission tracks. The particle in Figure 2a appears to have a uniform U distribution and has a calculated concentration of 76.7 ppm U. The particle is thus not a U mineral but is probably a zircon ($ZrSiO_4$) or a rare earth mineral such as monazite. The particle in Figure 2b shows a distinctly different distribution in which the U appears to be "zoned" or adsorbed on the surface of a mineral particle. Because of the non-homogeneous distribution, the U content was not calculated but is less than 10 ppm. Although several similar particles were observed in the SRC I analysis, the total number of "particulate" U fission tracks represented only 11.2% of the total U content of SRC I and that the bulk of the U (0.236 ppm) is uniformly distributed, either associated with the organic matrix or as minor constituents in more abundant minerals (e.g., clays) distributed throughout the matrix.

TABLE 4: Data on Uranium Concentrations in SRC I

Nature of Sample	Average Fission Track Density (tracks/cm^2)	U Concentration (µg/g)
Bulk SRC I	3.81×10^4	0.266
Particle - area 2.0×10^{-6} cm^2	1.1×10^7	76.7

SOLVENT REFINED COAL PRODUCTS

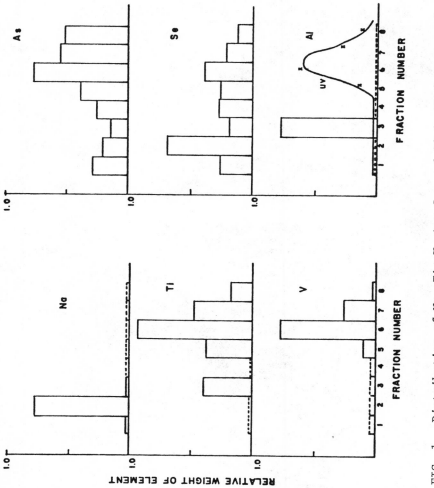

FIG. 1. Distribution of Na, Ti, V, As, Se, and Al in SM-2 Separated SRC I Chromatographic Fractions

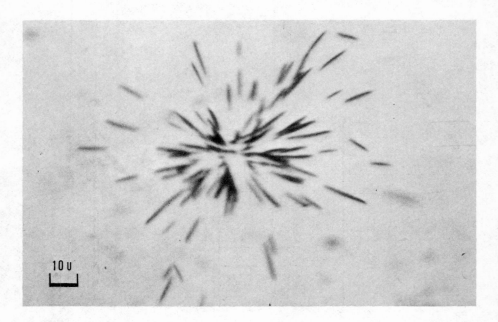

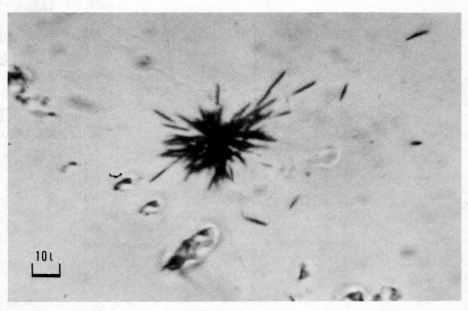

FIG. 2. Fission Track Patterns of Particulate Species in SRC I
(Fig. 2a Upper, Fig. 2b Lower).

CONCLUSIONS

Solid SRC I derived from a Kentucky No. 9/14 coal by non-catalytic hydrogenation is depleted in most trace metals but retains 88% of Ti present in the coal. Chromatographic separation of SRC I into molecular size fractions using SEC showed that Ti is associated predominantly with high molecular weight organic material (excluded) and that V shows a similar behavior. A SEC separation using SM-2 which exhibits greater adsorption of SRC than does SX-1 shows that the Ti present appears to be both particulate (TiO_2) and organically associated, the latter predominating. Vanadium appears to be associated only with the organic component of SRC I whereas Na and Al are associated only with non-UV absorbing material appearing with the solvent front which is probably inorganic particulate material.

Fission track studies of U in SRC I show that 88.8% of the U is uniformly distributed in the SRC I matrix but that 11.2% of the U is associated with mineral particles containing up to 76 ppm U and which are probably zircons or rare earth minerals.

REFERENCES

1. A. J. Forney, W. P. Haynes, S. J. Gaiser, R. M. Kornusky, C. E. Schmidt, and A. G. Sharkey. Trace Elements and Major Components Balances Around the Synthane PDU Gasifier. Aspects of Fuel Conversion Technology II, p. 67, EPA Report EPA-600/2-76-149 (1976).
2. J. C. Craun and M. J. Massey. Proc. Symp. Potential Health and Environmental Effects of Synthetic Fossil Fuel Technologies. U.S. DOE CONF 780903, p. 42 (1979).
3. D. W. Koppenaal, H. Schultz, R. G. Leff, F. R. Brown, H. B. Booher and E. A. Hattman. Preprints ACS Div. Fuel Chemistry 24, 299 (1979).
4. G. L. Anderson, A. H. Hill and K. K. Fleming. Proc. Symp. Environmental Aspects of Fuel Conversion Technology IV (April 1979, Hollywood, FL). U.S. EPA-200/7-79-217, p. 203 (1979).
5. W. D. Shults, ed., ORNL/NSF/EATC-18, 1976.
6. S. G. Hildebrand, R. M. Cushman and J. A. Carter. Proc. 10th Annual Conf. Trace Substances in Environmental Health, University of Missouri, Columbia, MO, p. 14 (1976).
7. P. M. Yavorsky and S. Akhtar. Proc. Symposium Environmental Aspects of Fuel Conversion Technology, St. Louis, MO, May 1974. EPA-650/2-74-118, pp. 325-30.

8. H. Schultz, G. A. Gibbon, E. A. Hattman, H. B. Booher and J. W. Adkins. PERC Report, PERC/RI-77/2, 1977.
9. R. G. Leff, C. E. Schmidt, R. R. DeSantis and A. G. Sharkey, Jr. PERC/RI-77/12, Pittsburgh Energy Technology Center (1977).
10. H. Schultz, E. A. Hattman, G. A. Gibbon, H. B. Booher and J. W. Adkins. Chapter 19 in <u>Analytical Methods for Coal and Coal Products</u>, C. Karr, Jr., Editor. Academic Press, Vol. I (1978).
11. C. E. Jahnig. Evaluation of Pollution Control in Fossil Fuel Conversion Processes: Liquefaction, Sec. 2, SRC. EPA-650/2-74-009-f (PB 243 694), 1975.
12. R. H. Filby, K. R. Shah and C. A. Sautter. J. Radioanal. Chem. <u>37</u>, 693 (1977).
13. W. M. Coleman, P. Perfetti, H. C. Dorn and L. T. Taylor. Fuel <u>57</u>, 612 (1978).
14. D. W. Koppenaal and S. E. Manahan. Env. Sci. and Technology <u>10</u>, 1104, 1976.
15. J. S. Fruchter, J. C. Laul, M. R. Petersen and P. W. Ryan. Preprints ACS Div. Petrol. Chem. <u>22</u>, 793 (1977).
16. R. R. Ruch and others. Final Report on Contract EY-76-21-8004, U.S. DOE report FF-8004-42 (1979).
17. R. H. Filby, K. R. Shah, M. L. Hunt, S. R. Khalil and C. A. Sautter. U.S. Department of Energy Report, FE/496-T-17, 1978.
18. R. H. Filby, K. R. Shah and C. A. Sautter. Proc. Environmental Aspects of Fuel Conversion Technology III, Hollywood, FL, Sept. 1977, EPA-600/7-78-063, p. 266, 1978.
19. R. H. Filby and S. R. Khalil. Proc. Symposium on Potential Health and Environmental Effects of Synthetic Fuel Technologies, (ORNL), Gatlinburg, TN, Sept. 1978 (in press).
20. R. H. Filby, S. R. Khalil and M. L. Hunt. Interim Report on Trace Elements in the SRC Processes (in press).
21. D. J. Swaine in <u>Recent Contributions to Geochemistry and Analytical Chemistry</u>, A. E. Tugarinov, Editor. Wiley, New York, 1976.
22. F. J. Fernandez. At. Abs. Newsletter <u>16</u>, 33 (1977).
23. M. O. Andreae. Anal. Chem. <u>49</u>, 820 (1977).
24. D. H. Maylotte, J. Wong, F. W. Lytle, R. O. Greegor and R. L. St. Peters. Submitted to Science.
25. G. A. Dailey and D. C. Michelson in <u>Environmental, Health and Control Aspects of Coal Conversion: An Information Overview</u>, Vol. 1, H. M. Braunstein, E. D. Copenhaver and H. A. Pfuderer, Editors, p. 51, ORNL/E15-94 (1977).
26. R. H. Filby. Preprints ACS Div. Petrol. Chem. <u>18</u>, 630 (1973).
27. R. H. Filby. Unpublished data.
28. R. N. Miller and P. H. Given in <u>Ash Deposits and Conversion Due to Impurities in Combustion Gases</u>, R. W. Bryers, Editor.
29. G. Eskenazy. Fuel <u>51</u>, 221 (1972).

COMPARISON OF MAINFRAME AND MINICOMPUTER SPECTRAL ANALYSIS CODES IN
THE ACTIVATION ANALYSIS OF GEOLOGICAL SAMPLES

James R. Vogt and Christopher Graham

Research Reactor and Department of Nuclear Engineering
University of Missouri
Columbia, Missouri 65211

ABSTRACT

In recent years, small laboratory computers for the acquisition and reduction of gamma-ray spectra have increased in capacity, speed and versatility to the point that they are comparable in many capabilities to the main frame computers of a few years ago. Several computer-based gamma-ray spectrometer systems using computers having memories of 256 kilobytes and hard disk bulk storage devices are now available. Generally supplied with these systems are applications software packages for gamma-ray spectra peak location and peak area determination. There has been a question among gamma-ray spectroscopists as to whether or not any of these codes are as accurate as the vintage main frame computer codes which have been the mainstay of gamma-ray spectroscopy for the past ten to fifteen years. This paper presents a comparison study of one of the most widely used main frame codes, GAMANL, and the applications software supplied with the Nuclear Data 6620 gamma-ray spectroscopy system.

DESCRIPTION OF CODES

Of the several more popular gamma-ray spectra reduction codes written for main frame computers, GAMANL and its variations have been very widely used. Many of the modifications or "improved versions" use the Massachusetts Institute of Technology (MIT) mathematical treatments in GAMANL and merely add sections for energy calibration and calculation of elemental concentrations from the generated peak areas. The original version of GAMANL was developed at MIT by Harper, Inouye and Rasmussen for the study of gamma-ray spectra obtained during (n,γ) reaction studies (1). This code was subsequently revised by the Nuclear Chemistry Group at MIT for use with data obtained from nuclear spectroscopy or neutron activation analysis experiments. It is this version that has been widely used and added to. The MIT modifications included replacing the peak locating routine with a cross-correlation function developed by Black (2). Black's method assumes the peaks are Gaussian in shape and indicates the presence of a peak when there is a positive correlation of the search profile with the spectral data in a given region. Another change was modification of the peak-area finding routine to subtract the background from the raw data and report that difference as the peak area. Subroutine FOURT was replaced with a much faster FOUR2 (from the MIT Lincoln Laboratory) which requires the number of data points to be a power of two. Subroutines LINEAR, ADJUST, PKALIB and sections on solid-angle correction and improved resolution were removed from the program.

A tape reading and conversion routine, TAPECON, was developed at the University of Missouri by Kay and Burton (3) and added to the above MIT GAMANL code. TAPECON corrects certain tape errors such as replacing missing F-fields, replacing missing data points with the average of points on each side, ignoring all EOF indicators and correcting certain types of parity errors. The total program with a plotting routine was named BARFF because of the voluminous nature of the output data. The program detects multiplets by an excessively large FWHM value. It analyzes up to triplets by fitting Gaussians to the data. Above triplets the code gives data on the three most intense components. This version of GAMANL has been in constant use by the Nuclear Analysis Program at the University of Missouri for the past ten years.

The peak search and peak area calculation code furnished with the Nuclear Data 6620 gamma-ray spectrometer system has been described in a recent release (4). The peak search and peak area code operates in three stages which are not necessarily in sequence. In the first stage a zero area rectangular fold-in function is passed over the spectrum. This function has a center wing twice the size of the left and right wings and compares the area of the three wings. This test is quite dependent on the Sensitivity parameter selection which is supplied by the user. When a region with positive slope is found it looks for a negative slope. If both are within appropriate limits it determines that the region contains a peak or peaks. In the second or peak definition stage of the code, the local minima on the left and right hand sides are located. On finding the minima, the peak shape criterion and minima are used to find the peak boundaries. If a Gaussian fit for all peaks has not been selected and if the peaks are not multiplets, the backgrounds on the left and right sides are calculated using 2 to 5 channels depending on expected peak width. The peak area is then calculated by subtraction. If a Gaussian fit for all peaks was selected or if the region contains a multiplet, the code moves to the third stage for Gaussian fitting. This stage is made-up of two sub-steps. The first step

ANALYSIS OF GEOLOGICAL SAMPLES

examines the second differential (three maxima for each peak) and uses it to determine the number of peaks and their appropriate centroids and boundaries. If the region contains a single peak it is passed through a Gauss-Jordon fitting routine. From the unique solution to the Gaussian equation for a specific peak, the FWHM, centroid and height is calculated. If the region contains a multiplet the code goes to the second sub-step for an iterative solution. For the largest peak of the multiplet, the rough estimates of peak height and width are used to subtract the largest peak from the region. The largest peak in the residual is then treated in the same manner. The whole process is repeated using the new estimates. A Chi-square test is then used to determine if the second iteration is significantly different than the first. If not, the process is terminated. If a divergent fit is detected the first iteration is used. Iterations may be continued up to a maximum of ten.

EXPERIMENTAL

In order to provide information on the comparative results for the two codes in real experimental situations, the study was conducted by performing analyses on U.S. Geological Survey (USGS) standard rocks. This approach has the advantage that the spectral details studied are those frequently found in complex samples and which may not be readily duplicated using pulsers. The approach has a disadvantage in that the "true" values are not known. This is particularly true with respect to USGS rocks as compared with certified standards such as those distributed by the National Bureau of Standards (NBS). In retrospect, it may have been better to use NBS standards, and such a study is planned. In this study 5 replicate samples of approximately 30 mg each of USGS rocks G-1, G-2, W-1 and BCR-1 were sealed in Supra-sil quartz vials and irradiated in a rotating position in the University of Missouri Research Reactor for 10 hours at a flux of 6×10^{13} n cm^{-2} sec^{-1}. No flux monitors were used since previous experiments had indicated that the flux variation from sample to sample in this rotating position is $< \pm 2\%$. Following irradiation, the samples were allowed to decay for 21 days and were then counted in the quartz vials over a period of two days using one of our two ND6620 systems (Figure 1). A Ge(Li) detector having a relative efficiency of 10% and a resolution of 2.5 keV was used for counting. The resolution of this detector was less than optimal due to a persistent line noise problem. The samples were counted for a live time of 4400 sec with a typical dead-time of $\leq 10\%$ on an automatic sample changer (Figure 2). The sample changer used was constructed using parts of a Nuclear Chicago Model 1085 gamma scintillation changer. The modifications were based on a design by Friedman and Tanner (5) and on additional changes suggested by Tanner (6). Decay corrections were made for Ce-141, Cr-51, Nd-147, Sc-46, Rb-86, Fe-59, and Ta-182. The gamma-ray spectra for G-1, G-2, W-1 and BCR-1 are shown in Figures 3 to 6, respectively. For the peak analyses using the ND6620 software, the Gaussian fitting

Fig. 1. The Two 256 Kbyte Gamma-ray Spectrometer Systems

Fig. 2. Automatic Sample Chamber With Ge(Li) and LEPS Detectors

ANALYSIS OF GEOLOGICAL SAMPLES

101

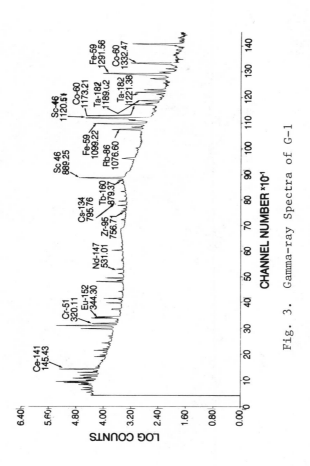

Fig. 3. Gamma-ray Spectra of G-1

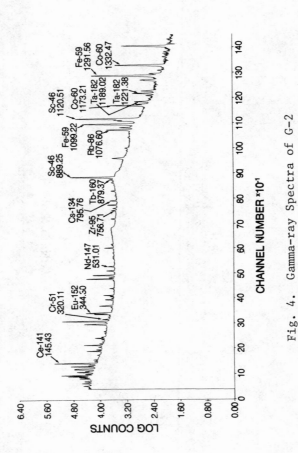

Fig. 4. Gamma-ray Spectra of G-2

ANALYSIS OF GEOLOGICAL SAMPLES

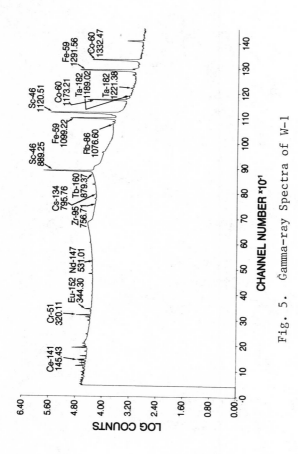

Fig. 5. Gamma-ray Spectra of W-1

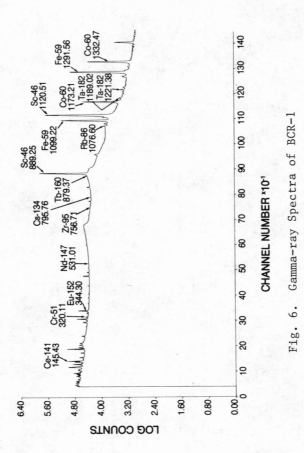

Fig. 6. Gamma-ray Spectra of BCR-1

routine was selected for all peaks, the Sensitivity was set at 2.5, the Shape Parameter was set at 5.0% and the FWHM was set at the measured resolution of the detector. The average peak area per mg of sample weight of each peak for the five replicate samples was calculated using GAMANL, the ND6620 software, and by hand from the spectral printout.

For determining elemental concentrations the values given by Flanagan (7) for G-1 were used as the "standard concentrations". G-1 was chosen as a standard since it has been analyzed by many laboratories over a number of years. However, of the 11 elements selected for the study, only 7 have "recommended" values. In addition, some of the "recommended" values are suspect. The concentrations used for G-1 are given in Table 1. Using G-1 as the standard; concentrations for Ce, Cr, Nd, Cs, Tb, Rb, Eu, Sc, Fe, Co and Ta were calculated for G-2 (Table 2), W-1 (Table 3) and BCR-1 (Table 4) and compared with reported values of Flanagan (7) and Baedecker (8). The values given by Flanagan in brackets are magnitudes and the values in parentheses are averages of reported results. Baedecker's results were obtained by activation analysis with reduction using his code SPECTRA3 (9,10). Baedecker's results are based on the analysis of 48, 7, and 56 replicate samples for G-2, W-1 and BCR-1 respectively. Zr-95 was originally included in the calculations and the 756.71 keV peak is shown in the figures. However, Zr-95 was dropped from consideration due to an interference from Eu-154. We presently determine Zr by measuring Nb-95.

Table 1. Concentration Levels of G-1 Used for Standards (7)

Element	Concentration (ppm unless % indicated)
Ce	[170]
Cr	[20]
Nd	56
Cs	1.5
Tb	0.54
Rb	220
Eu	1.3
Sc	(2.9)
Fe	1.36%
Co	[2.4]
Ta	1.5

Concentrations without brackets or parentheses are recommended. Values with parentheses are averages, and values with brackets are magnitudes.

Table 2. Comparison of Results for G-2 (Concentrations in PPM Unless % Indicated)

Energy (keV)	Nuclide	Flanagan	Baedecker	ND6620	GAMANL	Hand
145.43	Ce 141	[150]	157	170	173	176
320.11	Cr 51	[7]	7.8	5.7	12.2	13.1
344.30	Eu 152	1.5	1.25	1.51	1.9	1.66
531.01	Nd 147	60	54	60	58	60
795.76	Cs 134	[1.4]	1.3	1.3	1.4	1.4
879.37	Tb 160	0.54	0.51	0.50	0.39	0.49
889.25	Sc 46	(3.7)	3.29	3.53	3.63	3.61
1076.60	Rb 86	168	164	179	180	182
1099.22	Fe 59	(1.85)%	1.87%	1.97%	2.0%	2.0%
1120.51	Sc 46	(3.7)	3.29	3.42	3.38	3.45
1173.21	Co 60	5.5	4.2	4.53	4.9	4.9
1189.02	Ta 182	0.91	0.89	0.80	0.79	0.79
1221.38	Ta 182	0.91	0.89	0.77	0.76	0.75
1291.56	Fe 59	(1.85)%	1.87%	1.94%	1.98%	1.95%
1332.47	Co 60	5.5	4.2	4.4	4.5	4.6

Table 3. Comparison of Results for W-1 (Concentrations in PPM Unless % Indicated)

Energy (keV)	Nuclide	Flanagan	Baedecker	ND6620	GAMANL	Hand
145.43	Ce 141	[23]	22	24	50	24.1
320.11	Cr 51	(114)	113.7	100	152	177
344.30	Eu 152	1.11	0.99	1.0	1.1	0.93
531.01	Nd 147	15	12	25	17	<15
795.76	Cs 134	0.9	1.0	0.93	0.95	0.84
879.37	Tb 160	0.65	0.64	1.4	Missed	0.56
889.25	Sc 46	(35.1)	35.09	36.7	37.9	40.0
1076.60	Rb 86	21	27	25.4	46.1	29.7
1099.22	Fe 59	7.76%	7.76%	7.88%	8.1%	8.04%
1120.51	Sc 46	(35.1)	35.09	35.1	35.9	40.0
1173.21	Co 60	47	43.1	38.6	43.8	42.4
1189.02	Ta 182	0.50	0.47	0.36	0.35	0.38
1221.38	Ta 182	0.50	0.47	0.43	0.45	0.44
1291.56	Fe 59	7.76%	7.76%	7.79%	8.09%	8.04%
1332.47	Co 60	47	43.1	36.8	38	38.8

Table 4. Comparison of Results for BCR-1 (Concentrations in PPM Unless % Indicated)

Energy (keV)	Nuclide	Flanagan	Baedecker	ND6620	GAMANL	Hand
145.43	Ce 141	53.9	53	54.8	89	56.9
320.11	Cr 51	(17.6)	13.9	10.9	13.8	18.9
344.30	Eu 152	1.94	1.76	1.70	1.84	1.71
531.01	Nd 147	29	30	33	33	28
795.76	Cs 134	.95	1.2	1.1	1.1	0.98
879.37	Tb 160	1.0	1.1	1.7	Missed	1.0
889.15	Sc 46	[33]	30.15	33.8	34.4	34.6
1076.60	Rb 86	46.6	53	56	66	56
1099.22	Fe 59	9.37%	9.11%	9.88%	10.1%	10.1%
1120.51	Sc 46	[33]	30.15	32.1	33.6	33.8
1173.21	Co 60	[38]	34.5	32.9	37.0	36.0
1189.02	Ta 182	0.91	0.88	0.84	0.66	0.66
1221.38	Ta 182	0.91	0.88	0.49	0.75	0.76
1291.56	Fe 59	9.37%	9.11%	9.77%	10.1%	9.85%
1332.47	Co 60	[38]	34.5	31.4	32.4	33.2

DISCUSSION

The agreement of GAMANL and the ND6620 software with the values of Flanagan and Baedecker were generally best for G-2. This was to be expected since G-1 and G-2 have similar compostions so that the spectral shapes are comparable. In considering the overall results it should be noted that the sample sizes used were less than recommended for homogeneity. The spectra for W-1 and BCR-1 are dominated by Sc-46 which is a factor of ten higher than in G-1 and G-2. Overall both codes and the hand calculation gave satisfactory results, particularly since some of the peaks were chosen because of their complexity or difficulty. Both codes gave better than expected results for some of the more difficult peaks.

In the comparison of GAMANL with the ND6620 software, for the Ce-141 doublet at 145.43 keV, the ND6620 software did very well while GAMANL correctly determined G-2 but was very high for both W-1 and BCR-1. For the 320.11 Cr-51 peak neither code did particularly well. The reason for this is not apparent, although the Cr value for G-1 is not well known. The ND6620 results for Cr were self-consistent and GAMANL did well on BCR-1. For the 344.30 Eu-152 peak, which was a doublet in G-2 and BCR-1, both codes gave excellent results except for the GAMANL value for G-2. The 531.01 Nd-147 and 795.76 Cs-134 peaks were selected because of their low peak height to background ratio in W-1 and BCR-1. In fact, some of the peaks cannot be seen on the plots. The typical peak height to background ratios for Nd-147 were 0.016 for W-1 and 0.040 for BCR-1, and for Cs-134 were 0.068 for W-1 and 0.059 for BCR-1. The results for Cs-134 were excellent, the results for Nd-147 were acceptable except for W-1 where both codes were high. The 879.37 Tb-160 peak shows up well in G-2 but is a small tic on the lower energy side of the very large Sc-46 peak in W-1 and BCR-1. The ND6620 software did well on G-2 and was high on W-1 and BCR-1, while GAMANL was low for G-2 and missed the peak completely for W-1 and BCR-1. The 1076.60 Rb-86 peak is a small peak on the Compton edge of Fe-59. The ND6620 software did very well in this situation, while GAMANL was high on BCR-1 and very high on W-1. Both codes did well on 1099.22 Fe-59 peak which was clean and on the 1120.51 Sc-46 and 1173.21 Co-60 peaks which were doublets but where the peak of interest was greatly dominant. The shape of the 1189.02 Ta-182 peak was very poor in both the W-1 and BCR-1 spectra, indicating some interferences. Both codes gave reasonable results for G-2, but GAMANL was low for W-1 and BCR-1 and the ND6620 software was low for W-1. The results for the 1221.38 Ta-182 peak were acceptable except for the ND6620 value for BCR-1. Both codes gave reasonable results for the 1332.47 Co-60 peak. A summary of the agreement of the results is given in Table 5 and a summary of comments on the peaks is given in Table 6.

Table 5. Summary of Agreement (%) of Elemental Concentrations*

Energy (keV)	Nuclide	G-2 ND6620	G-2 GAMANL	G-2 Hand	W-1 ND6620	W-1 GAMANL	W-1 Hand	BCR-1 ND6620	BCR-1 GAMANL	BCR-1 Hand
145.43	Ce 141	+ 8.3	+10.2	+12.1	+ 4.3	+117	+ 4.8	+ 1.7	+65.1	+ 5.6
320.11	Cr 51	-18.5	+56.4	+67.9	-12.0	+ 33.3	+55.2	-21.6	- 0.72	+ 7.4
344.30	Eu 152	+ 0.6	+26.7	+10.7	0	0	- 6.1	- 3.4	0	- 2.8
531.01	Nd 147	0	0	0	+66.6	+ 13.3	N.D.**	+10.0	+10.0	- 3.4
795.76	Cs 134	0	0	0	0	0	- 6.7	0	0	0
879.37	Tb 160	- 1.9	-23.5	- 3.9	+115	N.D.**	-12.5	+35.3	N.D.**	0
889.25	Sc 46	0	0	0	+ 3.13	+ 8.0	+14.0	+ 2.4	+ 4.2	+ 4.9
1076.60	Rb 86	+ 6.5	+ 7.1	+ 8.3	0	+70.7	+ 2.7	+ 5.7	+24.5	+ 5.7
1099.22	Fe 59	+ 5.3	+ 7.0	+ 7.0	+ 1.5	+ 4.4	+ 3.6	+ 5.4	+ 7.8	+ 7.8
1120.51	Sc 46	0	0	0	0	+ 2.3	+14.0	0	+ 1.8	+ 2.4
1173.21	Co 60	0	0	0	- 4.5	0	- 1.6	- 4.6	+ 6.8	0
1189.02	Ta 182	-10.1	-11.2	-11.2	-23.4	-25.5	-19.1	- 4.5	-25.0	-25.0
1221.38	Ta 182	-13.5	-14.6	-15.7	- 8.5	- 4.3	- 6.4	-44.3	-14.7	-13.6
1291.56	Fe 59	+ 3.7	+ 6.4	+ 4.3	+ 0.38	+ 4.3	+ 3.6	+ 4.2	+ 7.8	+ 5.1
1332.47	Co 60	0	0	0	-14.6	-11.8	- 4.3	- 8.9	- 6.1	- 3.8
No. of Peaks Within ± 15%		14/15	12/15	13/15	12/15	10/15	12/15	12/15	11/15	14/15
No. of Peaks Within ± 25%		15/15	13/15	14/15	13/15	10/15	13/15	13/15	12/15	15/15

*Percent deviation from closest value of Flanagan or Baedecker if outside range. Defined as zero if within range. **Not Detected. Treated as > 25% error in summaries of no. of peaks within % error.

ANALYSIS OF GEOLOGICAL SAMPLES

Table 6. Summary Comments on Peaks

Energy (keV)	Nuclide	G-2*	W-1	BCR-1
145.43	Ce 141	Doublet	Doublet	Doublet
320.11	Cr 51	Clean	Clean	Clean
344.30	Eu 152	Doublet	Clean	Doublet
531.01	Nd 147	Clean	Low Pk. Ht. to Bkg.	Low Pk. Ht. to Bkg.
795.76	Cs 134	Clean	Low Pk. Ht. to Bkg.	Low Pk. Ht. to Bkg.
879.37	Tb 160	Clean	Low Pk. Ht. on Sc edge	Low Pk. Ht. on Sc edge
889.25	Sc 46	Clean	Clean	Clean
1076.60	Rb 86	Clean	On Fe-59 Compton Edge	On Fe-59 Compton Edge
1099.22	Fe 59	Clean	Clean	Clean
1120.51	Sc 46	Triplet	Doublet	Doublet
1173.21	Co 60	Doublet	Doublet	Doublet
1189.02	Ta 182	Clean	Poor Shape	Poor Shape
1221.38	Ta 182	Clean	Clean	Clean
1291.56	Fe 59	Doublet	Doublet	Doublet
1332.47	Co 60	Poor Baseline	Clean	Clean

* The comments on G-2 also apply to G-1

CONCLUSIONS

While the hand calculations gave more results that were within ±15% of Flanagan's or Baedecker's values than did GAMANL, the hand calculations were not better overall than those of the ND6620 software. Based on the number of peaks within ±15% the hand calculations gave better results for BCR-1, the ND6620 software gave better results for G-2, and the two methods were tied for W-1. If one examines the number of peaks for which one computational method had the lowest percent deviation of the three methods used, (Table 5) there is a very significant relative difference. Of the 45 gamma-ray peaks analyzed, the ND6620 software had the lowest percent deviation for 22, GAMANL had the lowest for 5 and the hand calculations had the lowest for 8. The other ten peaks were ties; two between GAMANL and the ND6620 software, one between hand calculation and the ND6620 software, and seven were three-way ties all at zero percent deviation indicating that the results were all within the range of Flanagan's and Baedecker's values. While the study showed that all three methods of peak area determination gave acceptable results for most of the peaks studied, it also showed that a good computer code is significantly better than hand calculations. This was true even for the large and well defined peaks of Sc-46 and Fe-59. In conclusion, the ND6620 software in general gave results closer to those of Flanagan and Baedecker than did GAMANL or the hand calculations. In particular, the ND6620 software did better on multiplets and odd peaks than did the other two methods. Our original question then, as to whether or not any of the current minicomputer codes are as reliable as the vintage mainframe codes, can be answered in the affirmative. In this case, the minicomputer code was superior.

REFERENCES

1. Harper, T., Inouye, T. and Rasmussen, N., "GAMANL, A Computer Program Applying Fourier Transforms to the Analysis of Gamma Spectral Data", Report No. MIT-3944-2, Massachusetts Institute of Technology, Cambridge (1968).
2. Black, W.W., Nucl. Instrum. Methods, $\underline{71}$, 318 (1969) and $\underline{82}$, 141 (1970).
3. Kay, M.A., and Burton, C. in "Nuclear Science Group Technical Report 1970-1971", Vogt, J.R., ed., University of Missouri (1971).
4. ND6620 Peak Search Program Algorithm, anon., Nuclear Data, Inc. Schaumburg, Ill. (1980).
5. Friedman, M.H. and Tanner, J.T., "An Automated Activation Analysis Data Acquisition System, J. Radioanal. Chem $\underline{25}$, 269 (1975).
6. Tanner, J.T., Personal Communications to J. Vogt (1978).

7. Flanagan, F.J., ed., "Descriptions of Eight New USGS Rock Standards", Geological Survey Professional Paper No. 840, p. 171 (1976).
8. Baedecker, P.A., Personal Comminication to J. Vogt (1979).
9. Baedecker, P.A., Proceedings of the Conference on Computers in Activation Analysis and Gamma-ray Spectroscopy, DOE Report No. CONF-780421, 373 (1979).
10. Baedecker, P.A., SPECTRA: Computer Reduction of Gamma-ray Spectroscopic Data, in Advances in Obsidian Glass Studies, Taylor, R.E., ed., 343, Noyes Press, Park Ridge, N.J. (1976).

APPLICATION AND COMPARISON OF NEUTRON ACTIVATION ANALYSIS
WITH OTHER ANALYTICAL METHODS FOR THE ANALYSIS OF COAL

R. A. Cahill, J. K. Frost, L. R. Camp, and R. R. Ruch

Illinois State Geological Survey

Champaign, IL

ABSTRACT

X-ray fluorescence, optical emission, atomic absorption, neutron activation, and standard ASTM chemical methods are used at the Illinois State Geological Survey (ISGS) to supply accurate and reliable data on the chemical composition of coal and coal-derived materials. Accuracy of the methods is demonstrated by examples from interlaboratory comparisons for many elements and by comparison of ISGS results with certified values and literature results from reference coal samples from the National Bureau of Standards.

Trace element mobility during coal pyrolysis, methods of coal beneficiation, chemical forms of elements in coal, and use of elemental distribution patterns to help interpret the geochemistry of the Illinois Basin coal fields are among the topics currently being investigated.

INTRODUCTION

Although sulfur content continues to be the most important factor limiting the use of coals, particularly those of the Illinois Basin, other inorganic elemental constituents will become factors as the use of coal increases. The assessment of the effect upon the environment of coal combustion for electrical power generation will be concerned with the potential release of arsenic, mercury, lead, selenium, and other toxic trace elements to the atmosphere. With the likelihood that coal liquefaction and gasification will become economically feasible, it is important to consider the effects of major, minor, and trace elements on conversion catalysts. Increased coal utilization will also produce larger amounts of wastes from coal cleaning operations, electrical power generation, and coal conversion processes. These wastes will need to be evaluated for both their effect on the environment and for their potential for economic recovery of valuable metals.

Fundamental geochemical studies are needed to investigate the distribution of inorganic species within individual coal seams and between different coal basins. These results can be used to help interpret the depositional environment and diagenetic history of a given coal seam. Knowledge from these studies can also be applied to predicting what elements can be easily removed by coal cleaning operations.

The purpose of this paper is to review the analytical approach to coal that has been used at the Illinois State Geological Survey (ISGS) to aid in answering the many questions regarding the concentration and distribution of inorganic elements in coal. Comparisons of analytical methods and results from intralaboratory studies will be discussed and comparisons of results of analyses of National Bureau of Standards reference coal samples by different laboratories will be made.

ANALYTICAL PROCEDURES

Instrumental methods—atomic absorption and optical emission spectroscopy, neutron activation analysis, and X-ray fluorescence spectroscopy—have been applied to the analysis of coal at ISGS. The analytical method chosen for each element depends primarily on a method's inherent precision and accuracy, but the cost, time, type of sample preparation required, and the need for trained personnel are also important factors.

ISGS has been involved in the analysis and investigation of the inorganic constituents in coal for many years. Initial studies by Ruch, Gluskoter, and Shimp (1974) reported data on 82 coals from the Illinois Basin. These authors, in their review of the literature on the chemical nature of coal ash, noted that prior to 1970 most investigations of trace elements in coal were based on an analysis of high temperature coal ash, which usually reported only semi-quantitative results obtained by optical emission spectroscopy. The analytical procedures used at ISGS for the 1974 report included atomic absorption spectroscopy for Cd, Cu, Ni, Pb, and Zn; optical emission spectroscopy for B, Be, Co, Cr, Cu, Ge, Mn, Mo, Ni, Pb, Sn, V, Zn, and Zr; X-ray fluorescence spectroscopy for Al, Ca, Cl, Fe, K, Mg, P, S, Si, and Ti, and also attempted for As, Br, Cu, Mn, Ni, Pb, V, and Zn; radiochemical neutron activation analysis for As, Br,

Ga, Hg, Sb, and Se; and instrumental neutron activation analysis using a 3 x 3 inch NaI(Tl) detector for Na and Mn. The details of these procedures can be found in Ruch, Gluskoter, and Shimp (1974).

Gluskoter et al. (1977) expanded on the 1974 report on coal by including 71 additional coals from eastern and western coal-producing areas and by reporting the concentrations of 23 other elements. Results were also reported for 64 washed coals and 40 coal bench samples. The analytical procedures used in this study are summarized in Table 1 and the details of the procedures can be found in Gluskoter et al. (1977). The addition of instrumental neutron activation analysis with high resolution Ge(Li) detectors permitted the routine determination of 38 elements.

A project for the determination of valuable metals in liquefaction process residues necessitated the adaptation and development of procedures for analysis for previously undetermined elements. Ruch et al. (1979) reported results on 18 sample sets from 6 different liquefaction processes for 71 elements. They report the details of treatment of the samples and of the analytical methods. The methods of analysis are summarized in Table 2. For this study radiochemical procedures were developed for the analysis of the rare earth elements Er, Gd, Ho, Nd, Tm, and also for Y, Pd, Pt, and Au, which could not be determined by instrumental neutron activation analysis. Energy-dispersive X-ray fluorescence spectroscopy (XES) was also added to the methods used.

In the course of these and other investigations at ISGS, results for several hundred coal and coal-related materials have been compiled. Two or more analytical techniques were often used for the determination of an element (Tables 1 and 2).

Agreement of results from different analytical techniques is one means of assuring that a large amount of data has few erroneous results. For example, if results for a single element determined by two independent techniques were in large disagreement, that particular element could be redetermined; however, if only one method was available, the errors might go unnoticed. In many cases, the choice of the better method of analysis was easily made; in other cases where the choice was unclear, the recommended values reported were the average of results by two or more methods. The criteria used to make these decisions was based on a method's demonstrated precision and accuracy relative to available reference standards and the frequency of erroneous or biased results. Examples of the results of comparisons of methods are shown in Figures 1 to 6. The comparison is shown graphically in scatter plots that include the calculated regression line and correlation coefficient.

TABLE 1. Analytical Procedures Used To Determine Trace Element Values In Whole Coal And Bench Samples

Element	Procedure
Rb, Cs, Ba, Ga, In, As, Sb, Se, I, Sc, Hf, Ta, W, La, Ce, Sm, Eu, Tb, Dy, Lu, Th, U, Yb, (Au)	INAA
Na, K, Br, Fe	INAA, XRF
Cl	INAA, XRF, ASTM
Mg, Ca, Al, Si, P, Ti	XRF
Be, Ge, Zr	OE-P, OED
Cr, Co, Mo	OE-P, OED, INAA
Ag, Sn	OE-P
Ni, Zn	OE-P, OED, AA, XRF
Hg	RNAA
B	OED
Pb	OE-P, AA
Sr	OED, INAA
F	ISE
V	OE-P, OED, XRF
Cu	OE-P, OED, AA
Mn	OE-P, INAA
Cd	AA, OED

Source: Gluskoter et al. (1977).

TABLE 2. Methods Used To Determine Elements In Coal Liquefaction Feed Coals And Residues

Element	Method
As, Br, Cl, Cs, Dy, Ga, Hf, I, In, Na, Rb, Sb, Sc, Se, Ta, Th, U, W	INAA
Au, Er, Gd, Hg, Ho, Nd, Pb, Pr, Pt, Tm, Y	RNAA
Ce, Eu, La, Lu, Sm, Tb, Yb	INAA, RNAA
Al, Ca, Mg, P, Si, Ti	XRF
Cd, Li, Pb, Zn	AA
Ag, Mn	INAA, OEP
B, Ge, Mo	OED
Ba, Sr	INAA, XES
Be, Bi, Tl, Zr	OEP
C, H, N, S	ASTM
Co, Cr,	INAA, OED, OEP
Cu, Ni	AA, OED, OEP
F	ISE
Fe	AA, XRF, INAA
K	XRF, INAA
Sn	OED, XES
Te	XES
V	OED, OEP

Source: Ruch et al. (1979).

Figure 1 compares results by X-ray fluorescence spectroscopy (XRF) with those by instrumental neutron activation analysis (INAA) for potassium in 70 whole coals. The results generally fall near the X = Y line and the coefficient of the regression is good at 0.93.

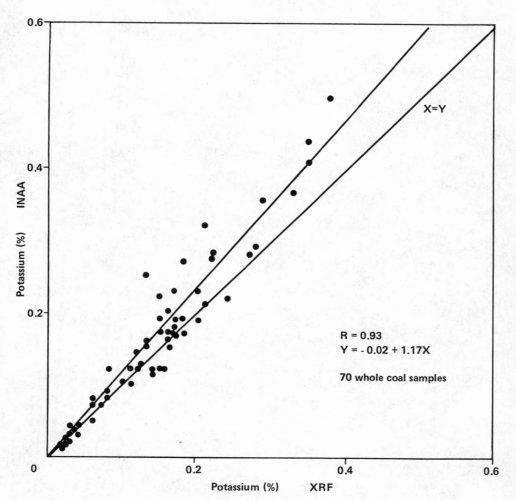

Figure 1. Comparison of potassium results by INAA and XRF.

Figure 2 compares results by XRF with INAA for the determination of iron in 68 whole coals. The results show more scatter than that observed for potassium in Figure 1. The correlation coefficient of the regression line is also lower (0.87) with the INAA results being systematically higher relative to XRF data. The scatter becomes significant at iron concentrations greater than 2 percent; this indicates that sample homogeneity could be a problem, or that a systematic error exists between the methods.

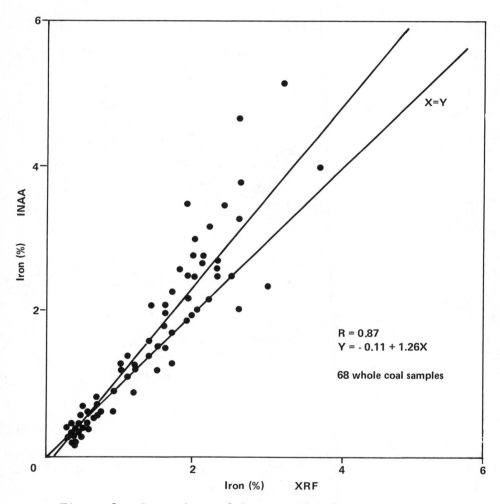

Figure 2. Comparison of iron results by INAA and XRF.

In Figure 3, INAA results are compared with those obtained by neutron activation with a radiochemical separation (RNAA) procedure for selenium in 25 whole coals. The selenium determination by INAA using the 264.6 keV γ-ray of ^{75}Se is subject to interferences by the 264.4 γ-ray of ^{182}Ta and the 279 keV γ-ray by ^{203}Hg. Also the 136 keV peak is often poorly resolved from the 133 keV line of ^{181}Hf. A high bias would then be expected by INAA over results from RNAA. This is the case for a majority of the samples; however, sometimes RNAA gave higher results. The agreement between the two methods is nevertheless good (i.e. in most cases within ±25%). After the introduction of INAA using Ge(Li) detectors, the radiochemical selenium procedure was used only for crosschecking.

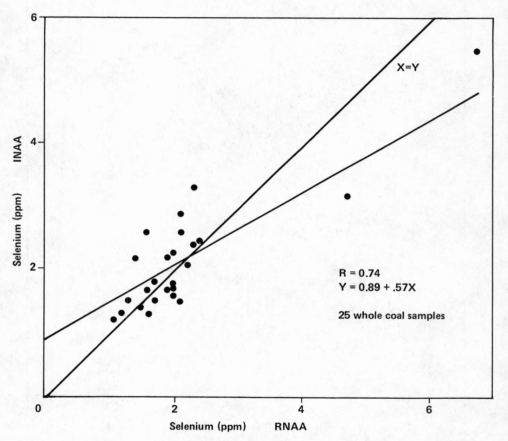

Figure 3. Comparison of selenium results by INAA and RNAA.

Nickel results are compared in Figure 4 as determined by INAA and atomic absorption spectroscopy (AA). In this case the scatter was probably due to imprecision of the INAA results because rather poor counting statistics are associated with the determination of nickel using the activity from ^{58}Co.

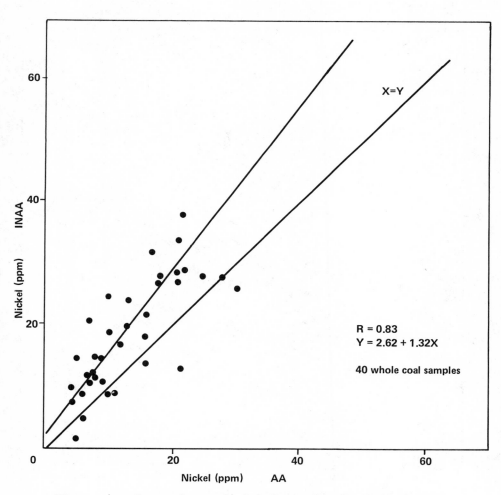

Figure 4. Comparison of nickel results by INAA and AA.

Figure 5 shows results for barium determined by INAA and energy-dispersive X-ray fluorescence spectroscopy (XES). The agreement of the data on the 39 whole coals is within ±25% for 30 of the 39 plotted values. The determination of barium by INAA is often subject to errors associated with poor counting statistics for ^{131}Ba in coal samples measured 28 to 32 days after irradiation, as is normally done at ISGS. Ideally, a count should be made after a decay of 7 to 10 days, but this was not practical because of the demand for the use of the counting equipment. Another important factor may be the presence in the coal samples of discrete particles of the mineral barite (BaSO$_4$); these may not be completely homogenized by grinding.

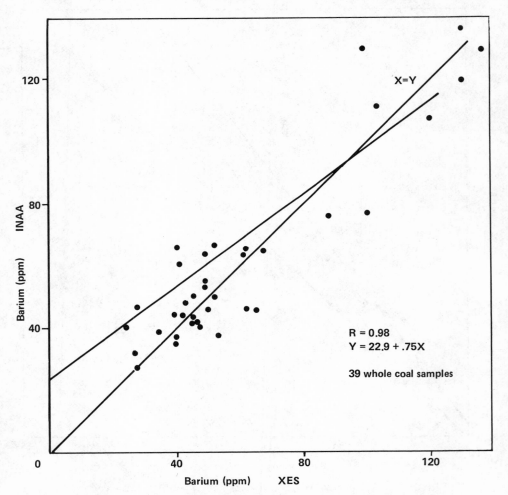

Figure 5. Comparison of barium results by INAA and XES.

Results of determinations of strontium by INAA and optical emission spectroscopy (OED) in 55 coal samples are compared in Figure 6. Strontium was difficult to determine by OED when concentrations were greater than 200 ppm; for 12 of the 95 samples reported on in Gluskoter et al. (1977), only INAA results were reported. The results by INAA using the short-lived ^{87m}Sr have a lower detection limit and better counting statistics than those determined for ^{85}Sr. Figure 6 shows a high degree of scatter which illustrates the difficulty in choosing a recommended value. In later studies, Sr results obtained by AA and XES were found to agree well with INAA; correlation coefficients of 0.98 and 0.97, respectively, were obtained.

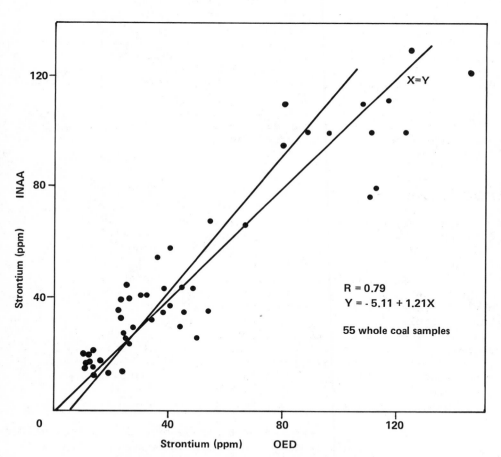

Figure 6. Comparison of strontium results by INAA and OED.

An excellent means of determining the accuracy and precision of the results of coal analysis is repeated analyses of standard reference materials during the course of an investigation. Replicated INAA results for NBS 1632 were published in Gluskoter et al. (1977) and Ruch et al. (1979). These data compared rather favorably to certified values and results from Ondov et al. (1975). However, NBS 1632 has been out of supply for some time. Table 3 compares results obtained in routine analysis by INAA of NBS 1632A with those reported by Germani et al. (1980).

The results compare well; however, in the case of Fe, Na, Ba, Cl, I, Mo, Rb, and Sr, the precision in this work is not as good as that reported by Germani et al. (1980). The poorer precision for Fe and Na is more difficult to account for than is that for the other elements which tend to have poorer counting statistics.

There has been wide application of INAA for the analysis of coal. This is illustrated in Table 4 in which the percentages by method of results reported in the literature on analysis of NBS standard reference coals are summarized (Gladney, 1980). This table does give a somewhat biased view because laboratories reporting INAA results often present results on more than 40 elements, and methods such as XRF or AA normally report far fewer elements. It should also be noted that AA and XRF have large usage in industrial and commercial laboratories, but INAA does not.

APPLICATIONS

The results reported in Ruch, Gluskoter, and Shimp (1974) and Gluskoter et al. (1977) provide a valuable data base that has been widely used to make decisions regarding utilization of coal. Results have been used for estimating possible emissions to the atmosphere from coal combustion, have provided ranges of elemental concentrations in coals in the major producing areas of the country, and have provided data on the potential removal of sulfur and other inorganic elements by specific gravity separation methods. From their washability curves an organic affinity index was calculated that could be used to predict the association of an element in coal and the degree of success one could expect to have in its removal. The elements As, Pb, Cd, Hg, Se, and Zn were consistently found in the inorganic fraction of the coals and hence should be relatively easy to remove by conventional coal-cleaning operations.

Ruch et al. (1979) reported on the fate of elements in 6 different coal liquefaction processes using 18 sets of samples that were thought to represent equilibrium conditions in the given process. Calculations were made based on the difference in concentration of an element in the residue and that in the starting coal to obtain losses or possible gains of the element in the liquefaction process.

TABLE 3. Summary of Results for SRM 1632A by INAA

Element	This Work	No. Of Determinations	Germani et al.*
Fe (%)	0.98 ± 0.20	20	1.16 ± .03
K (%)	0.41 ± 0.02	27	.42 ± .02
Na (ppm)	730 ± 90	30	850 ± 40
As (ppm)	9.1 ± 1.0	30	10.2 ± .5
Ba (ppm)	123 ± 25	15	122 ± 11
Br (ppm)	40 ± 8	28	41 ± 4
Ce (ppm)	29 ± 3	21	32 ± 4
Cl (ppm)	720 ± 100	2	790 ± 20
Co (ppm)	7.8 ± .8	22	6.5 ± .2
Cr (ppm)	37 ± 4	22	34 ± 2
Cs (ppm)	2.3 ± .4	21	2.6 ± .3
Dy (ppm)	2.2 ± .1	11	2.2 ± .3
Eu (ppm)	0.53 ± .02	26	0.55 ± .03
Ga (ppm)	8.5 ± .5	26	8 ± .8
Hf (ppm)	1.6 ± .2	21	1.55 ± .08
I (ppm)	1.6 ± .4	2	1.8 ± .2
In (ppb)	41 ± 7	7	36 ± 4
La (ppm)	16 ± 1	27	18 ± 2
Lu (ppm)	0.21 ± .05	20	0.18 ± .03
Mo (ppm)	≤5		
Mn (ppm)	29 ± 2	17	32 ± 3
Ni (ppm)	21 ± 5	19	26 ± 4
Rb (ppm)	30 ± 4	21	29 ± 1
Sb (ppm)	0.7 ± .1	24	0.6 ± .1
Sc (ppm)	6.4 ± .5	22	6.8 ± .6
Se (ppm)	3.0 ± .4	20	2.6 ± .3
Sm (ppm)	2.7 ± .1	25	2.1 ± .07
Sr (ppm)	92 ± 12	11	84 ± 9
Ta (ppm)	0.35 ± .05	20	0.40 ± .03
Tb (ppm)	0.34 ± .05	20	0.36 ± .12
Th (ppm)	4.7 ± .5	20	4.8 ± .2
U (ppm)	1.2 ± .2	22	1.2 ± .1
W (ppm)	0.8 ± .2	22	0.6 ± .2
Yb (ppm)	1.1 ± .1	23	0.98 ± .08
Zn (ppm)	27 ± 5	22	31 ± 6

*Germani et al. (1980).

TABLE 4. Distribution by analytical method of literature
results on NBS reference coal samples*

Sample	Total no. of analyses	% of results by method						
		NAA	AA	XRF	MS	PAA	OES	Other
SRM 1632	894	76	4	4	2	4	6	6
SRM 1632A	129	82	6	9				3
SRM 1633	996	61	8	12	1	10	7	1
SRM 1635	114	85	4	7				4

NAA - Neutron Activation Analysis
 AA - Atomic Absorption Spectroscopy
XRF - X-ray Fluorescence Spectroscopy
 MS - Mass Spectrometry
PAA - Photon Activation Analysis
OES - Emission Spectroscopy

*As listed in Gladney (1980).

Most of the 71 elements determined were generally retained in the
residues. As, B, Br, Cl, F, Hg, N, S, and Ti were partially lost
during the liquefaction process, i.e., the concentrations (when
normalized) were lower in the residue relative to concentration in
the starting coal.

The chemical state or form in which elements occur in coal ultimately affects their behavior and fate in coal utilization. Because
trace elements are often intimately associated with the coal material,
and because the concentrations are often quite low, the direct determination of chemical forms of elements in coal is difficult. Specific
gravity separation is one approach that has been used for many years
to determine organic/inorganic associations of various constituents
in coal. Results obtained by this approach have recently been compared to results obtained on 27 coals that had undergone selective
acid digestion to chemically remove the mineral fraction of coal
(Kuhn et al., 1980). Seven coals were also extracted with ammonium
acetate and results were compared with these data.

The method used to chemically remove minerals from the organic
fraction of the coal is a variation of the procedure for the determination of forms of sulfur in coal (American Society for Testing
and Materials, 1978), in which HCl and HNO_3 are used under prescribed
conditions to extract sulfate and pyritic sulfur. In addition, treatment with Hf is used to dissolve the silicate minerals present.

The ammonium acetate extraction procedure was designed to determine which elements were easily exchangeable in coal. Calcium, Na, Mg, Cl, and Sr were found to be exchangeable.

Elemental concentrations obtained by analysis of chemically demineralized coal were found to be comparable to the values predicted for the concentration of an element in the organic fraction of coal by extrapolation of values obtained from specific gravity fraction studies. Table 5 summarizes concentrations for 14 elements in chemically demineralized coal from 3 regions of the country. The levels of the major ash-forming elements Si, Al, Fe, and Ca are generally reduced below 100 ppm. Except for Br, most of the trace elements are also reduced by more than 50 percent. These elements were interpreted to be primarily associated with the mineral fraction of coal.

The regional variation of elements occurring in the Herrin No. 6 Coal of the Illinois Basin is currently under extensive investigation. Such variation can help to interpret the depositional history of the basin. An example of such a distribution is shown in Figure 7, which is a plot of the bromine distribution in coals in the Illinois Basin. The highest levels of bromine are found in the southeastern areas, which are the deepest part of the basin and which were influenced by brines. Detailed studies of chemical and

TABLE 5. Mean concentrations in chemically demineralized coals

Element	Illinois Herrin No. 6*	Eastern*	Western*
Si	56 ± 19	60 ± 12	56 ± 20
Al	61 ± 14	144 ± 160	76 ± 58
Fe	96 ± 53	107 ± 75	71 ± 77
Ca	38 ± 22	71 ± 72	61 ± 69
As	<1	<1	.5 ± .3
B	8 ± 2	14 ± 7	7 ± 6
Br	6 ± 4	7 ± 6	2 ± 2
Co	.4 ± .2	3 ± 4	.5 ± .5
Cr	6 ± 3	7 ± 4	1 ± 3
Cu	5 ± 3	4 ± 2	4 ± 3
Ni	6 ± 5	3 ± 2	--
Sb	.4 ± .3	.4 ± .3	.4 ± .2
Sr	4 ± 3	29 ± 19	--
Zn	<5	<1	--

*All values in ppm

Source: Kuhn et al. (1980).

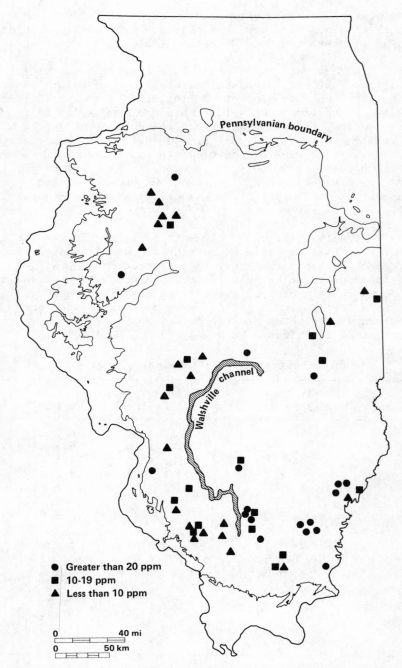

Figure 7. Regional distribution of bromine in the Herrin (No. 6) Coal Member (ppm, whole coal basis).

mineralogical variation within a given coal seam aid in the interpretation of the sequence of how, when, and where the inorganic elements are incorporated into coal.

REFERENCES

American Society for Testing and Materials, 1978, Test for forms of sulfur in coal, D 2492-77, *in*: 1978 Annual book of ASTM standards, Part 26, gaseous fuels; coal and coke; atmospheric analysis, ASTM, Philadelphia, p. 332-337.

Germani, M. S., Gokmen, I., Sigleo, A. C., Kowalczyk, G. S., Olmez, I., Small, A., Anderson, D. L., Failey, M. D., Guloval, M. C., Choquette, C. E., Lepel, E. A., Gordon, G. E., and Zoller, W. H., 1980, Concentrations of elements in the National Bureau of Standards bituminous and subbituminous coal standard reference materials, Analytical Chemistry, v. 52, p. 240-245.

Gladney, E. S., 1980, Elemental concentrations in NBS biological and environmental standard reference materials, Analytica Chimica Acta, v. 118, p. 385-396.

Gluskoter, H. J., Ruch, R. R., Miller, W. G., Cahill, R. A., Dreher, G. B., and Kuhn, J. K., 1977, Trace elements in coal: Occurrence and distribution, Illinois State Geological Survey Circular 499, 154 p.

Kuhn, J. K., Fiene, F. L., Cahill, R. A., Gluskoter, H. J., and Shimp, N. F., 1980, Abundance of trace and minor elements in organic and mineral fractions of coal, Illinois State Geological Survey, Environmental Geology Notes 88, 67 p.

Ondov, J. M., Zoller, W. H., Olmez, I., Aras, N. K., Gordon, G. E., Rancitelli, L. A., Abel, K. H., Filby, R. H., Shah, K. R. and Ragaini, R. C., 1975, Elemental concentrations in the National Bureau of Standards' environmental coal and fly ash standard reference materials, Analytical Chemistry, v. 47, p. 1102-1111.

Ruch, R. R., Gluskoter, H. J., and Shimp, N. F., 1974, Occurrence and distribution of potentially volatile trace elements in coal, Illinois State Geological Survey, Environmental Geology Notes 72, 96 p.

Ruch, R. R., Russell, S. J., Malhotra, R., Steele, J. D., Bhagwat, S. B., Dreher, G. B., Cahill, R. A., Frost, J. K., Harvey, R. D. and Ashby, J. F., 1979, Determination of valuable metals in liquefaction process residues, Illinois State Geological Survey Final Report to Department of Energy, FE-8004-42, 187 p.

AN AUTOMATED MULTIDETECTOR SYSTEM FOR INSTRUMENTAL NEUTRON
ACTIVATION ANALYSIS OF GEOLOGICAL AND ENVIRONMENTAL MATERIALS

Sammy R. Garcia, Walter K. Hensley, Michael M. Minor
Michael M. Denton, and Mary A. Fuka.
Los Alamos National Laboratory
P.O. Box 1663, M.S. 776
Los Alamos, N.M. 87545

ABSTRACT

An automated multidetector system for instrumental neutron activation analysis (INAA) of geological and environmental materials was constructed. The system was evaluated using NBS SRM's 1632a Coal and 1633a Coal Fly Ash. Data are presented to show the precision, accuracy, and stability of the system over a 3-month period. As a result of the evaluation, elemental concentrations for Dy, La, Eu, Tb, Yb, Lu, and Ta, which have not been reported in the literature, were determined.

INTRODUCTION

Several investigators have reported trace element concentrations in NBS SRM's 1632a Coal and 1633a Coal Fly Ash.[1-5] Because of the continuing need for standards that are reliable, independent data from other laboratories are of value to the scientific community.

The automated system was designed to process sediment samples from the National Uranium Resource Evaluation (NURE) program using the facilities at the Los Alamos National Laboratory Omega West Reactor (OWR).[6] Because of the nature of the NURE program, a large inventory of samples was the result. The system had to be capable of processing more than 60,000 samples annually.

System Description

There are two identical automated systems, each of which have a neutron detector and four Ge(Li) detectors. Standard ADS's, amplifiers and preamplifiers are used with the system and are temperature controlled to ensure long term stability. Figure 1 illustrates schematically one of the automated systems. Samples are placed in ethylene-butylene co-polymer irradiation vials 1.2 cm in diameter and 6.0 cm long and are weighed on a top-loading electronic balance which is interfaced to a PDP-11/34 computer.[7] The samples are loaded into clips which can hold 50 horizontally stacked samples. These clips are uniquely numbered and encoded with a wired socket so that the code can be read into the computer when a matching plug at the loader (see Figure 1) is attached to the clip. Each automatic loader was designed to hold four clips, corresponding to a maximum throughput of 400 samples per day for the two systems.

All data acquisition and sample movements are controlled by a PDP-11/34 computer utilizing standard CAMAC interface electronics. The gamma-ray spectra are automatically dumped to 9-tract magnetic tape for subsequent data reduction on a PDP-11/60 computer.

System Operation

It is necessary to run 400 samples daily during the 8 hours of irradiation time available. There are two reactor ports available allowing 126 seconds for each sample to be irradiated and moved to and from the reactor. To optimize the sensitivity for a maximum number of elements, the short and long irradiation times were adjusted accordingly. The following activation scheme was developed.

- Flux 6×10^{12} n/cm^{2}sec

- Irradiation Time 20 sec (short-lived activities)
  20 + 96 sec (long-lived activities)

- Decay Time 1274 sec (short-lived activities)
  14 days (long-lived activities)

Initially, detector energy calibration, resolution, and ADC linearity were measured with a NBS 4216 mixed gamma-ray source containing ^{109}Cd, ^{57}Co, ^{139}Ce, ^{203}Hg, ^{113}Sn, ^{85}Sr, ^{137}Cs, ^{88}Y and ^{60}Co. The resolution and characteristic peak shape as a function of energy were determined for each detector, and the resolution was found to be less than 2 keV FWHM at the 1332 keV line of ^{60}Co for all eight Ge(Li) detectors.

Each sample is counted three times: first for delayed-neutron assay in a high efficiency neutron detector and subsequently for short-lived and long-lived activities.[7] The samples are first irradiated for 20 sec followed by a 10 sec delay, then a 20 sec neutron count to determine uranium, and after a 1274 sec decay period, they are counted for 496 sec in a fixed geometry for the short-lived gamma-ray activities. The electronic dead-time for a normal sediment sample is generally under 15%. Following the "shorts" count, the

NEUTRON ACTIVATION ANALYSIS OF GEOLOGICAL MATERIALS

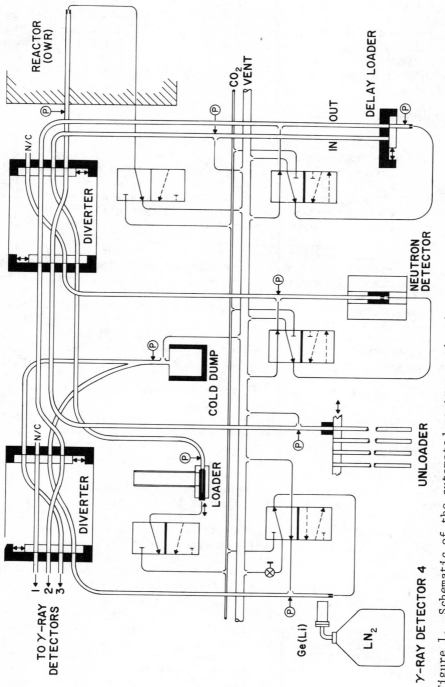

Figure 1. Schematic of the automated neutron activation system. P = Photocell, N/C = No Connection.

samples are returned to the reactor for an additional 96 sec irradiation to build up long-lived activities. The samples are then automatically unloaded and stored in pipes located underground for a 14-day decay period before being counted the third time. With each sample irradiation, a flux monitor reading is taken from a fission ion-chamber adjacent to the irradiation site and used for flux normalization.

EXPERIMENTAL

Pure element standards with a quoted metallic purity of 99.99% were used to prepare the trace element standards. The matrix used for the preparation of the pure element standards was silicon dioxide with a quoted metallic purity of 99.9% with a particle size less than 325 mesh. A Spex Industries 8000 Mixer/Mill was used for all mixing procedures. Polystyrene vials with methacrylate mortars were used for mixing the pure element standards.

A selected quantity of the stock pure element standard was pipetted onto 25 grams of silicon dioxide and evaporated to dryness under an infrared heat lamp. The standard was then mixed for 1 hour using the Spex Industries Mixer/Mill and triplicate splits taken to test for homogeneity. The standard deviation for the triplicate analysis was generally less than 2% for each standard prepared. Standards used for the "majors" were USGS standard rocks and NBS SRM's. Specifically the standards used were USGS QLO-1, STM-1, RGM-1, SCo-1, and NBS SRM's 1632a Coal and 1633a Coal Fly Ash.

The pure element and reference materials were irradiated following the activation scheme mentioned earlier in the text. The gamma-ray spectra are analyzed using an overlayed version of the computer code RAYGUN.[8,9] RAYGUN makes use of a library containing data on radionuclides of interest: half-life, cross-section, isotopic abundance, principal gamma-rays, and gamma-ray branching ratios. RAYGUN provides both qualitative and quantitative analysis of high resolution gamma-ray spectra. The code first determines a smooth background under a peak or peak-groupings then calculates whether the peak is too wide to be a singlet. If so, it attempts to split the peak according to the particular detector's characteristic peak shape at that energy. RAYGUN then calculates under each peak, or simply sums the area minus background for singlets or unsplittable groupings. Correction for known interferences is accomplished by extrapolating the intensity of the interfering line(s) from the intensities of the major lines of the interfering activity. The decay rate for each isotope is calculated from a weighted average of the various decay rates deduced from each of the gamma rays of the isotope. Elemental concentrations are then calculated from known cross sections, isotopic abundances, atomic weights, decay constants, neutron flux, and sample weight. There are 32 elements assayed from the automated analysis: 13 from the "shorts" analysis and 19 from the "longs" analysis. They include Na, Mg, Al, Cl, K, Ca, Ti, V, Mn, Sr, Dy and U (DNA) for the "shorts" and Sc, Cr, Fe, Co, Zn, Rb, Sb, Cs, La, Ce, Sm, Eu, Tb, Yb, Lu, Hf, Ta, Au and Th for the "longs". If one were to alter the counting scheme, additional radionuclides could be assayed at the expense of losing the high throughput achieved from the activation scheme described earlier. Other elements possible are Ga, As, W, Cd, Br, Au and Se.

RESULTS AND DISCUSSION

To evaluate the stability, accuracy and precision of the automated multidetector system, NBS SRM's 1632a Coal and 1633a Coal Fly Ash standards were used as the system evaluation materials. In-house and USGS rock standards were used to maintain a quality assurance program throughout the 3-month period this study was made. Long term stability was monitored using Na as a "shorts" reference element and Co as the "longs" reference element. These elements were chosen because there are no interfering peaks in their spectral analysis.

The results from the automated analysis of the NBS SRM's 1632a Coal and 1633a Coal Fly Ash are presented in Tables 1 and 2, respectively. The agreement of our elemental concentration with those of NBS and other values found in the literature is very good. From 50 data points accumulated over a period of 3 months, we find a standard deviation of 2.8% for the Na concentration and a 3.8% standard deviation for the Co concentration in 1633a Coal Fly Ash. It is important to recognize that these data points were collected from eight Ge(Li) gamma-ray detectors over the three month period.

In addition, we obtained values for several elements not previously assayed in NBS SRM 1633a Coal Fly Ash, including Dy, La, Eu, Tb, Yb, Lu, and Ta. We conclude that our systems have excellent long-term stability which is necessary for the production of high quality data.

Table 1. Multielement concentrations in NBS SRM 1632a Coal[a]. February 8 and 26, 1980 Activations

Element	Data Points	Average[b] ppm	Std Dev %	NBS[b] ppm	Literature[b] ppm
Na	24	884.	3.6		760[c], 850 ± 46[e]
Mg	3	1400	16		600-1300[c], 1300 ± 30[d]
Al(%)	24	3.07	4.1	(3.07)	2.97 ± 0.04[c], 2.9 ± 0.3[d], 3.01 ± 0.13[e]
Cl	24	897.	2.6		758 ± 50[c], 800 ± 70[d], 784 ± 17[e]
K(%)	18	0.43	15		0.41 ± 0.01[c], 0.42 ± 0.02[d], 0.42 ± 0.02[e]
Ca(%)	3	0.27	6.5		0.24 ± 0.02[c], 0.24 ± 0.02[d], 0.24 ± 0.02[e]
Ti	24	1720	9.8	(1750)	1650 ± 130[c], 1630 ± 70[d], 1620 ± 130[e]
V	24	46.9	5.4	44 ± 3	44[c], 44 ± 3[d]
Mn	24	33.7	3.6	28 ± 2	31 ± 2[c], 32 ± 3[d], 29 ± 5[e]
Ba	8	150	17		130[c], 122 ± 11[d]
Dy	22	2.56	10		2.2[c], 2.2 ± 0.3[d]
U	24	1.28	5.9	1.28 ± 0.02	1.2 ± 0.1[c], 1.21 ± 0.1[d]
Sc	24	6.56	3.5	(6.3)	6.5[c], 6.8 ± 0.6[d]
Cr	24	36	17	34.3 ± 1.5	35[c], 34 ± 2[d]
Fe(%)	24	1.12	8.2	1.11 ± 0.02	1.13 ± 0.03[c], 1.16 ± 0.03[d], 1.11 ± 0.06[e]
Co	21	6.6	17	(6.8)	7.0[c], 6.5 ± 0.2[d]
Cs	3	2.4	35	(2.4)	2.0[c], 2.0 ± 0.3[d]
La	14	15	17		16[c], 18 ± 2[d]
Ce	24	31.1	11	(30)	29[c], 32 ± 4[d]
Sm	24	2.5	14		2.3 ± 0.3[c], 2.8 ± 0.3[d], 2.10 ± 0.07[d]
Eu	19	0.51	16	(0.54)	0.55[c], 0.55 ± 0.03[d]
Lu	8	0.18	37		0.2[c], 0.18 ± 0.03[d]
Hf	11	1.8	18	(1.6)	1.6[c], 1.55 ± 0.08[d]
Th	24	4.8	12	4.5 ± 0.1	4.5[c], 4.8 ± 0.2[d]

[a]Data from two automated systems which represents eight Ge-(Li) gamma-ray detectors. [b]Values are in ppm unless otherwise specified. The values in parenthesis are not certified by the NBS. [c]Ref. 5. [d]Ref. 4. [e]Ref. 5. [f]Ref. 7.

Table 2. Multielement concentrations in NBS SRM 1633a Coal Fly Ash[a]
February 8,13,26 and May 16, 1980 Activations

Element	Data Points	Average[b] ppm	Std Dev %	NBS[b] ppm	Literature[b] ppm
Na	50	1720	2.81	1700 ± 100	2100 ± 600[c] 2100[d]
Mg	50	4500	11	4550 ± 100	
Al(%)	50	14.2	2.19	(14)	14.0 ± 0.2[c]
K(%)	50	1.84	7.37	1.88 ± 0.06	1.97 ± 0.04[c] 1.97[d]
Ca(%)	50	1.12	6.86	1.11 ± 0.01	1.29 ± 0.11[c] 1.29[d]
Ti	50	8060	4.58	(8000)	8400 ± 100[c] 8400[d]
V	50	301.	2.58	(300)	360 ± 40[c]
Mn	50	191.	2.25	(190)	190 ± 15[c] 31 ± 2[d]
Sr	50	819.	6.63	830 ± 30	
Ba	50	1500	5.95	(1500)	
Dy	50	16.6	7.69		
U[e]	50	10.2	2.12	10.2 ± 0.1	
Sc	50	40.6	3.16	(40)	
Cr	50	197.	6.44	196 ± 6	
Fe(%)	50	9.50	3.21	9.40 ± 0.10	9.7 ± 0.2[c] 9.7[d]
Co	50	46.2	3.82	(46)	
Zn	34	220	23	220 ± 10	
Rb	47	130	20	131 ± 2	
Sb	40	7.8	20	(7)	
Cs	50	10.6	10.4	(11)	
La	50	100	23		
Ce	50	183.	10.8	(180)	
Sm	50	20	22		16.0 ± 0.2[c]
Eu	50	2.98	11.2		
Tb	39	2.3	29		
Yb	50	10	18		
Lu	50	0.93	10.3		
Hf	50	7.78	10.9	(7.6)	
Ta	24	2.0	23		
Th	50	24.8	6.5	24.7 ± 0.3	

[a]Data from two automated systems which represent eight Ge(Li) gamma-ray detectors. [b]Values are in ppm unless otherwise specified. The values in parenthesis are not certified by the NBS. [c]Ref. 3. [d]Ref 4. [e]Ref. 7.

REFERENCES

1. G. A. Uriano, "Certificate of Analysis, SRM 1633a Trace Elements in Coal Fly Ash," NBS, National Bureau of Standards (1979).
2. J. Paul Cali, "Certificate of Analysis, SRM 1632a Trace Elements in Caol (Bituminous)," NBS, National Bureau of Standards (1978).
3. M. P. Failey, D. L. Anderson, W. H. Zoller, G. E. Gordon and R. M. Lindstrom, Anal. Chem., 51, 13 (1979).
4. M. Germani, I. Gokmen, A. C. Sigleo, G. S. Kowalczyk, I. Olmez, A. M. Small, D. L. Anderson, M. P. Failey, M. C. Gulovali, C. E. Choquette, E. A. Lepel, G. E. Gordon and W. H. Zoller, Anal. Chem., 52, 2 (1980).
5. E. S. Gladney, "Elemental Concentrations in NBS Biological and Environmental Standard Reference Materials-A Review," LASL, Los Alamos Scientific Laboratory, LA-UR 79-3158 (1980).
6. E. S. Gladney, D. B. Curtis, D. R. Perrin, J. W. Owens and W. E. Goode, "Nuclear Techniques for the Chemical Analysis of Environmental Materials," LASL, Los Alamos Scientific Laboratory, LA-8192-MS (1980).
7. M. M. Minor, W. K. Hensley, S. L. Stein, M. M. Denton, R. G. Martinez, J. W. Starner and M. E. Bunker, "An Automated Activation Analysis System for Trace Elements Assay of Stream Sediment Samples," LASL, Los Alamos Scientific Laboratory, to be published.
8. R. Gunnick, J. B. Niday, "Computerized Quantitative Analysis by Gamma-Ray Spectrometry," Description of the GAMANAL Program, Vol. 1 (1972).
9. RAYGUN, a derivative of GAMANAL, has been extensively modified by J. W. Starner of Los Alamos for use in the NURE Program.

MODES OF OCCURRENCE OF TRACE ELEMENTS AND MINERALS IN COAL:
AN ANALYTICAL APPROACH

Robert B. Finkelman*

U. S. Geological Survey

Reston, Virginia 22092

ABSTRACT

Recent analytical studies of coal have elucidated the modes of occurrence of many trace elements. Scanning electron microscope (SEM) work has shown that many trace elements can be present as micrometer-size accessory minerals. For example, Zn and Cd occur in sphalerite; Cu in chalcopyrite; Zr and Hf in zircon; rare earth elements (REE), Y, and Th in monazite and xenotime. Fission-track results indicate that U can occur as micrometer-size grains of uraninite, in zircons, or in organic combination. Most minerals in coals have either a detrital or an authigenic origin. More detailed studies are necessary to elucidate further the relationships among the trace elements, minerals, and organic components in coal.

INTRODUCTION

Atomic and nuclear analytical techniques have been widely used for trace element analysis of coals. Emphasis has generally been placed on determining the concentrations of the elements rather than on determining their mode of occurrence. However, it is the mode of occurrence that dictates the behavior of the element during the cleaning, combustion, conversion, or leaching of the coal or during the weathering of the coal or of its products upon disposal.

----
*Present address: Exxon Production Research Company, P. O. Box 2189, Houston, Texas 77001.

Recently various analytical techniques, including atomic and nuclear methods, have been used to determine the modes of occurrence of many trace elements in coal.[1] The intent of this paper is: to review briefly recent results obtained from several of these techniques; to suggest the most likely modes of occurrence for about 60 trace elements in coal; to discuss the sources of the minerals found in coal; and to suggest fertile areas for future research.

ANALYTICAL TECHNIQUES: REVIEW OF RECENT RESULTS

The SEM equipped with an energy-dispersive X-ray analyzer has proved to be an ideal instrument to study the inorganic constituents in coals.[2] This system was used to detect and analyze in situ, micrometer-size minerals in polished blocks of more than 100 coal samples from worldwide locations. The results indicate that many trace elements in coal can be present as micrometer-size accessory mineral grains scattered throughout the organic matrix (macerals).[3] For example, in many coals studied, Zn and Cd occur predominantly in the mineral sphalerite;[4] Cu in chalcopyrite; Zr and Hf in zircon; REE, Y, and Th in monazite and xenotime; Ba in barite; and Mn in siderite.

Electron microprobe analyses[5] have indicated that, in the Upper Freeport coal from Indiana County, Pennsylvania, As is present in solid solution in pyrite, probably having been emplaced by reaction of pyrite with epigenetic As-bearing solutions that pervaded the coal along fractures. Statistical analysis indicates that Hg has a similar mode of occurrence. The difference in the mode of occurrence between As and Hg and the other chalcophile elements is reflected in their behavior during sink-float separation of the coal. Those elements forming micrometer-size minerals within the macerals are concentrated in the lighter specific gravity fractions. Those elements associated with massive pyrite are concentrated in the heavier specific-gravity fractions.

The ion microprobe mass analyzer (IMMA) has recently been used in a quantitative mode on coal.[6] Clay-rich particles from the Upper Freeport coal were analyzed using a broadbeam (50 μm) approach. Recalculation of these data to a whole coal basis showed that these areas of concentrated detrital minerals contain virtually all of the following elements: Be, B, Mg, Al, Si, K, Cr, Mn, Rb, Zr, Nb, and Ba; substantial amounts of Li, Na, Ti, and V; and perhaps Cs. Qualitative IMMA results from pyrite indicate the presence of Tl and some Pb.

Recently,[7] an atomic absorption technique using a graphite furnace electrothermal atomizer was used to determine low-level concentrations of Pt, Pd, and Rh in coal. Most values were below 2 ppb on a whole coal basis. These low values argue against any

TRACE ELEMENTS AND MINERALS IN COAL

significant organic or sulfide complexing of Pt metals.

A variation of the fission-track technique[8] has been used to determine the mode of occurrence of U in coal. In one sample, virtually all the U occurred as micrometer-size grains of uraninite. In another sample, a significant amount of the U was associated with detrital accessory minerals, such as zircon. In other samples, the bulk of the U appeared organically bound.

Organic associations constitute a major mode of occurrence for several trace elements such as Se, Br, Cl, I, and Ge. X-ray diffraction and SEM results indicate that Ti-bearing minerals are common in many coals, however, no more than about 50 weight percent of the Ti can be accounted for in this manner[3]. The remainder may be bound to the organic constituents.

Evidence from analytical transmission electron microscopy[9] indicates the presence of a myriad of submicron crystallites, even in the cleanest coal. Perhaps some elements considered to be organically bound will be found in these crystallites.

SUGGESTED MODES OF OCCURRENCE

A list of the probable modes of occurrence of trace elements in coal is presented below. It must be emphasized that these suggestions are derived from broad generalizations and should be treated as such. It is likely that not all of a particular element in a coal will be in the mode or modes suggested. Nor will the element even occur in the suggested mode in every coal. Nevertheless, there appears to be sufficient evidence to indicate that substantial amounts of the element will be found in most coals in the modes indicated.

- Sb - probably as an accessory sulfide in the organic matrix
- As - solid solution in pyrite
- Ba - barite, crandallites, and other Ba-bearing minerals
- Be - organic association, clay
- Bi - accessory sulfide, perhaps bismuthinite
- B - generally organic association, illite
- Br - organic association
- Cd - sphalerite
- Cs - inorganic association, clays, feldspars, or micas
- Cl - organic association, NaCl
- Cr - clays
- Co - associated with sulfides, such as pyrite and linnaeite
- Cu - chalcopyrite
- F - unclear, probably several inorganic associations, such as apatite, amphibole, clays, and mica
- Ga - clays, organic association, sulfides

Ge – organic association, rarely in silicates, sphalerite
Au – native gold, gold tellurides
Hf – zircon
In – sulfides or carbonates
I   – organic association
Pb – pyrite, galena, PbSe, coprecipitated with Ba
Li – clays – illite, mixed-layer
Mn – siderite, calcite
Hg – solid solution with pyrite
Mo – unclear, probably with sulfides, or organic constituents
Ni – unclear, may be with sulfides, organic constituents, or clays
Nb – oxides
P   – various phosphates, some may be organically associated
Pt – perhaps native Pt alloys
REE – phosphates
Re – sulfides or organic constituents
Rb – probably illite
Sc – unclear, clays, phosphates or may have organic association
Se – organic association, as PbSe in Appalachian coals, pyrite
Ag – probably silver sulfides, but may be complex
Sr – carbonates, phosphates, organic association in low rank coals
Ta – oxides
Te – unclear
Tl – sulfides, probably epigenetic pyrite
Th – REE phosphates
Sn – inorganic, tin oxides or sulfides
Ti – titanium oxides, organic association, clays
W   – unclear, may have organic association
U   – organic association, zircon
V   – clays (illite)
Y   – REE phosphates
Zn – sphalerite
Zr – zircon

SOURCES OF MINERALS IN COAL

The sources of the minerals in coal may be as varied as the modes of occurrence of the trace elements.

Mineralization occurring after the formation of the coal is referred to as epigenetic mineralization. Epigenetic minerals, deposited in cleats and joints, are the most obvious and often the least troublesome of the minerals in coal. In most coals, these minerals are less abundant than minerals from other sources.[10] Epigenetic minerals are relatively easy to separate from the coal and are amenable to study by X-ray diffraction, petrography, and other conventional analytical techniques.

The sources of the non-epigenetic minerals in coal are often difficult to determine. Investigation of the sources included textural evidence, that is, the relationships between the minerals and macerals (Fig. 1), mineralogical evidence (Fig. 2), and chemical evidence such as REE distributions (Fig. 3) and trace element ratios. All three lines of investigation indicate that the bulk of the minerals in coals with ash contents greater than about 5 weight percent are detrital, that is, they had been physically carried by water or air into the depositional basin. The evidence also indicates substantial amounts of authigenic minerals in most coals. These minerals have formed in place before consolidation of the sediment. Virtually all the minerals in low-ash coals (5 weight percent) are authigenic; from 10 to 50 percent of the minerals in the higher-ash coals also are authigenic. The distinction between the detrital and authigenic mineral suites is quite sharp. Table 1 shows the mineral, maceral, and textural relationships of these two suites. In contrast to the epigenetic minerals, both mineral suites formed simultaneously with the enclosing coal and are termed syngenetic minerals.

A further complication is the evidence of widespread diagenesis, that is, post depositional alteration of the preexisting minerals, and of remobilization of some trace elements.[1]

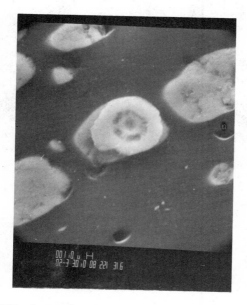

Figure 1. An SEM photomicrograph of an authigenic crandallite group mineral (circular) and authigenic kaolinite (light grey) filling pores in an inertinite (dark grey) particles. Scale bar is 1 micrometer.

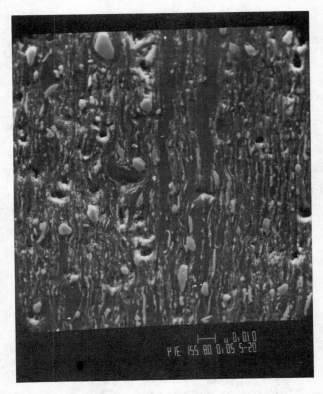

Figure 2. An SEM photomicrograph of detrital quartz (large, angular), rutile (bright grain in center), and illite (light grey) in a polished block of coal. Scale bar is 10 micrometers.

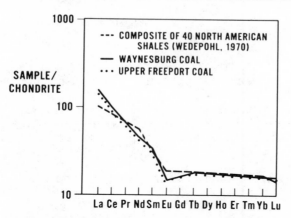

Figure 3. Comparison of REE abundances between ash-rich coals and shales.

CONCLUSIONS

The renewed interest in coal and the application to coal of new and sophisticated analytical techniques have allowed us to understand better the modes of occurrence of the trace elements. What more do we need to know about this subject?

First, we must understand the variation in mode of occurrence on all scales – within a coal bed, between coal beds in a single basin, and between coal basins.

Second, we must be able to quantify our impressions so that we can determine the proportion of each element in each of the various possible modes of occurrence. In order to do this we must refine our ability to separate physically and chemically the inorganic phases from the organic phases of coal. In one promising approach to this problem Kuhn et al.[11] progressively leached coals with mineral solvents and analyzed the "demineralized" residue. Finkelman and Simon,[12] using a similar leaching technique on a single coal, analyzed the inorganic residue at each stage. Another promising approach was taken by Palmer and Filby[13] who effected size-density separations of the fine-grained, low-temperature ash and then chemically and mineralogically analyzed these separates.

Finally, we must improve our ability to analyze smaller amounts of material, and to identify a broader array of elements and do so at lower concentrations. It is here that nuclear and atomic techniques can contribute most significantly contributions to fossil energy research.

TABLE 1
CHARACTERISTICS OF SYNERGENETIC COAL MINERALS

	AUTHIGENIC	DETRITAL
MINERAL RELATIONSHIPS	KAOLINITE SULFIDES CARBONATES CRANDALLITE GROUP APATITE BARITE	ILLITE MIXED-LAYER CLAYS QUARTZ ZIRCON RUTILE FELDSPAR
MACERAL RELATIONSHIPS	. INERTINITES . VITRINITE	. CARBOMINERITE[14] . FRAGMENTED MACERALS
TEXTURAL RELATIONSHIPS	. DISPERSED . PORES, PODS, OR ISOLATED GRAINS . EXHIBITS CRYSTAL FACES . INTIMATELY INTER-GROWN	. BANDED . ALIGNED . SUBANGULAR TO SUBROUNDED . INTERGROWTHS RARE (SOME RUTILATED QUARTZ

ACKNOWLEDGEMENTS

I would like to thank H. Gluskoter and Y. Sumartojo of Exxon Production Research Company and F. Simon of the U. S. Geological Survey for their helpful comments.

REFERENCES

1. Finkelman, R. B., Modes of occurrence of trace elements in coal. University of Maryland. Ph.D. dissertation, 301 p., 1980.

2. Finkelman, R. B., and Stanton, R. W., Fuel, v. 57, p. 763, 1978.

3. Finkelman, R. B., SEM 1978/I, p. 143, 1978.

4. Most mineralogical identifications have been confirmed by optical characterization or by X-ray diffraction analysis.

5. Minkin, J. A., Finkelman, R. B., Thompson, C. L., Cecil, C. B., Stanton, R. W., and Chao, E. C. T., Ninth International Carboniferous Congress, Compte Rendu, in press.

6. Finkelman, R. B., and Aruscavage, P. J., Coal Geology, v. 1, no. 2.

7. Finkelman, R. B., Simons, D. S., Dulong, F. T., and Stanton, R. W., Coal Geology, in press.

8. Finkelman, R. B., and Klemic, H., U. S. Geological Survey Journal of Research, 4, 6, 715, 1976.

9. Lin, J. S., Hendricks, R. W., Harris, L. A., and Yust, C. S., Acta Crystall., 11, 621, 1978.

10. Mackowsky, M.-Th., in Coal and Coal-bearing Strata. D. G. Murcheson and T. S. Westall, eds, 309-321, 1968.

11. Kuhn, J. K., Fiene, F. L., Cahill, R. A., Gluskoter, H. J. and Shimp, N. F., Ill. Geol. Survey, Environ. Geol. Notes 88, 67, 1980.

12. Finkelman, R. B., and Simon, F., unpublished data.

13. Palmer, C. A., and Filby, R. H. Personal Communication, 1980.

14. An association of coal with 20-60 volume precent mineral matter: International Committee for Coal Petrology, International Handbook of Coal Petrography: Supplement, Paris, France, 1971.

DEVELOPMENT AND CALIBRATION OF STANDARDS FOR PNAA ASSAY OF COAL*

Y. Nir-El, B. Director, T. Gozani, H. Bernatowicz,
E. Elias, D. Brown, and H. Bozorgmanesh

Science Applications, Inc.
Sunnyvale/Palo Alto, California

The applicability of the Prompt Neutron Activation Analysis (PNAA) method to large (100kg) coal samples has been demonstrated in a laboratory system developed by SAI, Sunnyvale[1-3]. The PNAA technique is based on irradiating the coal with neutrons and simultaneously detecting the characteristic prompt gamma radiation resulting from thermal neutron absorption in various elements in the sample. Detailed elemental analysis of the sample is obtained through careful analysis of the emitted gamma-ray spectrum. The accuracy of this method depends on the availability of a reliable set of coal standards. The elemental composition of this set should be well known and uncertainties of the order of ± 1% are desirable. The present available standards of coal are very small (1 g) and therefore not suitable for PNAA work. The aim of the present study was to develop and calibrate a new set of large coal standards which would be used in future PNAA field systems. This set has to cover a broad range of different types of coals, taking into account physical characteristics and elemental compositions. The present set of standards was calibrated against the sampling and analytical ASTM methods which were found to give reliable values.

Twenty-one different coal samples were contained in standard containers made of materials that do not interfere with PNAA measurements. Containers were filled to the top then the coals were bagged and sealed in polyethylene. Precautions were taken to ensure mechanical rigidity of the containers. Environmental effects, like moisture or density changes due to vibrations and shocks, were kept

---

*Work sponsored in part by Electric Power Research Institute.

to minimum. The coals used covered different particle size distributions, since the measuring systems were designed to minimize bulk density effects. The average bulk density of the coals was 0.90 ± 0.03 g/cm^3.

Measurements were carried out in a simulated rectangular belt geometry in which the broad sides of the container face the neutron source and the detector respectively. 120-350 µg ^{252}Cf in a bismuth holder was used as a neutron source. The gamma spectrum was measured by a Ge(Li) detector shielded by 5 inches of water and boric acid mixture, and a 2.5 inch thick cylindrical cup made of borated epoxy. The coaxial Ge(Li) detector has 17% efficiency and 1.9 and 6.5 keV energy resolution (FWHM) at 1.33 MeV and 7.640 MeV, respectively. After removal of the neutron source, a spectrum was measured for 7.5 hours to determine the ^{24}Na activity produced by thermal neutron capture in sodium.

A specially developed electronic system was used which is capable of handling high count-rates up to 200,000 cps while maintaining good energy resolution and spectral quality. The system is based on a specially designed time-variant filter with a trapezoidal weighing function[4]. Energy spectra covering the range 1-8 MeV were measured with a multichannel analyzer. Analysis of gamma peaks is done by a rapid program--GELI--which also corrects for interferences in peak areas.

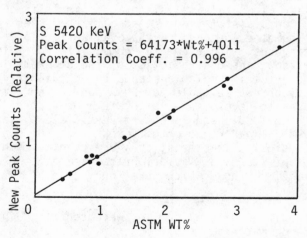

Figure 1. Calibration Curve for Sulfur in a Wide Variety of Coals.

Twenty-one coals which have a detailed ASTM chemical analysis were measured in the present work. Corrected peak areas were plotted versus ASTM weight percentages (wt%) for each coal. A least-squares fit of a straight line to the measured points gives a standard calibration curve for a specific γ line belonging to a certain element. Figure 1 shows a typical standard calibration curve for sulfur obtained by measuring its 5.420 keV prompt gamma transition. The correlation coefficient of the fitted straight line is 0.996, showing a very good linear correlation between the PNAA and the chemical ASTM values. The error of the slope is only 0.5% and the small offset may be explained by a possible small bias in the sampling required for the ASTM chemical analysis. The 5.420 keV peak areas of sulfur are regularly determined with high-precision--a typical value is ± 1% for a coal with sulfur content in the range of 1 to 2 wt%. The total error, including the calibration error, gives a final relative error of ± 1.2%.

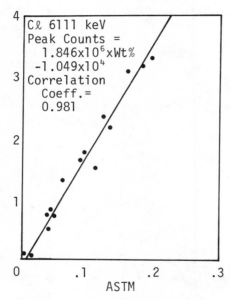

Figure 2. Calibration Curve for Chlorine in a Wide Variety of Coals.

The high sensitivity, selectivity and precision of the PNAA method are displayed in Figure 2 which presents the case of chlorine, a minor constituent of coal, having concentrations of 0-0.2%. Although the chemical analysis of Cl has inherently considerable uncertainties, the general good agreement with the present PNAA results is remarkable.

The coal samples used in these experiments establish the standard linear calibration curves and thus form a set of coal standards developed for practical application of the PNAA method. In an online system, an unknown coal is measured at identical experimental conditions. Peak areas calculated by the GELI program are readily converted into wt% by use of the standard calibration curves. The coal standards are now used for calibration of the Nucoalyzer-Sulfurmeter and CONAC being built by Science Applications, Inc.

REFERENCES:

1. T. Gozani, et al. "Prompt Neutron Analysis-Applications to Coal Analysis", ANS Trans 26, 160, (6/77). See also EPRI Reports RP 983-1 Vol. 1 through Vol. 8a 1979.

2. T. Gozani, et al. "Coal Steam Composition Analysis for Process Control Using Prompt Neutron Activation Analysis", ISA Symp. Instrumentation and Control for Fossil Demonstration Plants, June 1977.

3. T. Gozani, et al. "Advanced Techniques for Laboratory and On-Line Analysis of Coal", ISA Symp. Instrumentation and Control for Fossil Demonstration Plants, August 1979.

4. J. McQuaid, T. Gozani and D. Brown, "High Countrate High Resolution Electronic System for Ge(Li) Detectors". IEEE Trans. 19, Nuclear Science Symp. 1980.

ON-LINE NUCLEAR ANALYSIS OF COAL AND ITS USES[*]

D. R. Brown, H. Bozorgmanesh, T. Gozani

Science Applications, Inc. Palo Alto, California 94304

J. McQuaid

Lawrence Livermore Laboratory Livermore, California

ABSTRACT

On-Line Nuclear Analysis of Coal based on prompt neutron activation represents a technological breakthrough for real-time process control necessary for optimum efficiency in the use of coal. These analyzers are presently being installed to solve a variety of current problems in coal usage. This paper describes the general features of these instruments and gives a detailed discussion of a high counting rate spectroscopy system used with an analyzer based on a germanium detector. A brief discussion of various applications of these analyzers include control of coal blending, control of coal burning efficiency, and quality control in coal beneficiation and synfuel processes.

INTRODUCTION

In recent years the increased importance of coal in the national energy policy has brought to light many problems associated with traditional methods of coal processing and utililzation. The efficiency of new and costly coal beneficiation and synfuel processes can be greatly affected by changes in coal composition. The rise in operating costs of power generation due to compliance with stricter environmental standards for coal combustion emissions make it necessary for utilities to closely monitor the composition of the coal they receive. Traditional analytical and wet chemistry methods of monitoring coal composition are incapable of providing anything approaching real-time analysis

---

[*]Work supported in part by Coal Cumbustion Systems Division of the Electric Power Research Institute, Palo Alto, California

of coal. Typically, small samples of the coal stream are laboratory analyzed and the results made available between a day to a week later. By this time the coal is through the process stream, often already burned and no control is possible.

The need for real-time analysis of composition of bulk quantities of coal and the applicability of prompt neutron activation analysis (PNAA) to accomplish this has long been recognized[1,2]. Since 1975, Science Applications, Inc., under partial sponsorship by the Electric Power Research Institute (EPRI) has developed the prompt neutron analysis to a mature technology[3,4,5,6]. Two on-line nuclear analyzers of coal called Nucoalyzers, have been fabricated. These instruments which are based on lower resolution NaI detectors perform a limited compositional analysis of coal, mainly for its sulfur content. The first of these instruments is currently on-line at Detroit Edison's Monroe 3000 MW generating station and the second is scheduled to go on line at a TVA power station this winter. At present we are fabricating a second germanium detector based nuclear coal analyzer which will perform complete elemental analysis of coal in the process stream and will be installed in at TVA in summer, 1981.

The paper describes some general features of these instruments. More details are given on the high counting rate spectroscopy system developed for the high resolution on-line analyzer. Finally, a few of the areas to which these on-line nuclear coal analyzers are being applied are discussed.

System Description

A schematic of a Nucoalyzer is shown in Figure 1. The instrument is about 7.6m long, stands 4.3m high and weighs over 9 metric tons. Coal enters the system through an input hopper and is gravity fed and leveled onto a conveyor belt. A microwave transmission meter monitors the moisture[7] in the coal stream and a mass sensor is used to monitor the mass flow. The coal then enters the shielded nuclear interrogation region where it is irradiated by neutrons from a 300 mg source underneath the belt. The capture gamma rays are detected in a germanium detector system positioned above the source and coal. The signals are processed through an analog electronics module and fed into a microcomputer where the elemental abundances in the coal are deduced and output in standard engineering units. This microcomputer is also used for data logging and process control.

Figure 2 shows the Nucoalyzer which is installed at Detroit Edison. This instrument was calibrated using the PNAA coal standards discussed in a previous paper[8]. The photograph was taken during the acceptance testing at SAI's Coal Analysis Laboratory in Sunnyvale, California.

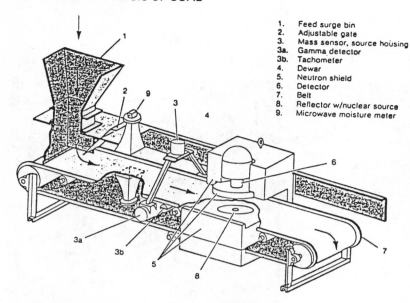

Figure 1. Schematic View of the Nucoalyzer

High Counting Rate Spectroscopy System

A fundamental problem at the outset of this program was to meet the conflicting requirements of good resolution and high counting rate. High counting rates are necessary to keep measurement times short enough, at a given level of statistical precision, to allow process control. The large dynamic range of the spectrum (1 to 10 MeV) means that the gamma peaks appearing in the spectrum span only a few channels. Degradation in resolution due to pulse pile-up can increase significantly the analysis uncertainties. Traditional pile-up rejection in commerical time invariant spectroscopic amplifiers are not useful here because they sacrifice throughput (in many cases up to 80 or 90%) for improved resolution. To overcome these problems, a high counting rate spectroscopy system based on time-variant shaping was developed.

This spectroscopy system includes a germanium (or Ge(Li)) detector with 20% to 30% efficiency coupled through a modified pre-amp to an RLC amplifier. The block diagram for this system is shown in Figure 3. The amplifier provides pole-zero cancellation, DC stabilization, and gaussian pre-filtering. The gated integrator convolutes the gaussian function with a rectangular function to produce a trapezoidal output wave form. This final wave form is about one-tenth that of the noise equivalent traditional time-invariant system. This large reduction in pulse width allows high counting rates at greatly reduced pulse pile-up.

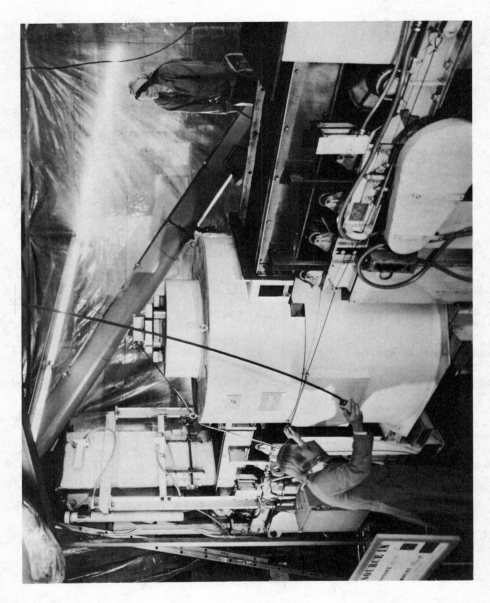

Figure 2. Nucoalyzer During Testing at SAI's Coal Analysis Laboratory, in Sunnyvale, CA

The time-varient gated integrator has a flat-top weighting function which is very important. It is this feature that gives the optimum weighting function for slow charge collection in a large Ge(Li) coaxial detector. These slow events result from defect traps in the germanium detector with collection times that are a function of gamma-ray energy. Therefore, a large number of slow events can be processed without degradation of the energy spectrum.

The system is capable of handling counting rates up to 150,000 counts per second while maintaining a good resolution (e.g, from <0.2% at 2 MeV to <0.1% at 8 MeV) and spectrum symmetry. This system allows us to maximize data throughput. Figure 4 shows a portion of spectra from three widely different coal types measured with this system. The rather clean separation of very closely spaced gamma ray lines illustrates the excellent resolution.

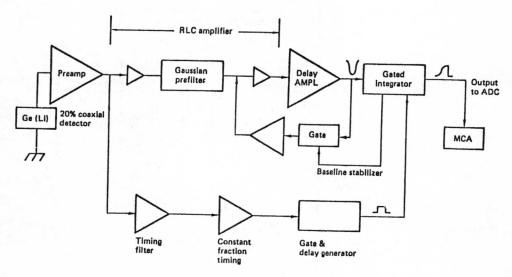

Figure 3. Block Diagram of Time-Varient System

These spectra also show the great differences in the capture gamma spectra from coal to coal. The compton background increase by a factor of five going from the low sulfur western coal to high sulfur east Ohio coal.

APPLICATIONS

Coal Blending

In many cases the most economical strategy for a coal burning utility to comply with sulfur dioxide emissions regulations is through blending of high sulfur coal with a low sulfur coal. This is the case for Detroit Edison's 3000 megawatt Monroe Power Plant where the first Nucoalyzer is presently in operation. The SO_2 emissions limits, within which this plant must operate, are 3.68 pounds of SO_2 per million BTU of heat input through 1984 and thereafter, 1.6 SO_2 per million BTU.

Although the blending strategy can assure compliance, a detailed economic study by Detroit Edison shows that a substantial savings is to be gained by using the Nucoalyzer to control the blending of the high and low sulfur coal[9]. The less expensive coal has a sulfur range of about 2%-4% whereas the more expensive coal has a sulfur range of 0.5%-1.1%. With the Nucoalyzer monitoring

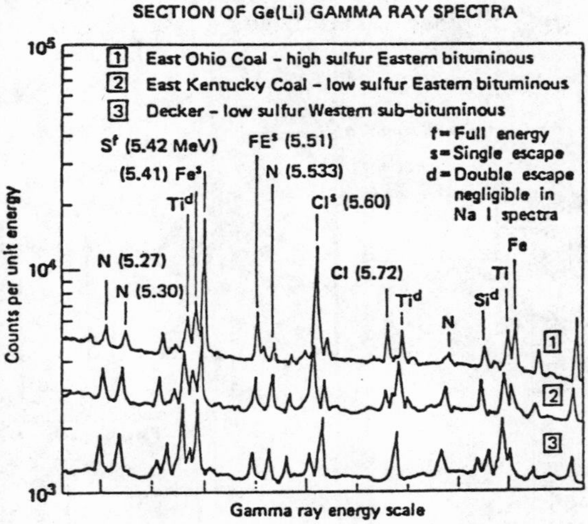

Figure 4. A Portion of the Energy Spectrum for Widely Different Samples of the Coal

the sulfur content of the blends, the more of the less expensive, high sulfur coal can be used, as shown:

Meter Control of Blending	Percent High Sulfur Eastern	Percent Low Sulfur Eastern
None	66%	34%
Nucoalyzer (Sulfurmeter only)	77.5%	22.5%
Nucoalyzer (Sulfurmeter/BTU)	78%	22%

Combustion Control

A Nucoalyzer in the process stream allows feed-forward plant control. The advantages that can be realized from ths application are potentially great. Through analysis of ash composition, and its slagging characteristics, the excess air fed to the boiler could be reduced. The same ash analysis could be used to reduce slag-related outages. Both these applications can increase the efficiency of coal burning. A rather different area of economic benefit can be realized from on-line BTU analysis. The heat value of the coal can be deduced from the elemental composition and moisture content. The accurate determination of the heat rate of all generating units in a large power system in real-time is necessary to optimize the system fuel input. Continuous BTU analysis of the coal into each unit can be used in the heat rate determination. From this knowledge, the load dispath equations can be optimized continuously.

Quality Control

Control of coal quality based on compositional analysis is presently important both for steam and metallurgical coal. The optimization of the coming synfuel production will also depend on careful control and process parameters based on variation in coal composition.

In all these applications the on-line nuclear analysis allows, heretofore impossible, real-time process control.

REFERENCES

1. R. F. Stewart, "Nuclear Measurements of Carbon Bulk Materials", ISA Transactions, Volume 6, Number 3, pp. 200-208, (1967).
2. T. Gozani, G. Reynolds, E. Elias, T. Maung, H. Bozorgmanesh, and V. Orphan, "Coal Stream Composition Analysis for Process Control Using Prompt Neutron Activation Analysis", 1977 Symposium on Instrumentation and Control for Fossil Demonstration Plants, Chicago ANL-78-7, July 1977.
3. T. Gozani, "The Development of Continuous Nuclear Analysis of Coal - A Review", ANS Trans., $\underline{28}$, 1978, 97.
4. T. Gozani, H. Bozorgmanesh, D. Brown, E. Elias, T. Maung, and G. Reynolds, "Coal Elemental Analysis by Prompt Neutron Activation Analysis", ANS Trans., $\underline{28}$, 1978, 88.
5. E. Elias, T. Gozani, V. Orphan, J. Reed, and D. Shreeve, "Prompt Neutron Activation Analysis - Applications to Coal Analysis", ANS Trans. $\underline{26}$, 1977, 160.
6. T. Gozani, et. al., "Nuclear Assay of Coal", Volumes 1 through 8, EPRI Report FP989. RP983-1, January 1979.

7. D. R. Brown, et. al., "Moisture Determination in Coal: Survey of Techniques", The 1978 Symposium in Instrumentation and Control for Fossil Demonstration Plants, ANL-78-62, pp. 514-539 (1978).
8. Y. Nir-El, B. Director, T. Gozani, E. Elias, D. R. Brown, and H. Bozorgmanesh, "Development and Calibration of Standards for PNAA Assay of Coal", paper given at ANS meeting on Atomic and Nuclear Methods in Fossile Energy Research, December 1-4, 1980, in Mayaquez, Puerto Rico.
9. O. J. Tassicker, T. Gozani, and R. Buckler, "Control of Coal Blending by On-Line Nuclear Analyzer of Coal".

ANALYSIS OF MINERAL PHASES IN COAL UTILIZING FACTOR ANALYSIS

Bradley A. Roscoe and Philip K. Hopke

Institute for Environmental Studies and
Nuclear Engineering Program
University of Illinois
1000 W. Western Ave.
Urbana, Illinois 61801

INTRODUCTION

There is much interest in the mineral phase inclusions of coal. The contribution of these mineral phases, i.e. calcite, pyrite, etc., to a coal sample may be determined utilizing several techniques: infrared absorption analysis (1), differential thermal analysis (2), electron spectroscopy (3), and x-ray diffraction (4). Other techniques are being utilized to learn how the minor and trace elements are distributed in these mineral phases. Neutron activation analysis in conjunction with coal washability studies have produced some information on the general trends of elemental variation in the mineral phases (5). These results have been enhanced by the use of various statistical techniques. One such statistical technique is target transformation factor analysis (TTFA) which has been previously demonstrated to be able to resolve sources of airborne particulate matter (6,7). Initial studies of TTFA have shown it to be able to produce elemental profiles of the mineral phases in coal (8). An advantage of TTFA is its ability to produce reliable quantitative results with a minimun of prior knowledge. In order to further explore the utility of this approach, a data set consisting of physically fractionated coal samples was generated. These samples were analyzed by neutron activation analysis and then their elemental concentrations examined using TTFA.

FACTOR ANALYSIS

The approach of the analysis is based on the assumption that a coal sample is a mixture of materials from a number of independent sources (mineral phases) that have relatively constant compositions. Thus, the amount of any given element present in a sample can be expressed as the sum of contributions from each of these mineral phases. Mathematically, the concentration of the ith element in the jth sample, x_{ij}, can be expressed by

$$x_{ij} = a_{i1}f_{1j} + a_{i2}f_{2j} + \ldots + a_{ip}f_{pj} \qquad (1)$$

where p is the number of mineral phases present. Each data point, x_{ij}, is represented as the product of a_{ik} and f_{kj}, where a_{ik} is the concentration of the ith element in the kth phase while f_{kj} is the amount of phase k in sample j.

The elemental balance described in equation 1 has been used in several fields of study: source resolution of environmental aerosols (6,7) and geological studies (9,10,11). In the geological area, most of the work consists of determining general trends of the data such as grain size distributions. The technique can also be applied to identify the mineral phases present and their composition (8). The goals of factor analysis are threefold: 1) to determine p, the number of mineral phases present in the sample, 2) to determine A, the elemental source profiles of those phases, and 3) to determine F, the contribution of each phase to each sample.

To perform a factor analysis, a set of n coal samples for which the concentration of m elements are known is required. The data set must include at least as many samples as there are mineral phases to be identified in the data. The more redundant data that are available, the better the factor analysis will be assuming that the data are reliable.

The data to be factor analyzed consists of multiple samples containing various amounts of different mineral phases. By forming a matrix of sample correlations, one can determine the number of phases present. This is done by diagonalizing the correlation matrix and looking at the eigenvalues and eigenvectors obtained. Several tests, which are described in the literature (6,12), can then be applied to the data so that the number of mineral phases present is determined. In addition, the eigenvectors of the largest eigenvalues can be related to the composition of the mineral phases that we are trying to determine.

Target transformation rotations are used to determine the mineral phase concentration profiles from the eigenvectors of the correlation matrix. Suspected mineral profiles are tested by rotating the eigenvectors toward them utilizing a least-squares fit (7,13). The error observed in this rotation tells us if the suspected mineral may possibly be in the sample. Using this process, suspected minerals may be either kept or deleted from further analysis. Further refinement of the retained test minerals may be accomplished by replacing the original mineral profile with the predicted profile found by the least-squares fit. Several iterations of this type may be used until the error of the least-squares rotation is acceptable.

The target transformation approach, as described above, may be modified to take into account the errors in the data by weighting the least squares fit (7). Several different weights may be used in this rotation including the elemental variance in the raw data, the squared average experimental error in the elemental concentrations, and the error-weighted variance in the raw data (14,15). These weighting techniques permit inclusion of the inherent error in the data and thus cause the iteration of the source profiles to converge more rapidly.

As has been recently shown by Roscoe and Hopke (15), a minimum of prior knowledge concerning the mineral phases is needed. Rather than start the target transformation with a suspected elemental profile of the mineral phase, it is possible to start with much simpler profiles. A unique test vector, one with zero concentration for all the elements except for one which is set to unity, may be used as an initial mineral phase. After the first target transformation rotation, the predicted vector can be substituted for the original unique vector, and the iteration continued. After several iterations, the initially unique vector will converge to a new vector with a complete elemental concentration profile. If all possible unique vectors are iterated in this manner and then normalized, a cluster analysis (16) will show specific groupings of similiar endpoints in the iterations which can be correlated to the elemental profiles of the mineral phases in the coal. In this manner, the mineral phases present in the coal can be identified with no prior knowledge.

PREVIOUS EXPERIMENTAL RESULTS

In a previous study (8), coal data reported by the Illinois State Geological Survey (5) were analyzed using target transformation factor analysis and no weighting of the rotation. Six mineral phases were identified as being present in the coal: organic fraction, organic sulfur, alumino-silicate, calcite, iron sulfate, and miscellaneous iron. The elemental profiles, including major, minor, and trace elements, of each phase were also determined. These general results were then compared to those of the Illinois State Geological Survey. As a next step, these samples were analyzed using a weighted rotation. The weighting used was that of the variance in the elemental data. Because of the nature of the data set, no additional information was obtained with the weighting.

The problem of the data set, with respect to factor analysis, is that the samples used were from a single coal seam covering three states. Since it is unrealistic to believe that the mineral phases of coal samples taken over three states have the same average elemental profiles, even though they are taken from the same coal seam, the initial assumption may have been violated. An ideal set of coal data for a factor analysis would be multiple samples from a specific mine and sub-dividing each sample on the basis of some physical property such as density. In this way, the variance of the mineral phases between samples is increased while the variances of the average mineral phase elemental profiles are decreased.

SAMPLE COLLECTION, PREPARATION, AND ANALYSIS

For this study, samples from an Illinois coal mine were obtained from the run-of-the-mill pile before washing. The ten samples acquired were dried so that all measurements would be made with respect to the dry weight of the coal. The samples were then crushed to a maximum particle size of approximately 1/4 inch by utilizing two stages of jaw crushers. Screening of the samples split the coal into two size fractions: <28 mesh and >28 mesh. The >28 mesh samples were then density separated using float-sink techniques. Appropriate mixtures of perchloroethylene and petroleum ether (Skelly B) were used to obtain the density fractions shown in Table I. As a result, 80 samples were obtained for analysis by neutron activation.

TABLE I
Coal Subsample Characteristics

Designation	Description
A	>28 mesh fraction before density separation
B	Density separated coal: 1.24 > s.g.
C	Density separated coal: 1.30 > s.g. > 1.24
D	Density separated coal: 1.35 > s.g. > 1.30
E	Denstiy separated coal: 1.40 > s.g. > 1.35
F	Density separated coal: 1.62 > s.g. > 1.40
G	Density separated coal: s.g. > 1.62
H	<28 mesh fraction

Using various combinations of sample irradiations and counts, the samples were analyzed for 54 elements. An additional sample was included in these analysis so that any elemental pick-up in the coal during the density separations could be identified. This sample was a >28 mesh sample that was soaked in perchloroethylene and petrolium ether and then dried. As a result of this extra sample, it was determined that only chlorine and bromine were absorbed by the coal during the density separations.

Of the 54 elements determined by neutron activation analysis, 30 had to be deleted from the data set before factor analysis could be applied. These elements were deleted since their concentrations in the samples were less than detectable. To keep from deleting too many elements, some samples which were deficient in many elements were deleted instead. As a result, a data set consisting of the concentrations of 24 elements in 58 samples was used.

DATA ANALYSIS

The first step of factor analysis is to determine the number of mineral phases present in the samples. Since determining the number of mineral phases is not always a easy choice, several tests are applied to the data. A large decrease in the magnitude of the eigenvalue found after diagonalization of the correlation matrix, will be an indication of the number of mineral phases present. As can be seen in Table II, there is a large decrease in the magnitude of the eigenvalue between 3 and 4 factors and again between 5 and 6 factors indicating the presence of either 3 or 5 mineral phases. For the RMS, chi-square, exner, and average percent error tests, which are calculated by reproducing the data with various numbers of factors, the number of factors is also determined by a large decrease in their magnitude. As can be seen

TABLE II
Results of Dimensionality Tests

Factors	Eigenvalue	RMS	Chi-square	Exner	% Error
1	55.3	138.	13600.	.229	120.
2	2.39	27.4	2840.	.0809	59.9
3	.315	5.99	1370.	.0230	49.0
4	.0208	2.68	1140.	.0116	43.5
5	.00701	.40	758.	.0012	25.6
6	.00003	.33	334.	.0009	18.5
7	.00002	.17	600.	.0007	22.4

in Table II, there is a large decrease in the magnitudes of each test between 2 and 3 factors and again between 4 and 5 factors indicating either 3 or 5 mineral phases. Since the number of phases present was not conclusively decided from these tests, the factor analysis was continued assuming either 3 or 5 mineral phases present.

The second step of factor analysis is to determine the elemental profiles of the mineral phases that are present in the coal using the iterative target transformation approach previously described. After 500 iterations of initially unique vectors, 24 elemental profiles for the 3 mineral phase solution were obtained. These 24 elemental profiles were normalized and then used in a cluster analysis (16). The results of the cluster analysis grouped similiar source profiles and 3 main groupings were found to occur. The same procedure was followed for a five mineral phase solution and 5 main groupings were found to occur after the cluster analysis. When the mineral phases were combined to see if they collectively reproduced the data, both the 3 and 5 mineral phase solution showed satisfactory results with the three factor solution being slightly better. Tables III an IV give the elemental profiles of the mineral phases determined for the 3 and 5 factor solutions while Tables V and VI show how well the data is reproduced for each solution.

It should be noted that the concentrations reported in Tables III and IV are relative values and not absolute. Normally a multiple linear regression analysis is performed on the final results to obtain appropriate scaling factors which will yield the absolute concentrations. Since data concerning the organic fraction of the coal were not available, this could not be done.

TABLE III
Elemental Profiles for Mineral Phases
for Three Factor Solution (ppm)

Element	a_1	a_2	a_3
Na	17800.	21700.	2500.
Al	416000.	176000.	0.00
K	75700.	28800.	7930.
Ca	1.77	358000.	42.7
Sc	105.	88.6	17.4
V	923.	983.	103.
Cr	596.	328.	110.
Mn	87.5	4360.	289.
Fe	10.9	119000.	460000.
Co	185.	110.	37.9
Ga	145.	134.	31.2
As	127.	335.	899.
Sb	32.5	10.6	14.8
Cs	54.7	.02	3.91
La	566.	226.	24.4
Ce	1140.	441.	62.1
Nd	543.	208.	43.3
Sm	64.9	35.0	7.06
Eu	15.5	8.47	1.90
Dy	43.5	22.8	11.7
Yb	27.5	14.0	2.69
Hf	21.4	4.69	3.88
Th	98.9	103.	5.89
U	27.7	33.4	6.18

TABLE IV
Elemental Profiles for Mineral Phases
for Five Factor Solution (ppm)

Element	a_1	a_2	a_3	a_4	a_5
Na	20600.	19800.	.66	220000.	5030.
Al	312000.	135000.	6500.	3.50	313000.
K	155000.	46800.	3690.	394000.	20300.
Ca	.00	384000.	.01	294000.	2.13
Sc	110.	76.5	7.59	897.	47.1
V	273.	655.	1.26	9280.	522.
Cr	679.	263.	53.7	4980.	257.
Mn	298.	4770.	335.	64.9	244.
Fe	26800.	98400.	460000.	158.	174000.
Co	246.	103.	21.3	1480.	74.8
Ga	196.	120.	8.36	1940.	32.0
As	798.	436.	851.	3700.	142.
Sb	63.9	9.89	7.82	553.	.35
Cs	72.7	.04	2.35	159.	28.3
La	131.	39.2	3.53	2260.	426.
Ce	278.	67.3	18.6	4640.	852.
Nd	46.9	.00	22.1	2260.	433.
Sm	.07	7.85	3.62	347.	51.3
Eu	2.77	2.55	.81	103.	10.9
Dy	34.1	12.0	6.96	410.	22.2
Yb	24.8	9.66	.59	192.	13.9
Hf	25.0	3.04	2.70	109.	11.3
Th	57.4	85.5	2.29	438.	66.8
U	24.6	26.3	.84	451.	6.67

TABLE V
Summary of Elemental Contributions to Average Sample
Composition for Three Factor Solution (ppm)

Element	a_1	a_2	a_3	Avg Pred Contrib	Avg Obs Contrib	Avg % Error
Na	461.	167.	185.	813.	566.	59.5
Al	10800.	1350.	.07	12200.	12200.	1.0
K	1970.	222.	586.	2770.	2670.	25.4
Ca	.01	2760.	3.38	2770.	2780.	.6
Sc	2.73	.68	1.29	4.70	3.39	53.2
V	24.0	7.58	7.61	39.2	24.4	93.3
Cr	15.5	2.53	8.13	26.2	19.0	52.7
Mn	2.27	33.6	21.4	57.3	54.9	70.6
Fe	.02	918.	34000.	34900.	35000.	.0
Co	4.81	.85	2.80	8.46	5.86	74.7
Ga	3.77	1.03	2.30	7.10	4.77	67.2
As	3.30	2.58	66.4	72.3	64.8	36.9
Sb	.84	.08	1.09	2.02	1.15	115.
Cs	1.42	.00	.29	1.71	1.53	22.4
La	14.7	1.74	1.80	18.3	13.8	47.5
Ce	29.5	3.40	4.58	37.5	27.9	70.1
Nd	14.1	1.61	3.20	18.9	13.3	56.6
Sm	1.69	.27	.52	2.48	1.63	68.3
Eu	.40	.07	.14	.61	.39	84.9
Dy	1.13	.18	.86	2.17	1.59	56.9
Yb	.72	.11	.20	1.02	.75	48.1
Hf	.56	.04	.29	.88	.63	66.3
Th	2.57	.79	.43	3.80	2.92	58.4
U	.72	.26	.46	1.43	.85	88.2
				Average Error		54.9

TABLE VI

Summary of Elemental Contributions to Average Sample Composition for Five Factor Solution (ppm)

Element	a_1	a_2	a_3	a_4	a_5	Avg Pred Contrib	Avg Obs Contrib	Avg % Error
Na	177.	135.		123.	132.	566.	566.	.1
Al	2670.	917.	415.	.00	8180.	12200.	12200.	.0
K	1360.	318.	236.	221.	532.	2670.	2670.	.0
Ca		2610.	.06	165.	.12	2780.	2780.	.0
Sc	.11	.52	.49	.50	1.23	3.68	3.39	22.9
V	.94	4.46	.08	5.20	13.7	25.7	24.4	36.0
Cr	2.34	1.79	3.44	2.79	6.72	20.6	19.0	21.7
Mn	5.82	32.4	21.5	.04	6.39	62.8	54.96	67.9
Fe	2.56	669.	29400.	.23	4550.	34900.	34900.	.0
Co	230.	.70	1.36	.83	1.96	6.96	5.86	53.0
Ga	2.11	.82	.54	1.08	.84	4.95	4.77	22.6
As	1.68	2.96	54.5	2.07	3.71	70.1	64.8	35.0
Sb	6.85	.07	.50	.31	.01	1.43	1.15	77.5
Cs	.55	.00	.15	.09	.74	1.61	1.53	18.2
La	.62	.27	.23	1.27	11.2	14.0	13.8	27.6
Ce	1.13	.46	1.19	2.60	22.3	28.9	27.9	42.6
Nd	2.39	.00	1.41	1.26	11.3	14.4	13.3	31.2
Sm	.40	.05	.23	.19	1.34	1.82	1.63	35.2
Eu	.00	.02	.05	.06	.29	.44	.39	40.9
Dy	.02	.08	.45	.23	.58	1.63	1.59	27.6
Yb	.29	.07	.04	.11	.36	.79	.75	21.5
Hf	.21	.02	.17	.06	.30	.76	.63	51.4
Th	.21	.58	.15	.25	1.75	3.22	2.92	38.8
U	.49	.18	.05	.25	.17	.87	.85	28.3
	.21					Average Error		29.2

DISCUSSION AND CONCLUSIONS

The three mineral phases found for the three factor solution can be labeled as alumino-silicate, calcite, and pyrite for a_1, a_2, and a_3, respectively. The high iron contribution in the calcite phase may be due to siderite ($FeCO_3$) which was not separated in the analysis. If this is true, a more realistic name for this phase may be carbonate.

For the five mineral phase solution, labeling of the minerals is a little more difficult. The a_2 and a_3 minerals appear to be calcite and pyrite for the five factor solution. This is substantiated by a cluster analysis of the 3 minerals from the 3 factor solution and the 5 minerals from the 5 factor solution. The results of the cluster analysis indicate similarity between the calcite and pyrite of both solutions, a_2 and a_3 respectively. The cluster analysis also shows similarity between the alumino-silicate of the 3 factor solution and the a_1 and a_5 mineral phases of the 5 factor solution. Both of these minerals appear to be alumino-silicates of different composition. The final mineral of the 5 factor solution, a_4, did not cluster with any other minerals. This phase could possibly be any of several less abundant minerals that are usually found in coal such as sphalerite (ZnS). The naming of this phase will have to wait until more data are available.

As this work has demonstrated, information concerning the mineral phases in coal can be acquired from factor analysis even with limited data. Additional data may permit the resolution of additional mineral phases as well as refinement of those already identified.

REFRENCES

1) P.A. Estop, J.J. Kovach, C. Karr, Anal. Chem. 40, 358(1968).

2) S. St. J. Warne, J. Inst. Fuel, 38(292), 207(1965).

3) H.J. Gluskoter, R.R. Ruch, Geol. Soc. Amer. Abstr., 3, 582(1971).

4) C.R. Ward, "Mineral Matter in the Springfield-Harrisburg (No. 5) Coal Member in the Illinois Basin", Illinois State Geological Survey, Circular 498, 35 pp. (1977).

5) H.J. Gluskoter, R.R. Ruch, W.G. Miller, R.A. Cahill, G.B. Dreher, and J.K. Kuhn, "Trace Elements in Coal: Occurances and Distributions", Illinois State Geological Survey, Circular No. 499, (1977).

6) D.J. Alpert and P.K. Hopke, Atmos. Environ., 14, 1137(1980).

7) D.J. Alpert and P.K. Hopke, "Modified Factor Analysis of Selected RAPS Aersol Data", Final Report to EPA, Order No. D6004NAEX, (1980).

8) B.A. Roscoe and P.K. Hopke, "Determination of Mineral Matter in Coal by Target Transformation Factor Analysis", presented Fourth International Conference on Nuclear Methods in Environmental and Energy Research, (1980).

9) A.T. Miesch, "Q-mode Factor Analysis of Geochemical and Petrologic Data Matricies with Constant Row-sums", Geological Survey Professional Paper 574-G, (1976).

10) J. Imbrie, "Factor and Vector Anaysis Programs for Analyzing Geologic Data", Report No. 6 of ORN Task No. 389-135, Office of Navel Research (1968).

11) R.P. Glaister and H.W. Nelson, B. Canadian Petroleum Geology, 22, 203(1974).

12) P.K. Hopke, R.E. Lamb, and D.F. Natusch, Environmental Science and Technology, 14, 1, (1980).

13) E.R. Malinowski and M. McCue, Anal. Chem., 48, 284(1977).

14) P.R. Bevington, Data Reduction and Error Analysis for the Physical Sciences, Chap. 5, McGraw-Hill Book Co. (1969).

15) B.A. Roscoe and P.K. Hopke, "Comparison of Weighted and Unweighted Target Tranformation Rotations in Factor Analysis", Computers & Chemistry, (in press).

16) P.K. Hopke, E.S. Gladney, G.E. Gordon, W.H. Zoller, and A.G. Jones, Atmos. Environ. 10, 1015(1976).

THE ANALYSIS OF INORGANIC CONSTITUENTS IN THE GROUNDWATER

AT AN UNDERGROUND COAL GASIFICATION SITE

        R.R. Ireland, W.A. McConachie, D.H. Stuermer,
        F.T. Wang and R.F. Koszykowski
        Lawrence Livermore National Laboratory
        University of California
        Livermore, California 94550

ABSTRACT

    A primary environmental consequence of underground coal gasification may be degradation of groundwater quality. This report presents the sampling, analytical and computational techniques used to study the inorganic constituents in the groundwater following the Hoe Creek II field study conducted in the fall, 1977. The physical and chemical processes controlling the observed concentrations are discussed with the aid of graphic displays of the analytical data and with the use of a chemical equilibrium modeling program.

INTRODUCTION

    Underground coal gasification shows increasing promise as a feasible technology for production of domestic synthetic fuels. One of the major environmental issues is the potential for contamination of groundwater aquifers from the residual reaction products of the gasifying process. The consequences must be considered if contaminated groundwater is to be used for human or livestock consumption or for agricultural purposes.

    One experimental in situ or underground coal gasification (UCG) project is located at LLNL's Hoe Creek site in northeastern Wyoming. As part of a cooperative effort between the Earth

---

    Work performed under the auspices of the U.S. Department of Energy by the Lawrence Livermore National Laboratory under contract number W-7405-ENG-48.

Sciences Division and the Environmental Sciences Division at LLNL, we have sought to determine the environmental consequences of UCG and to develop appropriate control technology options (Mead, et al., 1980).

Over the past three years, we have analyzed the physical properties and the organic and inorganic composition of groundwater samples from the Hoe Creek II field experiment. We report here on the sampling, analytical and computational methods used to study the trace element concentrations and to understand the physical and chemical processes involved.

The Hoe Creek II field experiment was conducted during the fall of 1977, with gasification ending on December 25, 1977. Approximately 2400 tons of coal from the Felix II coal seam (roughly 40m underground and 8m thick) and Felix I coal seam (about 5-10m above Felix II), both well below the static groundwater level, were gasified using air as the feed gas (Aiman, et al., 1978). (Fig. 1)

Groundwater samples were collected from the feed gas injection well, the two production wells, three wells located within the gasification burn cavity, and six wells outside the burn zone located in a roughly linear fashion up to 80m from the injection well (Fig. 2).

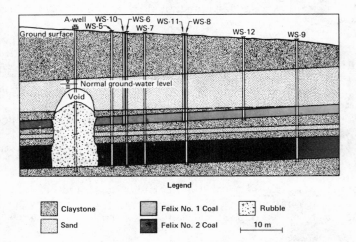

Fig. 1. Cross section of Hoe Creek II underground coal gasification site.

ANALYSIS OF INORGANIC CONSTITUENTS

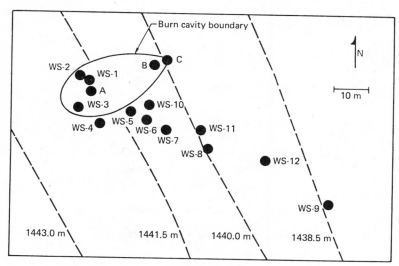

Fig. 2. Well placement at Hoe Creek II. A is the injection well, B and C are production wells. Wells WS-4 through WS-9 are in the Felix No. 2 coal seam, WS-10 to WS-12 are in the upper Felix No. 1 coal seam.

A recently developed well-head instrument package was used to measure the pH, conductivity, temperature, and redox potential of the groundwater at the time of sampling (Garvis and Stuermer, 1980). The concentrations of sulfate, cyanide, and phenol were measured by wet chemical methods on-site. Water samples were filtered immediately after collection through 1μ Nucleopore filters and stored in polyethylene bottles at 0-4°C until analyzed at LLNL. Samples destined for analyses by either ICP-OES or INAA were acidified after field filtration with 10 ml/l concentrated nitric acid (Ultrex, Baker Chemical).

ANALYTICAL

The instrumental analyses performed on the groundwater samples were ion chromatography (IC), inductively coupled plasma optical emission spectroscopy (ICP-OES), and instrumental neutron activation analysis (INAA).

The concentrations of seven anions were determined by IC using a Dionex Model 10 Ion Chromatograph with an anion exchange separator column followed by a strong acid resin column as a suppressor for the eluent (0.003N $NaHCO_3$ and 0.0024N Na_2CO_3) ions. A conductivity meter acted as the detection system and a Hewlett-Packard 3350 Laboratory Automation System aided in the data handling.

In the groundwater samples, F^-, SO_4^{-2}, and Cl^- were routinely detected by IC with PO_4^{-3}, NO_3^-, NO_2^-, and Br^- falling below the detections limits. The minimum detectable concentrations are 30, 450, 50, 290, 300, 170, and 270 ppb, respectively.

ICP-OES, a highly automated multi-element trace level analysis technique utilized a Jarrell-Ash Model 975 Atomcomp spectrometer (Peck et al., 1979). The elements that were routinely analyzed and their detection limits are listed in Table 1.

Table 1. Detection Limits of Elements in Water for ICP-OES.*

Element	Detection Limit (ppb)	Element	Detection Limit (ppb)	Element	Detection Limit (ppb)
Al	24	K	90	Pb	31
As	100	Li	0.4	Si	4
B	16	Mg	1	Sr	4
Ca	2	Mn	0.4	Ti	1
Cd	2	Mo	11	U	82
Co	3	Na	7	V	4
Cu	1	Ni	10	Zn	1
Fe	2	P	200	Zr	2

* Detection limit = four times the standard deviation of water blank.

INAA used gamma-ray spectrometry to determine the amount of each radioisotope that was formed when samples were irradiated in a nuclear reactor (Heft, 1977). Pellets were made for irradiation by drying 50 to 200 mls of filtered, acidified sample on 200mg of a micro crystalline cellulose. After an hour of irradiation (in a thermal neutron flux of 3.6×10^{13} neutrons/cm^2sec) and a cooling period of 10 days, the samples were counted on a Ge(Li) spectrometer system for 20,000 secs. The system can detect: Ag, Au, Ba, Cd, Ce, Cl, Co, Cr, Cs, Eu, Fe, Gd, Hf, Ir, Lu, Mo, Na, Nd, Rb, Re, Sb, Sc, Se, Sm, Sr, Ta, Tb, Th, Tn, U, Yb, Zn, and Zr. The detection limits vary with the volume of sample used in making the pellets.

ANALYSIS OF INORGANIC CONSTITUENTS

GRAPHIC DISPLAY

A computer based data integration program was used to assemble the analytical results and provide a graphical display of the concentration of each element as a function of distance from the feed gas injection well. Each data point for an element was individually expressed as a weighted mean concentration dependent upon the mode of analysis. Separate displays were made for each element and each sampling time i.e. 10, 51, 106, 170, 283, 452, and 660 days after termination of gasification. This display technique allowed us to quickly observe both changes with time and distribution with distance. This was especially valuable considering the large volume of data. (Fig. 3).

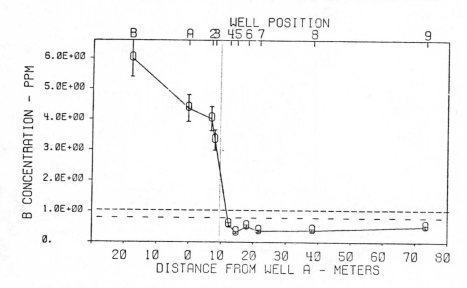

Fig. 3. Example of the computer generated graphics display of the boron concentration in the groundwater 10 days after termination of gasification. Well positions are plotted on the top boundary and distance from the injection well A on the bottom boundary. The vertical dotted line indicates the position of the burn boundary - data to the left is within the burn cavity and data to the right is outside the burn cavity. Recommended maximum permissible values (RMPV) for drinking water and irrigation water are indicated.

GEOCHEM

To gain insight into the possible chemical processes governing the distribution of the inorganics, we employed the computer modeling program GEOCHEM (Sposito and Mattigod, 1980; Mattigod and Sposito, 1979). GEOCHEM is a chemical equilibrium program based on thermodynamic associations and solubility product constants. The equilibria that can be calculated by the program are complexation, precipitation, oxidation-reduction, cation exchange, and metal ion adsorption. Thermodynamic data at 25°C and 1 atm pressure are stored in the data file of the program for combinations between 36 metals and 69 ligands. Corrections for ionic strength up to $3\underline{M}$ can be made. An example of one of the GEOCHEM output routines $\overline{\text{is}}$ shown in Fig. 4.

Trial runs were made with the model using average representative baseline data. Altered values of the inorganic input allowed us to assess the response of the goundwater system to various hypothetical elevated concentrations. The redox potential was also varied from a hypotheical pE of -9.0 to the observed value of 0.34. The pH was held constant at the measured baseline value of 7.8.

The use of GEOCHEM does not imply that the authors presumed the system to be at equilibrium. Rather, the model serves a valuable function as a tool for predicting the distribution and tendencies of chemical species in a dynamic and complex mixture.

DISCUSSION

Following gasification, groundwater that had been excluded from the burn cavity either by gas pressures or dewatering returns and leaches the inorganic constituents from the coal ash. Normal groundwater migration from the cavity can then transport contaminants through the aquifer, depending upon the solubility and sorption processes.

In general, the concentrations of most inorganic constituents increase within the burn zone due to this leaching. However, some constituents do not, e.g. Ba, Fe, Na, and HCO_3^-. The increases in concentrations are also observed up to 10m outside the cavity, but these initial values gradually diminish with time and distance, and are probably due to dilution and sorption. A notable exception again is Ba which always exists at lower than baseline levels outside the burn zone Baseline levels are generally represented by well WS-9, which in most circumstances remains relatively unchanged over the course of the sampling period.

ANALYSIS OF INORGANIC CONSTITUENTS

In contrast, some elements – Ca, Fe and Mg – are observed at increased concentrations further away, up to 30m from the cavity. This phenomenon is difficult to explain by groundwater migration alone. One hypothesis (Mead, et al., 1980) suggests that pressurized gases, (CO_2, H_2, CH_4) emanating from the burn zone travel significant distances through the aquifer. These gases react to produce a more reducing and acidic environment for the solubilization of Ca, Fe and Mg in place from the natural coal seam.

Primary Distribution of Metals and Ligands*

CA	AS A FREE METAL/	35.3 PERCENT
	BOUND WITH CO3-/	0.4 PERCENT
	IN SOLID FORM WITH CO3-/	54.0 PERCENT
	BOUND WITH SO4/	9.7 PERCENT
	BOUND WITH CL/	0.5 PERCENT
MG	AS A FREE METAL/	80.7 PERCENT
	BOUND WITH CO3-/	0.6 PERCENT
	BOUND WITH SO4/	17.7 PERCENT
	BOUND WITH CL/	1.0 PERCENT
K	AS A FREE METAL/	97.9 PERCENT
	BOUND WITH SO4/	2.1 PERCENT
NA	AS A FREE METAL/	96.6 PERCENT
	BOUND WITH CO3-/	0.1 PERCENT
	BOUND WITH SO4/	3.3 PERCENT
CD	AS A FREE METAL/	46.6 PERCENT
	BOUND WITH CO3-/	1.9 PERCENT
	IN SOLID FORM WITH CO3-/	19.4 PERCENT
	BOUND WITH SO4/	13.2 PERCENT
	BOUND WITH CL/	18.7 PERCENT
	BOUND WITH OH/	0.2 PERCENT
CO3-	AS A FREE LIGAND/	0.2 PERCENT
	BOUND WITH CA/	0.5 PERCENT
	IN SOLID FORM WITH CA/	69.6 PERCENT
	BOUND WITH MG/	0.4 PERCENT
	BOUND WITH NA/	0.3 PERCENT
	BOUND WITH H/	29.0 PERCENT
SO4	AS A FREE LIGAND/	83.3 PERCENT
	BOUND WITH CA/	6.7 PERCENT
	BOUND WITH MG/	5.3 PERCENT
	BOUND WITH K/	0.3 PERCENT
	BOUND WITH NA/	4.3 PERCENT
CL	AS A FREE LIGAND/	99.4 PERCENT
	BOUND WITH CA/	0.3 PERCENT
	BOUND WITH MG/	0.2 PERCENT

Fig. 4   Typical GEOCHEM Output from the Subroutine OUT138.

The graphic displays of the data quickly support this view. Both Eh and pH are initially lower than baseline values out to 50m and gradually increase with time. Additional supporting evidence comes from increased values of HCO_3^- and conductivity at these distances. (Fig. 5)

The GEOCHEM program provides insight into the interpretation of the visual data. Ca^{+2} is increased initially both inside and outside the burn cavity and in both areas the values decrease with time to less than baseline. The model suggests that Ca^{+2} is released inside the burn zone from $CaCO_3$ when the high temperatures drive off the gaseous CO_2. In the outlying regions, as mentioned previously, a lowered pH and Eh also solubilize $CaCO_3$. With time, as the pH and Eh recover and rise, reprecipitation occurs and Ca^{+2} values diminish. (Fig. 6)

The model also helps explain the behavior of Fe. Concentrations of Fe outside the burn zone parallel those of Ca, but internally Fe never reaches any elevated values. GEOCHEM predicts that outside, the reduced Fe^{+2} will eventually oxidize to Fe^{+3} and reprecipitate as $Fe_2(CO_3)_3$ and $Fe(OH)_3$. Inside the cavity, substantially increased concentrations of Si and S lead to the rapid removal of liberated leached Fe^{+2} from solution as solid $FeSiO_2(OH)_2$ and FeS_2 (Fig. 7).

GEOCHEM also predicts that Mg^{+2} is subject to the same precipitation pressures outside the cavity in time as Fe and Ca. However, inside the burn zone, the concentration of free Mg^{+2} is somewhat buffered by precipitating with $B(OH)_4^-$ so that internal concentrations of Mg^{+2} are neither significantly elevated nor depressed.

Boron, which is produced in high concentrations within the burn cavity and appears to persist, is of environmental concern in that it is known to be phytotoxic to some vegetation at values above 1 ppm. However, the graphic display reveals that B rarely migrates beyond the edge of the cavity (Fig. 3, 8). The equilibruim model suggests that, as above, free $B(OH)_4^-$ ligands are readily removed from solution by available Mg^{+2} ions. Figure 8 provides a good illustration of the mirror image data which is consistent with the GEOCHEM prediction.

GEOCHEM also offers a plausible explanation for the behavior of three other constituents - Ba^{+2}, SO_4^{-2} and AsO_4^{-3} - shown in the graphical displays. Ba^{+2}, as mentioned previously, is the only element which occurs at less than baseline values both inside and outside the cavity following gasification. The model suggests that the primary reason is the substantial increase in SO_4^{-2} following gasification with resultant $BaSO_4$ precipitation. Furthermore, the model also predicts that enough

ANALYSIS OF INORGANIC CONSTITUENTS

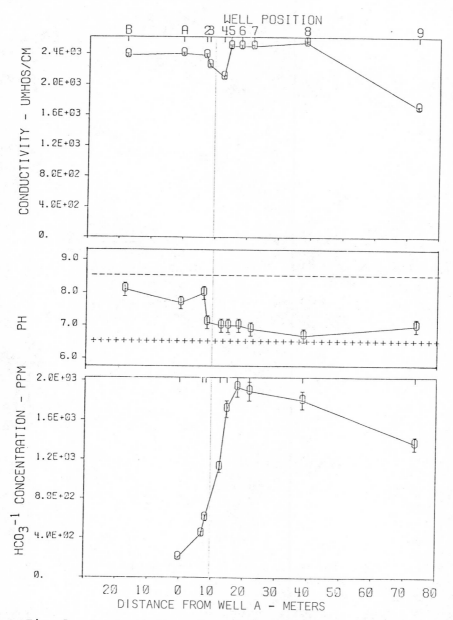

Fig. 5   Conductivity, pH, and HCO_3^- graphic displays.

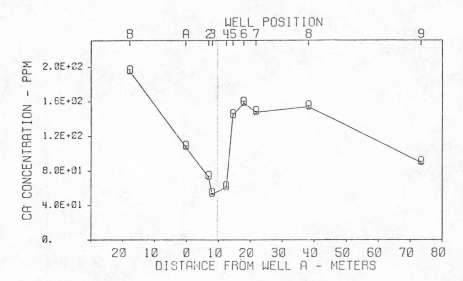

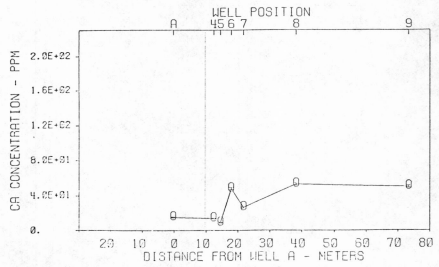

Fig. 6 Initial increased values of Ca^{+2} are seen as a result of gas induced lowering of pH and pE. In time, degassing of H_2 and CO_2 leads to higher values of these paramters.

ANALYSIS OF INORGANIC CONSTITUENTS

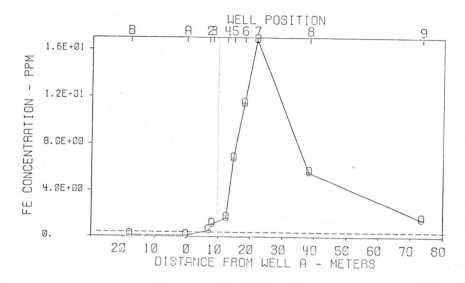

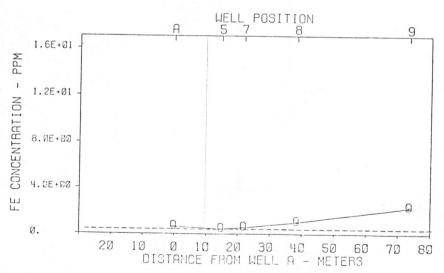

Fig. 7 Fe concentrations are always depressed within the burn zone due to precipitation with silicates and sulfides. Outside, Fe behaves similarly to Ca and Mg.

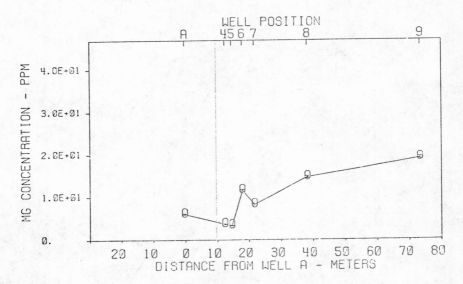

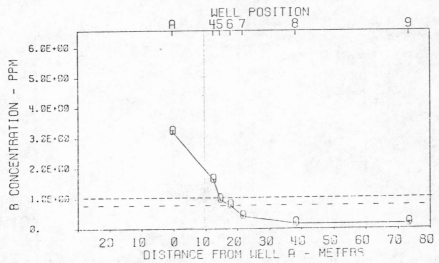

Fig. 8 Graphic display indicating that $[Mg^{+2}]$ and $[B(OH)_4^-]$ appears to be inversely proportional outside the burn cavity. This is consistent with the GEOCHEM prediction that all boron is precipitated as magnesium borate.

free Ba^{+2} is still available to effectively control the AsO_4^{-3} concentrations as solid Ba_3AsO_4. The display and the model tend to confirm this theory as AsO_4^{-3} remains generally near baseline concentrations.

The GEOCHEM chemical equilibrium model serves as a valuable adjunct in interpreting the graphical data. If the two tools are consistent and the explanations plausible, they may significantly alleviate our concerns about the environmental consequences of certain potentially toxic elements. For example, all Hg^{+2} is predicted to be removed from solution as $Hg(OH)_2$ and almost all Pb^{+2} as $PbCO_3$. But the models also suggest that some elements will remain dissolved. Trace species such as Cu^{+2}, Zn^{+2}, Ni^{+2} and Cr^{+3} are predicted to be bound in soluble form with HCO_3^-, Se^{+4} as soluble $MgSeO_3^0$, and CN^-, MoO_4^{-2}, and NH_3 all exist as essentially free ligands.

By adjusting the inorganic input to the model, we can test the response of the system to hypothetical elevated concentrations. But when observed levels do not behave appropriately, we know that other physico-chemical forces are at work. One factor not mentioned and known to exist is the self-cleansing property of the coal itself (Wang, 1979). Other factors would be absorption on other particle surfaces, colloidal agglutination, and organic interactions.

Nonetheless, the computer generated graphic display and GEOCHEM model have facilitated our interpretation of the inorganic data in the UCG groundwater by increasing our understanding of the inorganic buffering processes at work.

SUMMARY

We have presented the sampling methodology and the analytical tools used to produce data on the inorganic constituents present in the groundwater at Hoe Creek II. A graphic display plotting linear well sites with elemental concentration values allows for a quick grasp of the trends for either increasing or decreasing concentrations as functions of both time and distance. A computer equilibrium program GEOCHEM has been used as a tool for elucidating some apparent reasons for natural chemical buffering. By and large, the equilibrium program predictions and the observed concentration values over distance and time have been consistent.

ACKNOWLEDGEMENTS

The authors acknowledge the support received from the Department of Energy, Division of Pollutant Characterization and Safety Research for this work. We also sincerely thank Darrel

Garvis for the collection of the samples as well as Warren Mead for cooperation in the sampling efforts. Furthermore, for their expertise in the analytical areas, we thank Bob Heft (INAA), Art Langhorst (ICPOES) and Bill Steele (AA). Appreciation is also extended to Walt Martin for development of the plotting routines. Finally, we thank Shas Mattigod and Gary Sposito for their continuing support in aiding us with the GEOCHEM program.

REFERENCES

Aiman, W.R., C.B. Thorsness, R.W. Hill, R.B. Rozsa, R.J. Cena, D.W. Gregg and D.R. Stephens, 1978, The Hoe Creek II Field Experiment on Underground Coal Gasification, Preliminary Results, Lawrence Livermore National Laboratory, Rept. UCRL-80592.

Garvis, D.G, and D.H. Stuermer, 1980 "A Well-head Instrument Package for Multiparameter Measurement During Well Water Sampling", Water Research, 14:1525-1527.

Heft, R.E., 1977, Absolute Instrumental Neutron Activation Analysis at Lawrence Livermore Laboratory, Lawrence Livermore National Laboratory, CA. UCRL-80476.

Mattigod, S.V. and G. Sposito, 1979, Chemical Modeling of Trace Metal Equilibria in Contaminated Soil Solutions Using the Computer Program GEOCHEM. in: "ACS Symposium Series, No. 93, Chemical Modeling in Aqueous Systems," E.A. Jenne, ed., American Chemical Society, Washington D.C.

Mead, S.W., F.T. Wang, D.H. Stuermer, E. Raber, H.C. Ganow, R. Stone, 1980, Implications of Ground-Water Measurement at the Hoe Creek UCG Site in Northeastern Wyoming, Sixth Annual Underground Coal Conversion Syumposium Afton, Oklahoma, July 13-17, 1980. Also published as Lawrence Livermore Laboratory, Rept. UCRL-84083.

Peck, E.S., A.L. Langhorst, D.W. O'Brien, 1979, Analysis of Natural Waters with Automated Inductively Coupled Plasma Spectrometer System, Lawrence Livermore Laboratory, UCRL-81043.

Wang, F.T., 1979, "The Sorptive Property of Coal," in Proc. of the 5th Underground Coal Conversion Symposium, Alexandira, Virginia, June 18-21, 1979, pp. 403-407. Also published as Lawrence Livermore Laboratory, Report UCRL-82703.

APPLICATION OF PIXE, RBS AND HIGH ENERGY PROTON MICROBEAMS TO THE ELEMENTAL ANALYSIS OF COAL AND COAL WASTE

H. W. Kraner, A. L. Hanson and K. W. Jones

Brookhaven National Laboratory
Upton, New York 11973

S. A. Oakley, I. W. Duedall and P. M. J. Woodhead

Marine Sciences Research Center
State University of New York
Stony Brook, New York 11794

INTRODUCTION

Several developing techniques of elemental analysis based on nuclear technology are becoming more actively applied to research in the earth science and fossil fuel research.[1] This paper will present a brief description of several relatively new analytical methods which, together with more established techniques such as the scanning electron microprobe, provide a comprehensive array of methods for materials analysis. These methods, which are being used in a continuing study of the elemental constituency and integrity of coal waste products, include proton and x-ray induced fluorescent x-ray analysis (PIXE and XRF), alpha particle backscattering (Rutherford backscattering, RBS), proton microbeam analysis and electron microprobe analysis.

The role of nuclear-based techniques in materials and earth science research is yet to be fully defined. For example,

---

*Research supported by the U. S. Department of Energy, Division of Basic Energy Sciences, under Contract No. DE-AC02-76CH00016 and New York State Energy Research and Development Authority as manager for Department of Energy, Environmental Protection Agency, Electric Power Research Institute, and Power Authority of the State of New York.

Rutherford backscattering is a well-used tool in surface and coat-analysis in the semiconductor industry but little recognized as a simple technique for major elemental identification in other fields. This technique was, however, the basis for the lunar surface material measurement carried out early in the lunar research program in the surveyor landers.[2] Earth scientists are major proponents of electron microprobe analysis and yet are only beginning to appreciate the merits of the proton microprobe which, in many respects, complements the electron beam in terms of elemental sensitivity, reduced background and differences in spatial resolution and response. Complementarity is a key concept in defining a comprehensive set of analytical capabilities.

Also to be discovered is the most effective area of application for these techniques in fossil fuel research. With regard to the study presented here, it may evolve that broad surveys of balance studies on loss and composition of coal waste materials exposed to the marine environment may be most useful. Alternatively, detailed probing of the trace element constitution of microscopic areas of specific samples will be most useful and give basic information on solubilities, leaching rates, etc. which cause material changes with time.

This description of nuclear based analytical techniques will not include direct comparison with other, more established, analytical techniques, and neutron activation analysis will not be discussed. It is routinely used elsewhere, fully appreciated by the earth scientists and readily available. Many methods are certain to find preferred roles and selection by experimenters with particular familiarities. There are, however, some general aspects of the methods to be presented which are distinct and worth noting out of the specific context. Nuclear based analytical methods often require little, if any, sample preparation. For the most part, matrix or chemical states of the sample do not affect the measurement. In the case of the proton microprobe, analysis with good spatial resolution and high trace element sensitivity can be accomplished in the laboratory outside of the vacuum with the ability to position and observe the sample during the measurement. These features point to a general feature of overall convenience and integrity of measurement. The use of particle beam from a Van de Graaff accelerator in several methods often elicits the comment that the method is therefore not generally available. However, there are literally hundreds of these small accelerators in existence, many with applied programs and outside user capability. The technology and access is certainly no more difficult than in neutron activation analysis, a well-accepted analytical tool.

ELEMENTAL ANALYSIS OF COAL AND COAL WASTE

ANALYTICAL METHODS

XRF and PIXE (including the proton microprobe) identify elements by exciting atoms resulting in the emission of characteristic x-rays. Protons of several MeV and electrons of $\sim$ 20 keV (typical of electron beam probes) move at velocities close to K and L shell orbital velocities of light elements and therefore have a high probability ($\sim 10^3$ barns) to ionize and produce a characteristic x-ray. Similarly, a 20-keV photon has high probability for an atomic photoelectric effect in which a K (or L) electron is ejected from a heavy atom. Johansson, et al.[3] first showed that proton-induced x-ray emission was sufficiently sensitive to measurement of trace quantities in various matrices. Johansson et al. have more recently reviewed the field in detail[4] which has grown and matured to the extent that several international conferences have resulted.[5]

The applications of highly collimated proton beams which approach micron dimensions have been increasing dramatically over the last several years. Cookson[6,7] first showed the usefulness of a proton microprobe and several active groups[8,9,10,11] have emerged. Clarke has reported[1] on several specific microprobe measurements on coals with the Harwell machine. Shroy, et al.[12] have shown that a simple highly collimated proton beam may be used in the laboratory ambient which allows good specimen cooling, ease of specimen manipulation, the investigation of large or vacuum unsuitable samples (e.g., gassy) and good spatial resolution and system sensitivity. The ion microprobe clearly complements the more familiar scanning electron microscope (SEM) or microprobe.

The ability of a heavy charged particle to discern the atomic weight of a scattering particle has been appreciated since the early discoveries of Rutherford, which inspires the common description: Rutherford backscattering. Beside the application to lunar materials evaluation, the technique is well used in semiconductor materials and surface analysis and a comprehensive book by Chu, Mayer and Nicolet is available.[13] The ion beam commonly used is a 2-3 MeV alpha beam from a low energy accelerator.

COAL WASTE MATERIALS

In addition to studying several representative samples of coal of different ranks and NBS standard reference samples, coal waste materials comprise the major application of the techniques that have been described. Justification for the application of our research to coal waste is based on the following: There is urgency to convert from oil to coal burning at northeastern power plants and conversion has begun. An important obstacle to utilizing coal is the large volume of coal combustion products which include mainly fly ash and flue gas desulfurization sludge (scrubber sludge) if scrubbers are used. Both these wastes will require disposal.

The dumping of either the untreated scrubber sludge or fly ash in the sea would be quite unacceptable, probably having deleterious environmental effects.[15] IU Conversion Systems, Inc. (IUCS), Horsham, Pa., has developed a marketable stabilized coal waste by combining the scurbber filtercake with the fly ash. Basically this system treats sludge and fly ash with lime;cementitious reactions convert the mix to a stable material that can range from a clay-like substance to hard blocks. These blocks are being used to fabricate an artificial reef.[16]

The stabilization reactions in the formation of the coal waste blocks are similar to those of concrete but do not give the yield strength of concrete and are more porous and permeable.[17] The bulk density of the blocks is about 80% that of concrete, due to the lighter fly ash used and the absence of high density aggregate materials. In some instances, bottom ash can be added, as an aggregate, to the coal waste blocks.

The coal wastes investigated in the present study case were from three coal-fired power plants: (1) Conesville power station, Ohio, a 820-MW plant using bituminous coal; (2) the Elrama, and (3) Phillips power stations both near Pittsburgh, Pennsylvania. The coal waste (Conesville and Elrama) were stabilized by IUCS as cubic foot blocks. Using the Poz-o-Tec^R process (IUCS), lime was added to a fly ash and scrubber sludge mix which was wetted, compacted in a plywood mold and cured for 28 days to stabilize the mixture into block form. The ratio (by dry weight) of fly ash to scrubber sludge was near 1:1. (For the purpose of this paper, scrubber sludge refers to the scrubber sludge filtercake and includes any carry over of fly ash from particle collectors; the carry over fly ash was a significant problem only at the Elrama power station.) The bulk mineralogy[18] of the cured blocks was determined by x-ray diffraction analysis of unoriented mounts of powdered coal waste materials. Major components, based on mineralogy and elemental composition by wet and instrumental methods[16] are given in Table 1. Smaller blocks were cut from the 1 ft^3 blocks and reacted in seawater tanks for a period of about 5-6 months, depending upon the mix. The seawater, changed approximately bi-weekly, was monitored for major and minor constituents leached from the blocks. The results of the leaching study have been reported elsewhere.[19] After this exposure samples from the surface and interior of representative blocks were taken, crushed and powdered and prepared for analysis by re-forming as pellets 10 mm in diameter, 1 mm thick. Thin sections for proton and electron microscopy were also prepared. Identical samples were pre-

Table 1. Composition, expressed as percent dry weight, of major components in blocks of stabilized fly ash and scrubber sludge from the Elrama and Conesville power plants. (*nominal ratio)

Power Plant Fly Ash: Scrubber Ratio*	Fly Ash	$CaCO_3$	$CaSO_3 \cdot \tfrac{1}{2}H_2O$	$CaSO_4 \cdot 2H_2O$	$Ca(OH)_2$	SUM
Elrama 1:1	53	17.4	18.4	4.3	--	93
Conesville 1:1	48	2.0	33.6	--	5.9	90

pared from blocks not exposed to seawater (unreacted). Samples of the sludge and fly ash from the Conesville plant were also available as well as representative bituminous coals. The coal waste project has continued with large production runs of 18,000 8 in. × 8 in. × 16 in. In September 1980 these blocks were dumped via barge in the Atlantic Ocean approximately three miles south of Fire Island just east of Fire Island Inlet. They are shown on shore and in place in Fig. 1. The coal waste reef will be studied in situ for a period of three years.[20]

Figure 1. Stabilized scrubber sludge/fly ash blocks, left, and in-situ, above.

RESULTS

The results presented will first emphasize the techniques of PIXE, XRF, RBS and microprobes as applied to the analysis of coal waste block material and associated biota. A brief later section will describe measurements of anthracite, bituminous and lignite coal.

1A. PIXE

Most measurements of proton induced x-ray spectra from the several samples of interest reported here were made with an external proton beam in the microprobe configuration. For increased sensitivity, a relatively large .006 in. diameter aperture was used which required a 0.001 in. aluminum window to maintain vacuum. This window caused a relatively small extraneous background.

The higher energy portion of the PIXE spectra from samples of Conesville blocks from the surface of reacted blocks, the unreacted blocks and the NBS SRM1632a standard coal are shown in Fig. 2. The measurements were made for a determined integrated beam charge and used a .030 in. Mylar absorber on the detector to reduce the overall count rate by eliminating copious low energy x-rays. Ni, Cu, Zn, Ga, Ge, As, Rb and Sr can be observed in the block spectra, both reacted and unreacted. Ga and Ge are known trace elements in coal[14] and their concentrations can be estimated from the Zn and As in the NBS standard coal shown below which contains 28 ± 2 and 9.3 ± 1 µg/gm, respectively. Arsenic is also prominent in the block spectra and would be expected from the fly ash component.

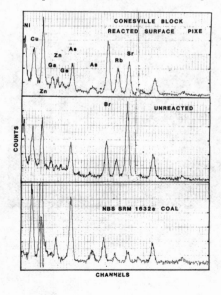

Figure 2. PIXE x-rays spectra from 7 to 17 keV of samples of Conesville block, reacted and unreacted and NBS SRM 1632a Coal.

Fig. 3 shows the full energy range spectra of the Conesville samples from reacted (surface), reacted (interior) and input bituminous coal without the low energy x-ray absorber. Prominent x-rays include S, Ca, Ti, Mn and Fe in addition to those in the higher energy portion. Some reduction of Ca and S relative to the large Fe peak may be observed in the reacted surface spectrum. The major components of these spectra, together with an interior sample of the reacted Conesville block, were analyzed quantitatively by a computer fitting program to the K_α lines of interest. It is assumed the iron minerals present are not mobilized so that the iron intensity should be presumed constant in reacted and unreacted blocks. Therefore it is of interest to compare the ratios of observed elemental x-ray intensities to the iron intensities in the several samples. These ratios are shown in Table 2A.

Fig. 4 is a composite of PIXE spectra of the Conesville sludge, fly ash and bituminous coal. It is clear that the fly ash is the principal source of trace elements for the blocks.

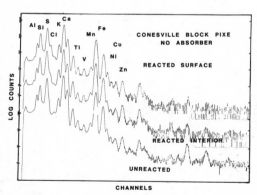

Figure 3. PIXE x-ray spectra of Conesville block: reacted surface, reacted interior and unreacted material. No low energy x-ray absorber was used.

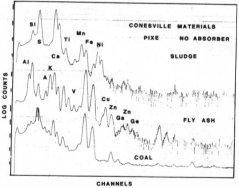

Figure 4. Conesville block materials: sludge, fly ash and representative bituminous coal. PIXE, no low energy absorber

Table II. Iron Ratios to Observed Elements in Conesville Block

Ratio	A. PIXE			B. XRF		C. RBS	
	Unreacted	Reacted Interior	Reacted Surface	Unreacted	Reacted Surface	Unreacted	Reacted Surface
Fe/O						0.08	0.17
Fe/S	3830	5050	8260	90	220	0.12	0.35
Fe/K				280	220		
Fe/Ca	14	31	34	6	11	0.06	0.164
Fe/Ti	71	109	67	92	94		
Fe/Mn	165	230	160				
Fe/Ni				1270	1510		
Fe/Cu	1510	2270	1510	660	580		
Fe/Zn	1170	1180	950	100	100		
Fe/Ga	7400	6630	6640	880	1160		
Fe/Ge	11400	7600	9950	880	640		
Fe/As	1540	1520	1840	160	175		
Fe/Pb				450	330		
Fe/Rb				100	75		
Fe/Sr				5.4	12.6		

1B. XRF

Spectra of unreacted and reacted surface samples of the Conesville blocks are shown together with the NBS SRM1633a fly ash standard for reference in Fig. 5. Many elements are evident including higher Z trace elements, e.g. As, but it is difficult to qualitatively judge if leaching has occurred in the reacted sample. Therefore quantitative analysis of the K_α peaks have been carried out and iron to element ratios were calculated and are listed in Table 2B. It should be noted that a clear Pb L_β is observed at 12.62 keV which implies that a Pb L_β is mixed with the As K_α at 10.52 keV. A Pb standard was measured to determine the Pb L_α/L_β ratio and thus remove the Pb L_α interference from the As K_α based on the Pb L_β intensity. In Table 2B we notice that Ca and S and Sr as well have been leached from the reacted block.

1C. RBS

Coal and coal waste samples were measured by Rutherford backscattering using 2.8 MeV alpha particles from the 3.5 MV Van de Graaff accelerator in a standard scattering chamber geometry. As implied by name, the scattering angle with respect to the incoming beam was 165° and the billiard ball kinematics apply. A ^{241}Am alpha source together with an electronic pulser were also present during the runs but they are off scale between channels 3800 and 4096. The pulser assures continuous monitoring of the system gain and count rate stability and the 5.486-MeV α source provides an absolute energy calibration. It is possible to calibrate the machine energy and energy scale to a few keV and thus measure the elemental edge energies accurately with high assurance of the particular specie.

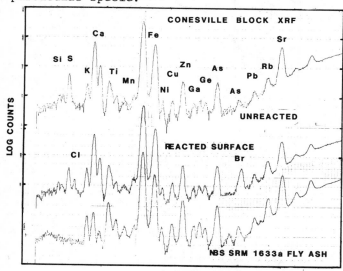

Figure 5. X-ray spectra from XRF of Conesville unreacted block, reacted surface material and a NBS SRM 1633a Fly Ash standard.

Fig. 6 is a composite of several RBS spectra plotted on a log scale of samples of NBS S RM 1632a and SRM 1633a coal, fly ash, unreacted and reacted block material, from the lower to the upper spectrum respectively. On this scale, quantitative analysis is difficult; however, the major elements and minor heavy elements observed are: C, O, Al, Si, S, Ca, Fe, As, Ba and very heavy elements (Hg or Pb). It is noteworthy that this technique allows measurement with a continuous relative efficiency of light elements down to carbon. Fig. 7 is an enlargement of the edges of interest for the reacted surface and unreacted block material. The height of each edge is proportional to the relative amount of each element present, and it is of interest to compare the ratio of O, S and Ca to that of Fe as has been done with the other techniques in Table 3C. We find again that sulphur and calcium are reduced in the reacted surface material by factors between two and three. A reduction of over two for oxygen is also observed which would be expected because $CaSO_3 \cdot \frac{1}{2}H_2O$, and unreacted $Ca(OH)_2$ are the major oxygen-containing phases (Table 1) leached from the blocks.

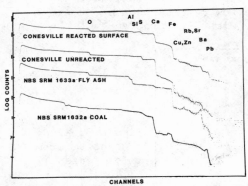

Figure 6. Backscattered alpha particle spectra (RBS) of pressed samples of Conesville reacted surface and unreacted material and NBS SRM 1633a and 1632a Fly Ash and Coal standards.

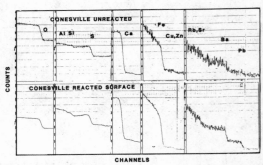

Figure 7. Enlargement of portions of RBS spectra from unreacted and reacted surface samples of Conesville material

ELEMENTAL ANALYSIS OF COAL AND COAL WASTE

It is also of interest to note that barium (or perhaps cesium) is observed in the RBS spectrum as a minor constituent. The barium L x-rays at 4.457 and 4.828 keV, if excited by PIXE or XRF, would be obscured by the more abundant, ubiquitous titanium K x-rays at 4.508 and 4.931 keV. Thus a third independent method is helpful to resolve interferences.

1D. Proton and Electron Microprobe Scans

Because the collimated proton beam is available externally in the laboratory, it is possible to scan samples in either a continuous line scan or to position the beam on particular areas of interest. Scans of thin, 30 μm sections were made with protons at 4 mm intervals across a midsection of sample of ~ 4 cm in length. The sample was essentially homogeneous in the sludge/ash mixture and the particulates were smaller than the .006 in. aperture used which determines the spatial resolution. Several positions in the midsection were found to be relatively lower in iron and correspondingly higher in Mn and Ni. The reduction in Fe could be due to the fact that scan positions included (randomly) fewer small Fe particles which were found to be the major iron source.

Figure 8 is a montage of scanning electron microscopic results showing the micrograph with a variety of particles and a slight secondary emittance difference in areas of the matrix. The x-ray scans shown below confirm that the different areas of the matrix are Ca-rich and others Si-rich. The ubiquitous Fe is found in particles up to ~ 20 μm in size. This scan of major elements suggests why the proton scans showed very uniform results although some abnormally high concentrations of <u>trace</u> constituents (in fly ash particles, for example) would be expected in the proton scans.

1E. Organic Remnants

Several samples of biota, taken from the surfaces of blocks placed in the Atlantic for about one month, were dried and analyzed by XRF for possible uptake of elements that might be identified with the blocks themselves. These included calcareous tube building worms (<u>Sabellaria vulgaris</u>), a hydroid (Obelia) and worm casings. A group of mud-tube worms (Asabellides occulata) was also run. Figure 9 is a XRF spectrum of the worm casing typical of all measurements except for the mud-tube worms. We find Si, K, Ca, Ti, V(sl), Mn, Fe, Ni(sl), Cu, Zn, Ga(sl), Pb, Br, and Sr. Only the very minute suggestion of Ga has a possible identification with coal or coal waste products. Although the Pb L_α interferes with the observation of As, observed in the fly ash component of the bricks, the Pb L_β is also prominent and of approximately sufficient intensity to assign the 10.5 keV peak as the accompanying Pb L_α. The larger mud-tube worms and their preparation exhibited a large amount of Pb, high Mn and Ca.

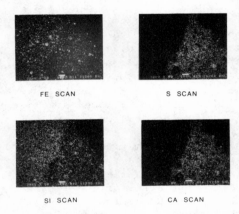
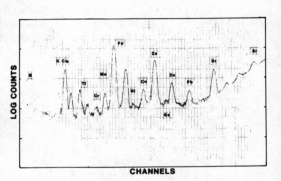

Figure 8. Scanning electron micrograph and associated x-ray scans of thin section of Conesville reacted block material.

Figure 9. XRF x-ray spectrum of a mud-tube worm (<u>Sabellaria vulgaris</u>) casing taken from blocks after one month in the ocean.

2. Coal

Samples of the several ranks of coals were studied by PIXE, XRF and RBS. Fig. 10 shows a collection of XRF spectra taken for the same x-ray fluence on samples (from the top) of anthracite, bituminous (representative of the Conesville input material) and North Dakota lignite. Sulphur is evident in all materials as are other major elements, Ti, Fe, Zn, Br and Sr. However, As is observed in the anthracite and is present in greater quantitites in the Conesville bituminous sample, and it is not observed in the lignite. Arsenic was strongly observed in Conesville fly ash and the composite bricks. Lignite is significantly different, showing very little calcium, iron and strontium but high titanium and zinc.

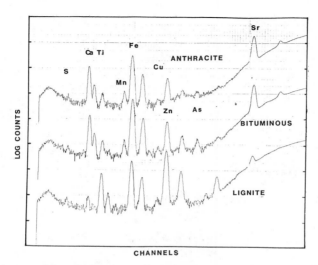

Figure 10. XRF x-ray spectra of samples of Anthracite, Bituminous and Lignite coals.

Rutherford backscattering results have shown in Fig. 6 the ability to discern minor and trace quantities of heavy elements with a continuous (Z^2) sensitivity. Further, it has been previously demonstrated[21] that RBS provides a simple means for finding major element relative abundances with agreement between measured and certified values of pressed NBS standard coal (SRM 1632a) within 15%.

In itself, this is not an extraordinary result except that it should be realized that the RBS measurement can proceed concurrently with PIXE and be used to directly analyze a thick target matrix. The characterization of the sample matrix is essential for accurate self-absorption corrections that must be applied to the x-ray spectra for trace analysis.

CONCLUSIONS

Proton and x-ray induced x-ray emission have proved to be sensitive and convenient methods to measure major trace element concentrations in bulk quantities of coal and coal waste materials. These techniques are complementary in their sensitivities as a function of atomic number, and both require little sample preparation. The PIXE measurements were made with the proton beam in air in a microprobe configuration. Collimated proton beam scans were made on several thin sections of fly ash/sludge block materials and good trace sensitivities were observed for small specific volumes; SEM scans showed a high degree of material homogeneity which precluded significant elemental variations at the $\sim$ 100 µm spatial resolution used. An excellent study of the trace elemental distribution in the mineral components of macerals of bituminous coals has been carried out by Chen, et al.[22] Many of the trace elements in bulk observed in this study, such as Ge, Ti and As, were observed in different distributions within these minerals. A spatial resolution of $\sim$ 2 to 4 µm was required for these observations.

Rutherford backscattering was used to directly observe major and minor elemental concentrations in coal waste materials and in several representative ranks of coals. RBS is useful for only trace concentrations of heavy elements, but it does provide a method independent of fluoresced x-rays for detection of possible middle Z interferences. Some information as to the organic ("volatile") or inorganic (less volatile under the beam) nature of the elemental matrix may be deduced from the shape of the surface or near full energy edge of the spectrum. Highly peaked, surface-augmented spectra are often observed for elements such as Ca which may better withstand beam surface heating effects as small refractory particles.

The observed depletion, demonstrated in this study, of Ca and S (and also O) in the reacted block is due primarily to the dissolution of $CaSO_3 \cdot \frac{1}{2}H_2O$ in seawater. Duedall et al.[19] in an independent study, using the same blocks, determined the flux of Ca^{2+} and SO_3^{2-} (reported as SO_4^{2-}) from the blocks to seawater to range between 10^{-7} mole/mm^2/day during the first 20-30 days to 2-3 $\times$ 10^{-8} mole/mm^2/day at the end of the leaching experiment (148-170 days, depending upon the mix). Using these fluxes, a one-dimensional model based on diffusion was developed. The model predicts diffusivities (for Ca^{2+}) of 1.2-3 $\times$ 10^{-9} cm^2/sec and the depth (x_c) of penetration of the diffusion process: in 10 days x_c = 0.06-0.10 cm; in 365 days, x_c = 0.39-0.61 cm; and in 30 years, x_c = 2.1-3.3 cm. In future work PIXE, XRF, and RBS will be used to confirm x_c for a variety of blocks exposed to seawater for different periods of time.

Arsenic, present in trace amounts in coal, is an element of concern and is enriched in fly ash.[14] The form of As in fly ash is unknown. However, because of its volatility most of the As probably becomes attached to the surfaces of the fly ash particles during and subsequent to combustion processes. This view is supported by the fact that As is rapidly mobilized in aqueous solutions whose pH > 9. Groenwold et al.[23] reported As concentrations in excess of 600 µg/ℓ in ground water (below a fly ash disposal field) whose pH was 11.6; for ground water whose pH was 7.9 (e.g., the pH of seawater), the reported As concentration derived from another fly ash field was 35.3 µg/ℓ. The Fe/As ratios in Table II suggest that very slight As had leached from the reacted blocks. Further research will focus on the calibration of PIXE, XRF, and RBS methods in order to calculate the rate of loss of As from the blocks. The proton microprobe is expected to play a unique role in further characterizing fly ash particles as to their arsenic and accompanying trace elemental constituency.

ACKNOWLEDGEMENTS

It is a pleasure to thank Ms. Myrna Jacobson for supplying the samples of reef-associated organisms. Mr. Frank Roethel and Dr. Ramesh Dayal supplied samples of the block material, and their efforts are well appreciated. Mr. Edward Reilly was as usual very helpful in Van de Graaff operation. Sincere thanks are offered to Dr. John Warren for excellent electron microscopy results.

REFERENCES

1. G. Clark, "The Future of Ion Beam Analysis in Earth Science," Presented at the 6th Conference on the Application of Accelerators in Research and Industry, Denton, Texas, November 3-5, 1980. To be published in the IEEE Transactions on Nuclear Science, February (1981).
2. A. Turkevich, Science 134:672 (1961) and A. Turkevich, et al. Surveyor Project Final Report, Part II Scientific Results. NASA Tech. Rep. 32-1265, JPL, CIT, Pasadena, California, p. 303.
3. T. B. Johansson, R. Akselsson and S. A. E. Johansson, Nucl. Instrum. Methods 84:141 (1970).
4. S. A. E. Johansson and T. B. Johansson, Nucl. Instrum. Methods 137:473 (1976).
5. c.f. Nucl. Instrum. Methods 137 (1976) and Proceedings of the 2nd International Conference on Particle Induced X-Ray Emission and Its Analytical Applications, Lund, Sweden, June 9-12, 1980.
6. J. A. Cookson, A. T. G. Ferguson and F. D. Pilling, J. Radioanal. Chem. 12:39 (1972).
7. J. A. Cookson, Second International COnference on Particle Induced X-Ray Emission and Its Analytical Applicattions, Lund, Sweden, June 9-12, 1980.

8. F. Bosch, A. El Goresy, B. Martin, B. Povh, R. Nobiling, D. Schwalm and K. Traxel, Science 199:765 (1978).
9. P. Horowitz, M. Aronson, L. Grodzins, W. Ladd, J. Ryan, G. Merrian and C. Lechene, Science 194:1162 (1976).
10. G. J. F. Legge, C. D. McKenzie and A. P. Mazzolini, J. of Micros. 117:Pt.2, 185 (1979).
11. C. J. Maggiore, Materials Analysis with a Nuclear Microprobe, Scanning Electron Microscopy, 1:(1980), Om Johari, ed., IITRI and LA-UR 80-390.
12. R. E. Shroy, H. W. Kraner and K. W. Jones, Nucl. Inst. & Meth. 157:163 (1978).
13. W. K. Chu, J. W. Mayer and Marc-A. Nicolet, Backscattering Spectrometry, Academic Press, N. Y. (1978).
14. Trace Contaminants from Coal, S. Torrey, ed., Noyes Data Corp., Park Ridge, N. J. (1978).
15. R. R. Lunt, C. B. Cooper, S. L. Johnson, J. E. Oberholtzer, G. R. Schimke and W. I. Watson (1977), Technical Report EPA-600/7-77-051, Arthur D. Little, Inc. Acorn Park, Cambridge, Mass.
16. P. M. J. Woodhead and I. W. Duedall, Technical Report FP-1252, Electric Power Research Institute, Palo Alto, Cal. (1979).
17. J. D. Seligman and I. W. Duedall, Environmental Science and Technology, 13:1082 (1979).
18. R. Dayal, Personal Communication.
19. I. W. Duedall, J. S. Buyer, M. G. Heaton, S. A. Oakley, A. Okubo R. Dayal, M. Tatro, F. J. Roethel, R. J. Wilke, J. P. Hershey, Second International Ocean Disposal Symposium, April (1980), Woods Hole Oceanographic Institution, Woods Hole, Mass.
20. P. M. J. Woodhead, J. H. Parker and I. W. Duedall, Artifical Fishing Reef Symposium (1980), D. Aska, ed., Florida Sea Grant Program, Daytona Beach, Fla. (in press).
21. H. W. Kraner and K. W. Jones, Bull. Am. Phys. Soc. 24:No.7, 823 (1979).
22. J. R. Chen, H. Kneis, B. Martin, R. Nobiling, K. Traxel, E. C. T. Chao and J. A. Martin, Presented at the Second International Conference on Particle-Induced X-Ray Emission and Its Analytical Applications, (1980), Lund, Sweden.
23. G. H. Groenewold, J. A. Cherry, O. E. Manz, H. A. Gullicks, D. J. Hassett and B. W. Rehm, EPA Synposium on Flue Gas Desulfurization, October (1980), Houston, Texas.

INVESTIGATIONS OF MECHANISMS AND
STRUCTURES IN FOSSIL FUELS

INVESTIGATION OF COAL HYDROGENATION USING

DEUTERIUM AS AN ISOTOPIC TRACER*

R. P. Skowronski,[†] J. J. Ratto,[§] and L. A. Heredy[†]

[†]Rockwell International
Energy Systems Group
8900 De Soto Avenue
Canoga Park, California 91304

[§]Rockwell International
Corporate Science Center
1049 Camino Dos Rios
Thousand Oaks, California 91360

ABSTRACT

Mechanisms of coal hydrogenation were investigated by using a deuterium tracer method. This method makes it possible to determine which structural positions in the coal react with hydrogen during liquefaction. A hydrogenation index (HI) and exchange index (EI) were formulated to measure the amount of deuterium incorporated due to hydrogenation and exchange reactions, respectively. In the coal-deuterium system, deuterium incorporation was found to vary both with product fraction and with structural position. In contrast, the deuterium contents of the fractions from donor solvent experiments were essentially uniform. The donor solvent experiments did, however, show preferential deuterium incorporation with respect to structural position. Important information with regard to the reaction mechanisms in the donor solvent system was obtained by analyzing the spent solvent mixture that was recovered from the reaction products. The results indicate that not only hydrogen donation but also hydrogen exchange involving the α-positions of

---

*This work was conducted under the auspices of the U.S. Department of Energy by Rockwell International under Contracts EF-77-C-01-2781 and DE-AC22-77ET11418.

tetralin can have a significant role in stabilizing the fragments that form during the thermal decomposition of the coal. In addition, evidence was obtained that there is also a direct route for deuterium incorporation into the coal products from the gas phase without the participation of tetralin.

INTRODUCTION

Hydrogenation of coal to obtain liquid fuels is receiving increasing attention in the United States. Although procedures for hydrogenating coal to form liquids have been known and practiced for decades, they have not been based on detailed fundamental knowledge of coal chemistry. In fact, relatively little is known about the mechanisms that occur during the hydrogenation of coal. Certainly such knowledge is needed for a fundamental understanding of coal chemistry, and may very well help to improve industrial coal hydrogenation processes.

The conversion of coal to liquids by hydrogenation necessitates increasing the hydrogen content of the coal by several percent. The high cost of hydrogen production and compression, however, is a prominent factor in the overall cost of obtaining liquid products from coal. Consequently, the efficient use of this gas is of significant importance in practical coal hydroliquefaction processes.

In light of these facts, a method which would trace hydrogen incorporation into the coal during hydrogenation would be of interest for the development of coal conversion processes and for a fundamental understanding of coal chemistry. Hydrogenation mechanisms are not readily elucidated by routine experiments with natural hydrogen which is primarily protium. Since the reactant coal originally contains protium, analysis of the hydrogenated products does not indicate where hydrogen incorporation occurred. One method which can trace hydrogen incorporation involves the use of deuterium as an isotopic tracer.

In our research, mechanisms of coal hydrogenation are being investigated by using a deuterium tracer method. This method makes it possible to determine which structural positions in the coal react with hydrogen during liquefaction. Hydrogenation experiments are conducted with deuterium gas and/or a fully deuterated solvent. In selected experiments, nitrogen, protium, or tetralin-h_{12} is used to distinguish sources of hydrogen. The soluble hydrogenation products are analyzed by using high resolution deuteron NMR spectrometry to determine into which structural positions the deuterium has been incorporated.

The use of deuterium as an isotopic tracer makes it possible to follow hydrogen incorporation during coal hydrogenation without significantly altering the course of the reactions or the chemical nature of the products. Isotope effects are most apparent when the isotope is directly involved in the rate-determining step. It is generally accepted that in the first phase of coal hydrogenation, the rate-determining step is the formation of radicals by the rupture of weak chemical bonds. After this step has occurred, the highly reactive fragments achieve stabilization by any of several reaction pathways such as addition of an atom or small group, rearrangement of atoms within the fragment, or polymerization. If this general concept of the reaction mechanism is correct, then the isotope effect will not be of great significance when using the deuterium tracer method.

When using an isotope such as deuterium to trace the progress of a chemical reaction, it is important to consider the possibility of the deuterium undergoing isotopic exchange with protium in different structural positions in the coal. For example, Gudkov et al.,[1] who studied both ^1H-^2H isotopic exchange and the hydrogenation of cyclohexane on a nickel catalyst, determined that the ^1H-^2H exchange proceeded more slowly than either hydrogenation or dehydrogenation. Recently, King and Stock[2] have investigated the exchange reactions between tetralin-d_{12} and diphenylmethane, a coal-related compound. They found that deuterium exchange is accelerated by the components of Illinois No. 6 coal and by the asphaltene-preasphaltene product fraction. There do not appear to be any relevant studies of thermal ^1H-^2H exchange in coal in the literature.

Relatively little use of deuterium as an isotopic tracer for coal research is found in the literature before 1976.[3,4] Recently, however, several research groups have used deuterium to investigate the mechanisms of coal conversion.[5-10] In our research, an intensive effort has been underway to obtain a better understanding of the hydrogenation reaction mechanisms and thereby improve hydrogen utilization and the efficiency of coal hydroliquefaction.

In the coal-deuterium gas system, deuterium incorporation was found to vary with structural position and was found to increase in the order oil < asphaltenes < preasphaltenes < residue. In the coal-tetralin-d_{12}-deuterium gas system, deuterium incorporation varied with structural position but not with product fraction. The results indicate that the α-tetralinyl radical is an important intermediate in hydrogen transfer to the coal. In addition, evidence was found that considerable direct interaction of the gas-phase hydrogen with the coal occurred in the donor solvent system.

EXPERIMENTAL

Materials and Apparatus

High-volatile A Loveridge Mine, Pittsburgh Seam bituminous coal (80.1 % C, 5.1% H, 1.6% N, 3.6% S, 9.6% O, by weight daf, 7.7% ash), -200 mesh, was dried at 115°C for 4 hours before use in each experiment. Technical-grade deuterium (>98 atom % deuterium, <1 ppm total hydrocarbons), high-purity protium (99.95% min.), and high-purity nitrogen were used. Tetralin-d_{12} was prepared in our laboratories[11] from naphthalene-d_8. The naphthalene-d_8 and tetralin-1-^{13}C were purchased from Aldrich Chemical Company. The deuterium contents of the naphthalene-d_8 and tetralin-d_{12} were both >99% of the hydrogen. Batch experiments were conducted using a 1-liter stirred autoclave (Autoclave Engineers) or a 0.25-liter rocking autoclave (Parr). Further details are available elsewhere.[11,12]

Experimental Procedure

In the coal-deuterium gas system, the 1-liter autoclave containing 25 g of dried coal was raised to the appropriate cold pressure with deuterium. Heating and stirring were begun. After ~100 min, the coal had reached reaction temperature. The coal was held at this temperature ±8°C for the selected reaction time. The autoclave and contents were then allowed to cool to ambient temperature at which point the stirring was terminated. The time required for the autoclave to cool to 300°C was ~40 min. Gas volumes were measured using a wet test meter, and product gas samples were collected for analysis. The system was disassembled, and the solid and liquid products were collected.

The experimental procedure for the experiments in which a hydrogen donor solvent was used consisted of the following steps. The system was assembled with the 25 g of solvent in the autoclave and the 25 g of coal in the injection vessel. The system was leak-tested and then filled to a cold pressure of 50 psig (6.9 MPa = 1000 psig) with the desired gas. The autoclave was then heated. When the temperature of the autoclave reached 340°C, stirring (100 rpm) was begun. At 440°C, the coal was injected, and the pressure raised to ~1900 psig. The temperature of the autoclave decreased to about 425°C within ~10 min. The reaction was conducted at 425°C for the desired reaction time. The pressure was ~2000 psig during the reaction period.

After the scheduled reaction time had elapsed, the furnace was lowered from the autoclave, and the external surface was air-cooled. Also, water was circulated through the autoclave's internal cooling coils. Injection of the coal at temperature and a cooldown rate of 100°C/min (425°C to 325°C) reduced the uncertainty in the reaction time. After the autoclave had cooled to ambient temperature, the gas volume was measured using a wet test meter, and product gas samples were collected for analysis. The system was disassembled, and the liquid and solid products were collected.

Product Analyses

Gaseous products were analyzed by gas chromatography coupled with mass spectrometry (GC-MS). In the experiments in which a solvent was used, the spent solvent mixture was distilled from the coal products. A portion of the spent solvent was then analyzed by GC-MS, and a separate portion was separated by GC and analyzed by nuclear magnetic resonance (NMR) spectrometry. GC-MS analyses were conducted by Shrader Analytical Labs, Inc., using a Pye-Unicam Model 105 chromatograph equipped with a flame ionization detector. A 7-ft x 1/4-in. OD glass column with 2% OV-17 was used and temperature programmed from 100-150°C at 4°C/min. The GC was interfaced with an AEI Model MS-30 mass spectrometer operating at maximum sensitivity.

The solid and liquid coal products were solvent fractionated into oil (hexane soluble, benzene soluble), asphaltene (hexane insoluble, benzene soluble), preasphaltene (THF soluble, benzene insoluble), and (THF) insoluble residue fractions. Product fractionation was carried out using three ACS reagent-grade solvents: hexane isomer mixture, benzene, and tetrahydrofuran. Samples of the fractions were combusted, and the resulting water was analyzed by MS (Shrader Analytical Labs, Inc.) to determine the protium and deuterium atom % compositions. Each fraction also was analyzed to determine its elemental composition.

Proton, deuteron, and carbon-13 NMR spectra of soluble fractions and recovered solvent mixtures were recorded by using a JEOL FX60Q FT NMR spectrometer. A 45° pulse was used which corresponds to 14 µs for ^1H, 75 µs for ^2H, and 6 µs for ^{13}C. The pulse repetition times were 6.0, 9.0, and 2.0 s, respectively. Chloroform-d was used as the ^1H and ^{13}C NMR solvent, and chloroform was used as the ^2H NMR solvent. Integrations of NMR spectra were carried out with software supplied by JEOL, Inc.

RESULTS AND DISCUSSION

Assessing the Sites of Hydrogen Incorporation

Proton NMR deconvolution spectrometry was compared with deuteron NMR spectrometry to assess its potential for determining into which structural sites the deuterium was incorporated. Proton NMR deconvolution spectrometry is an indirect technique that involves the subtraction of proton NMR spectra. Deuteron NMR spectrometry is a direct technique for detecting the deuterium in the hydrogenated coal products. The major advantage of the first technique is that proton NMR spectra are of very high quality with respect to resolution and signal/noise ratio. On the other hand, the major advantage of using the direct technique is that any error associated with the final data arises from only one spectrum.

For the purpose of comparing these two techniques, three pairs of experiments were conducted. These experiments were carried out in the temperature range of 380-400°C and at ~3000 psig for reaction times of 1/4 to 1 h. Since the deuterium incorporated into the coal products does not absorb at the protium frequency, the difference between the proton NMR spectra of corresponding fractions gives a quantitative measure of the hydrogen that was taken up at different structural positions during hydrogenation. For example, the proton NMR spectrum of the oil from an experiment carried out with deuterium may be subtracted from the proton NMR spectrum of the oil from a parallel experiment carried out with protium, to determine into which structural sites the hydrogen was incorporated during each of the two hydrogenations.

Such spectrum deconvolution is shown in Figure 1. The outer edge of the black area is the proton NMR spectrum of the oil fraction from one of the experiments that was conducted with protium. The inner edge of the black area is the proton NMR spectrum of the oil fraction from the corresponding experiment that was conducted with deuterium. The difference between the two spectra, shown by the area in black, is a measure of the distribution of deuterium which was taken up during hydrogenation in chemically different positions in the oil.

Deuteron NMR spectrometry directly measures the quantity of deuterium in each structural position in the coal products. In most cases, there was very good agreement between the values obtained by proton NMR deconvolution spectrometry and deuteron NMR spectrometry. The only poor agreement was observed for an oil fraction that contained only 4 atom % deuterium, based on total hydrogen. In those cases in which the ^{2}H content of the sample was over 10-15 atom %, there was good agreement between the two techniques. Details of this comparison have been presented elsewhere.[11] It was decided, on the basis of the results, that deuteron NMR spectrometry would be used to assess the structural sites of deuterium incorporation in all subsequent experiments.

Coal-Gas System

Much of the coal hydrogenation research that has been reported in the literature has been conducted with an added catalyst, a hydrogen donor solvent, or both. Relatively little has been done to investigate the reactions of coal in the presence of gas-phase

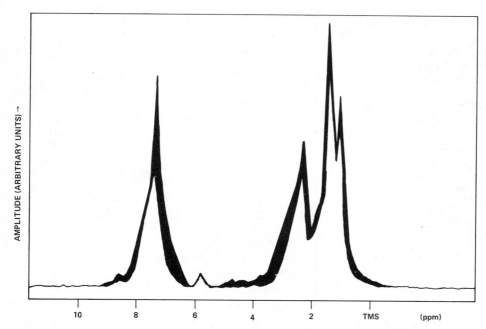

Fig. 1. Determination of Deuterium Incorporation by Proton NMR Deconvolution Spectrometry

hydrogen without a donor solvent and without an added catalyst. The following research was conducted specifically with a view toward elucidating the mechanisms of hydrogenation that occur in the coal-gas system.

In order to investigate hydrogen incorporation as a function of contact opportunity, a set of three experiments designed to vary the extent of hydrogen contact with the coal was carried out. All three experiments were conducted for 1 h with a pressure of ~3200 psig at 380°C. The first experiment (Experiment A) was performed using nitrogen as the gas-phase reactant. This was a control experiment which was designed to prohibit any possibility of reaction with gas-phase hydrogen.

The two other experiments in this set (Experiments B and C) were conducted using gas-phase deuterium. In Experiment B, a deuterium atmosphere was used together with the standard experimental procedure. In Experiment C, a deuterium atmosphere was used together with a vacuum pretreatment procedure which allows the deuterium to penetrate the pores of the coal to a greater extent.

This procedure consists of evacuating the pores of the coal for 24 h and then filling the autoclave with deuterium and allowing it to enter the pore structure for 24 h before the experiment is begun.

Figure 2 depicts the two general reactions that are of interest in this set of experiments. In Experiment A, only Mechanism 1 was possible because a nitrogen atmosphere was used. In Experiments B and C, both Mechanisms 1 and 2 were possible. The vacuum pretreatment procedure that was used in Experiment C was designed to enhance the availability of deuterium at the reaction sites. It may improve the contact opportunity of the deuterium and coal in an indirect as well as a direct manner. For example, an indirect improvement may result from the removal of adsorbed H_2O or O_2 from the coal surface.

(1) $\quad {}^1H:R'\quad +\quad .R\quad \longrightarrow\quad {}^1H:R + R'.$
(hydrogen from coal)   (coal radical)

(2) $\quad {}^2H_2\quad +\quad .R\quad \longrightarrow\quad {}^2H:R + {}^2H.$
(hydrogen from gas-phase) (coal radical)

Fig. 2. Schematic Fragment Stabilization Mechanisms

The solvent fractionation and gas-phase data gave an indication of the extent of occurrence of mechanisms that involve bond cleavage. As coal is heated to temperatures above $350°C$, thermal bond scission within the coal structure increases which produces highly reactive radicals. The coal fragments that are produced can be stabilized by such means as addition of an atom or small group, rearrangement of atoms within the fragment followed by the loss of a small group, or polymerization. If a fragment is stabilized by hydrogen, the hydrogen must necessarily be very near to the site of bond scission.

In the experiment in which a nitrogen atmosphere was used (Experiment A), reaction pathways in which fragments are formed in the coal by thermal bond scission and are stabilized by gas-phase hydrogen were eliminated. All of the hydrogen that stabilized such fragments must have come from the coal itself--autostabilization. The solvent fractionation data indicate that little soluble product was formed under these conditions. It appears that the autostabilization mechanism cannot stabilize a high percentage of the fragments formed and so relatively little of the low molecular weight product is obtained.

Polymerization may have occurred, which could account for the large percentage of insoluble material (89%) formed. Perhaps the high inert gas pressure lowered the yield of soluble products by decreasing the mobility of the intermediate radicals and reducing the probability of their reaction with hydrogen-donating sites in the coal. This would enhance the probability of recombination of these radicals to form higher molecular weight products.

When gas-phase hydrogen ($^2H_2$) was used (Experiments B and C), more low molecular weight products were formed than when nitrogen was used. In Experiment B (17% conversion), a deuterium atmosphere was used together with the standard experimental procedure. The gas-phase hydrogen apparently reacted with some of the radicals that were formed. Such a mechanism reduces the number of radicals available for polymerization and in this manner reduces the extent of contribution of the polymerization mechanism.

In the experiment in which the pores of the coal were evacuated and $^2H_2$ was allowed to enter the pore structure for 24 h before the experiment was begun (Experiment C), the highest yield of soluble products (31%) was obtained. This procedure was designed to increase the contact between the $^2H_2$ and the internal surface area of the coal.

Coal has a complex internal pore system. In order for gas-phase hydrogen to react with the considerable internal surface area of the coal, it must first travel into the pore structure. The higher concentration of hydrogen within the pore structure of the coal in Experiment C, as compared with Experiment B, apparently provided more hydrogen near to the sites of bond scission. In this manner, more of the coal fragments were stabilized by hydrogen, and the extent of the polymerization reaction was reduced. Reagent access appears to be quite important.

The deuterium distributions in the product fractions and product yields (Table 1) gave further insight into the mechanisms that occurred. In the experiments in which a deuterium atmosphere was used, considerable deuterium incorporation occurred. The atom % deuterium content increased in the order oil < asphaltenes < preasphaltenes < residue. The values for the hydrocarbon gases in each experiment are higher than those of the corresponding oil. Although the deuterium content in grams was larger in Experiment B than in Experiment C, the yield of soluble products was smaller. A very large fraction of the deuterium is in the residue in the products from Experiment B. Evidently, the deuterium incorporated in Experiment B was not as effective in producing soluble products as the deuterium incorporated in Experiment C.

The NMR spectrometric data reveal into which structural sites the deuterium incorporation occurred. The α-aliphatic (benzylic) position in Experiments B and C shows preferential deuterium uptake in all three soluble fractions. This may be partially due to exchange which is more facile in the benzylic positions. Preferential incorporation is indicated in some other fractions, such as in the aromatic positions in the oil fractions in Experiments B

Table 1. Deuterium Distributions in Product Fractions and Yields From Experiments B and C

Fraction	Atom % ^{2}H		Yield (%)	
	Experiment B	Experiment C	Experiment B	Experiment C
Gas	54 (0.16)*	41 (0.24)	7	10
Oil	35 (0.08)	35 (0.13)	6	10
Asphaltenes	48 (0.03)	43 (0.07)	2	5
Preasphaltenes	51 (0.03)	46 (0.06)	2	5
Insoluble Residue	60 (1.03)	50 (0.66)	83	70
Total Product	56 (1.33)	45 (1.16)	100	100

*Numbers in parentheses are deuterium contents in grams.

and C. However, it is difficult to propose a mechanism on the basis of this information alone.

The product gas data indicate that bond cleavage mechanisms occurred in all three experiments. Hydrocarbons, presumably formed by thermal bond scission, are present in the gas phase. The isotopic compositions of the hydrocarbon gases from Experiments B and C were determined. The hydrocarbons are so highly deuterated that it suggests that considerable ^{1}H-^{2}H exchange occurred.

The formation of H_2S increased from Experiment A to Experiment B and then to Experiment C. It appears that the availability of hydrogen at the sites of sulfur bond cleavage controls the extent to which the sulfur is removed via the H_2S formation mechanism. The data indicate that the hydrodesulfurization mechanism occurred at a particularly high rate in Experiment C. Considerable CO and CO_2 are present in Experiment A, less in Experiment B, and still less in Experiment C. This may be due to increased amounts of deuterium available to remove oxygen by water formation instead of via CO or CO_2 formation.

Several other experiments with coal and deuterium were conducted. The details of all of the coal-deuterium experiments are given elsewhere.[11,12] In this system, preferential deuterium incorporation was found in several sites, as shown in Table 2. In this table, words "strong" and "weak" refer to numerical values that have been calculated for preferential incorporation.[12] The values 1.1 to 1.3 were considered weak, and values >1.3 were considered strong. The absence of any notation in the column indicates that no preferential incorporation occurred. In some experiments, insufficient sample precluded NMR analysis of certain fractions. Therefore, the percentages given in Table 2 refer to the percentage of the total number of fractions analyzed.

Table 2. Preferential Incorporation in the Coal-Deuterium Experiments

Functional Region (ppm)*	Oil	Asphaltenes	Preasphaltenes
γ-aliphatic (0.0-1.0)		Strong in 50%	Weak in 11%
β-aliphatic (1.0-1.9)			
α-aliphatic (1.9-4.5)	Strong in 100%	Strong in 100%	Strong in 100%
Aromatic (4.5-10.0)	Strong in 38%		

*Parts-per-million from tetramethylsilane.

The α-aliphatic position consistently shows preferential deuterium incorporation in the oil, asphaltenes, and preasphaltenes. In some experiments, preferential incorporation is shown in the aromatic positions of the oils or in the γ-aliphatic positions of the asphaltenes. In one experiment, weak preferential incorporation was found in the γ-aliphatic positions of the preasphaltenes. This experiment was one of those which showed preferential incorporation in the γ-aliphatic positions of the asphaltenes.

A trend was found with respect to atom % deuterium content of the fractions when coal was reacted with $^2H_2$ without a donor solvent present. Table 3 lists the deuterium contents in the products of a typical experiment conducted without a solvent, compared with those of a typical experiment in which a solvent was present. Deuterium content was found to increase in the order oil < asphaltenes < preasphaltenes < insoluble residue for the experiments in which only coal and $^2H_2$ were used. EPR measurements indicate that unpaired electron density increases in the same order.[13] The values of the deuterium contents were essentially uniform when a donor solvent was used. We believe that this difference in results between the two systems reflects a difference in the hydrogenation mechanisms. Results from the donor solvent experiments are discussed in detail in a subsequent section.

Table 3. Deuterium Contents in Coal Products

Products	Atom % ^2H	
	$^2H_2$ Gas Only	$^2H_2$ Gas + Deuterated Solvent
Oil	29	38
Asphaltenes	31	45
Preasphaltenes	42	35
Insoluble Residue	59	37

Assessing the Extent of Hydrogenation and Isotopic Exchange

The extent of hydrogenation and isotopic exchange are calculated from a hydrogen isotope mass balance of the coal products, donor solvent (if used), and gas-phase hydrogen. The net amount of hydrogen added to the coal, H, is

$$H = n_H - n_{1H}^o$$

where n_H equals the amount of hydrogen in the coal products and n_{1H}^o is the amount of hydrogen in the starting coal. The amount of exchange is given by

$$E = n_{2H} - H$$

where n_{2H} equals the amount of deuterium incorporated in the coal products by hydrogenation and exchange. The calculations are conducted on an all-protium basis. A sample calculation from one of our donor solvent experiments that was conducted at 400°C for 1 h is given below:

$$H = n_H - n_{1H}^o = 5.5 - 4.7 = 0.8 \text{ (moles of H/100 g coal)}$$

$$E = n_{2H} - H = 3.0 - 0.8 = 2.2 \text{ (moles of H/100 g coal)}$$

We define:

$$\text{Hydrogenation Index (HI)} = \frac{H}{H + E} = \frac{0.8}{0.8 + 2.2} = 0.3$$

$$\text{Exchange Index (EI)} = \frac{E}{H + E} = \frac{2.2}{0.8 + 2.2} = 0.7$$

These values, HI and EI, indicate that ~30% of the deuterium incorporation was due to hydrogenation, and ~70% was due to isotopic exchange. Table 4 presents the experimental conditions and hydrogenation index values from four typical experiments. The higher the hydrogenation index, the clearer the assessment one obtains of the sites of hydrogen incorporation due to liquefaction. We are correlating these indexes with experimental parameters in an effort to determine what conditions lead to a higher relative amount of hydrogenation.

Table 4. Hydrogenation Index Values From Typical Experiments

Experimental Conditions	Conversion (%)	HI
360°C, 3000 psig, 1 h	15	<0.1
400°C, 3200 psig, 1 h	55	0.5
400°C, 2200 psig, 1 h (tetralin-d_{12} used)	31	0.3
400°C, 3000 psig, 1 h (tetralin-d_{12} used)	55	0.4

Coal-Donor Solvent-Gas System

Donor solvent coal hydrogenation was investigated using tetralin-d_{12} as a model donor solvent. High isotopic purity (~99 atom % ^{2}H) tetralin-d_{12} was prepared in our laboratories from naphthalene-d_8 for these experiments. Donor solvent coal hydrogenation is important because several of the processes that are in advanced stages of development employ donor solvents. Examples are the Solvent Refined Coal (SRC)[14,15] and Exxon Donor Solvent (EDS)[16] processes. The hydroaromatic structures in these solvents are of primary importance because some of them can release hydrogen to stabilize the coal fragments that are formed by thermal cleavage and thereby aid the liquefaction process.

Because one is able to differentiate between sources of hydrogen by using the deuterium tracer method, the method is very well suited for investigating the mechanisms of donor solvent coal hydrogenation. Table 5 presents a set of three experiments in which tetralin or tetralin-d_{12}, in a deuterium, protium, or nitrogen atmosphere, was used to elucidate the roles of gas-phase and solvent hydrogen in donor solvent hydrogenation.

Table 5. Experimental Matrix for Donor Solvent Hydrogenation Experiments

Experiment*	Coal +				
	Gas			Solvent	
	Protium	Deuterium	Nitrogen	Tetralin-h_{12}	Tetralin-d_{12}
D/P		X		X	
P/D	X				X
N/D			X		X

*Experiment name indicates gas isotopic composition/solvent isotopic composition (i.e., D/P refers to deuterium gas/tetralin-h_{12}).

In Experiment D/P, the only source of ^2H was the gas phase. Therefore, any deuterium found either in the spent solvent or in the coal products originated from the gas phase. In Experiments P/D and N/D, the only source of ^2H was the solvent, tetralin-d_{12}. In Experiment P/D, gas-phase hydrogen (as $^1H_2$) was present; in Experiment N/D, it was not. Any deuterium found in the coal products or gas phase came from the solvent. Similarly, in Experiment N/D, any protium found in the spent solvent originated from the coal. In this manner, the deuterium tracer method provides a clear accounting of the hydrogen during coal liquefaction.

Product Yields. The product yields from these experiments are presented in Table 6. The data from these experiments show that >25% conversion was obtained in each. Of these three, the lowest conversion occurred in the experiment in which the nitrogen atmosphere was used. Perhaps this is because gas-phase hydrogen was not present to react with the radicals formed by bond cleavage.

Experiment D/P, in which deuterium gas and tetralin were used, resulted in a 49% conversion. In contrast, Experiment P/D, in which protium gas and tetralin-d_{12} were used, yielded only 37% conversion. This is a considerable difference and may be indicative of an isotope effect in the solvent. The possibility of an isotope effect in the solvent is an important point which will be addressed in future experiments.

Table 6. Product Yields From Donor Solvent Hydrogenation Experiments (wt %)

Experiment	D/P	P/D	N/D
Gas	10	11	9
Oil	24	19	16
Asphaltenes	8	4	1
Preasphaltenes	7	3	3
Residue	51	63	71

Isotopic Composition of Gas-Phase Products. Each of these experiments yielded 2.0-3.0 g of gas-phase products, of which approximately half was methane. A detailed list of the products is given elsewhere.[17] The isotopic compositions of the methane, ethane, and propane are presented in Table 7. Comparing the results from Experiments D/P and P/D, the experiment in which the deuterium was in the solvent (P/D) shows more highly deuterated methane, ethane, and propane species than the experiment in which the deuterium was in the gas phase (D/P). The isotopic compositions of the propane in these two experiments were comparable.

The isotopic composition of the product gases from the experiment with nitrogen in the gas phase (N/D) is similar to that in which protium was present in the gas phase (P/D). The methane and propane results are similar. The presence of ethylene interferred with the interpretation of the ethane data from Experiment N/D. The overall similarity of the isotopic composition data from Experiments P/D and N/D is an indication that relatively little exchange between the hydrocarbons and hydrogen occurs in the gas phase.

The concentrations of the D_0 species can be correlated with the deuterium source used in the experiment. The D_0 species can form from the corresponding radicals by protium abstraction from the coal (in any experiment), the protium gas (in Experiment P/D), or from tetralin-h_{12} (in Experiment D/P). The amount of D_0 species is reduced if the same radicals abstract deuterium from the tetralin-d_{12} (Experiments P/D and N/D) or $^2H_2$ gas (Experiment D/P).

It appears that $^2H_2$ gas reacts more effectively than tetralin-d_{12} with the hydrocarbon gas precursors which may be small alkyl radicals. This tentative conclusion is supported by previous experimental data obtained on the hydrogenation of coal with deuterium gas at 380°C without added donor solvent (Experiment B). The D_0 components in that experiment were 3% for CH_4, 0% for C_2H_6,

Table 7. Isotopic Compositions of the Product Gases (%)

Component (Experiment)	D_0	D_1	D_2	D_3	D_4	D_5	D_6	D_7	D_8
CH_4									
D/P	28	29	30	13	0	–	–	–	–
P/D	36	25	13	14	12	–	–	–	–
N/D	40	24	14	11	11	–	–	–	–
C_2H_6									
D/P	8	31	32	19	10	0	0	–	–
P/D	23	17	19	15	10	10	6	–	–
N/D	*	*	*	11	9	12	13	–	–
C_3H_8									
D/P	27	24	17	10	10	6	4	2	0
P/D	35	31	16	7	3	2	3	2	1
N/D	34	27	14	9	3	4	5	2	2

*Interference from other species causes uncertainty in these values.

and 12% for C_3H_8, which indicates that the $^2H_2$ gas is a more effective deuterium source than tetralin-d_{12} for the conversion of the hydrocarbon precursors to the corresponding hydrocarbons. This may be due to the greater access of the hydrogen gas molecules to the pore structure of the coal, compared with that of the solvent molecules. Hydrogen exchange with the solvent probably contributes to a lesser extent to the formation of these species, because even after 1 h, no more than one-third of the solvent's hydrogen in any position has been exchanged.

Isotopic Composition of Liquid and Solid Products. The deuterium contents of the solvent-fractionated products are presented in Table 8. These data show that considerable deuterium incorporation occurred in each experiment. The weighted-average deuterium incorporation in Experiment D/P is 32%. In Experiment P/D, it is 27%. That is, there is greater overall deuterium incorporation in the products from the experiment in which deuterium was in the gas phase (D/P). Experiment P/D, in which fully deuterated solvent was used, shows less deuterium incorporation. This result indicates that a mechanism occurs which accounts for considerable interaction of the gas-phase hydrogen with the coal.

Table 8. Deuterium Contents of Products
(Atom % of Total Hydrogen)

Fraction	Experiment D/P	Experiment P/D	Experiment N/D
Gas	29	33	29
Oil	26	26	12
Asphaltenes	42	21	31
Preasphaltenes	34	15	28
Residue	46	20	36
Total Product	32	27	28

The deuterium contents of the asphaltenes, preasphaltenes, and residue depend principally on the isotopic composition of the gas phase. These fractions have high atom % deuterium contents in Experiment D/P, in which $^2H_2$ and tetralin-h_{12} were used, and low atom % deuterium contents in Experiment P/D, in which $^1H_2$ and tetralin-d_{12} were used. The values for these fractions from Experiment N/D, in which tetralin-d_{12} was used under N_2 pressure, were between these two extreme values. This indicates that the use of the inert gas, which precludes reactions with molecular hydrogen, allows pathways which are otherwise less significant to become relatively more important. No obvious trend with regard to the isotopic composition of either the oil or the gas fraction was observed. The elemental analysis data for these fractions are presented elsewhere.[18]

Isotopic Composition of Spent Solvents. The spent solvents from these three experiments were analyzed by GC-MS and NMR to determine the deuterium content of each solvent component. The GC-MS data for the recovered tetralin and naphthalene are listed in Table 9. The isotopic distribution of the tetralin and naphthalene from Experiment D/P are remarkably similar. Both show only small amounts of species more highly deuterated than the D_4 species.

In Experiments P/D and N/D, tetralin-d_{12} was used, and the recovered tetralin and naphthalene contain considerable amounts of deuterium. The isotopic distributions of the tetralin and naphthalene are similar in that the most highly deuterated species (D_8-D_{12} in tetralin and D_4-D_8 in naphthalene) are predominant. The isotopic composition of the tetralin in Experiment P/D is very similar to that from Experiment N/D. This indicates that isotopic exchange between the coal and the solvent is greater than it is between the hydrogen gas and solvent. This conclusion is further substantiated by the NMR results (vide infra).

Table 9. Isotopic Distributions in Tetralin and Naphthalene in the Recovered Solvents

Component (Experiment)	D_0	D_1	D_2	D_3	D_4	D_5	D_6	D_7	D_8	D_9	D_{10}	D_{11}	D_{12}
Tetralin													
(D/P)	28	32	22	11	5	1	1	0	0	0	0	0	0
(P/D)	1	1	3	3	4	3	3	4	6	11	21	23	17
(N/D)	1	1	2	3	3	2	3	1	4	9	23	30	18
Naphthalene													
(D/P)	30	32	21	10	4	2	1	0	0	-	-	-	-
(P/D)	4	6	7	11	14	20	18	13	7	-	-	-	-
(N/D)	1	2	4	7	11	17	21	22	15	-	-	-	-

The naphthalene from Experiment P/D, however, shows more of each of the species of low and intermediate deuteration (D_0-D_5) than that from Experiment N/D. This indicates that a significant amount of the gas-phase protium exchanged with the naphthalene. It should be noted that these GC-MS data are in good agreement with the NMR data which were obtained independently. They are presented in Table 10.

Table 10. Isotopic Compositions of Recovered Tetralin

Experiment	Incorporated Isotope	Atom % Isotope	Isotopic Distribution (%)		
			H_α	H_β	H_{ar}
D/P	^2H	13.5	76	14	10
P/D	^1H	23.4	75	17	8
N/D	^1H	18.6	74	17	9

In Experiment D/P, tetralin-h_{12} was used as the starting solvent, and only 13.5 atom % deuterium was found in the recovered tetralin. The deuterium distribution is H_α = 76%, H_β = 14%, and H_{ar} = 10%. In Experiments P/D and N/D, fully deuterated tetralin was used. The protium incorporation in the tetralin-d_x recovered from these experiments was 23.4% in P/D and 18.6% in N/D. That is, the amount of protium incorporated in the tetralin-d_x was larger in the experiment with the protium atmosphere than in the experiment with the nitrogen. This supports the finding in Experiment D/P that gas-phase hydrogen is incorporated into the solvent.

The data from Experiment D/P also indicate that there is direct incorporation of deuterium gas into the coal. At the end of Experiment D/P there was only 13.5% deuterium in the tetralin, and the α-aliphatic position of tetralin contained 31% deuterium (13.5 x 0.76 x 3) at the end of the experiment. Nevertheless, some of the coal products (asphaltenes, preasphaltenes, and residue), which formed throughout the duration of the reaction, contained considerably more deuterium. This large amount of deuterium incorporation into the coal products strongly indicates that there is also a direct route for deuterium incorporation into the coal from the gas phase without the participation of tetralin.

Experiments P/D and N/D display protium ratios of approximately 75% in the α-aliphatic positions, 17% in the β-aliphatic positions, and 8% in the aromatic positions. Whether or not protium is present in the gas phase, this ratio of incorporation is the same. It is essentially the same as the ratio of deuterium incorporated into tetralin. These three results were obtained at 425°C. The N/D system was also tested using reaction times of 1/4 and 1/2 h. It was found[19] that the extent of isotopic exchange with the solvent increases with time; however, the ratio of exchanged hydrogen in the three different structural positions of the solvent remains constant within this reaction time range (1/4 to 1 h).

With regard to the solvent, both the GC-MS and NMR data quantitatively indicate the importance of the α-positions in tetralin for the transfer of hydrogen from the gas phase to the coal (Experiment P/D) and from one site in the coal structure to another (Experiment N/D). The GC-MS data suggest that one structural position may be the most active, and the NMR data confirm that, indeed, one position is the most active and that it is the α-position(s). Perhaps the α-tetralinyl radical acts to channel gas-phase hydrogen into the coal. We intend to conduct experiments at shorter reaction times to investigate this.

<u>Preferential Incorporation in the Coal - Donor Solvent - Gas System</u>. In experiments conducted using the coal - donor solvent - gas system, preferential incorporation was found in several structural positions.[11,12,18,19] The clearest indication of such incorporation occurs when the only source of protium is the coal. This is so because then no other protium source can exchange with the deuterium that is incorporated into the coal products and in this manner obscure the results. The coal - donor solvent - gas system was investigated using tetralin-d_{12} and nitrogen and using tetralin-d_{12} and deuterium.

The preferential incorporation that occurred is summarized in Table 11. Strong preferential incorporation was found in the α-aliphatic positions in all fractions. Weak incorporation was found in the aromatic positions of the preasphaltenes from the tetralin-d_{12} and nitrogen experiments. The other instances of weak preferential incorporation are from the tetralin-d_{12} and deuterium experiments. Perhaps the weak preferential incorporation in the aromatic positions of the preasphaltenes which occurred in the tetralin-d_{12} and nitrogen system may have been obscured by stronger incorporation in other positions in the tetralin-d_{12} and deuterium system.

Table 11. Preferential Incorporation in the Coal - Donor Solvent - Gas System

Functional Region (ppm)	Oil	Asphaltenes	Preasphaltenes
γ-aliphatic (0.0-1.0)		Weak in 17%	Weak in 17%
β-aliphatic (1.0-1.9)			Weak in 17%
α-aliphatic (1.9-4.5)	Strong in 100%	Strong in 100%	Strong in 100%
Aromatic (4.5-10.0)	Weak in 75%		Weak in 50%

SUMMARY

Mechanisms of coal hydrogenation were investigated using a deuterium tracer method. This method makes it possible to determine which structural positions in the coal react with hydrogen during liquefaction. The experiments were conducted using bituminous coal with gas-phase deuterium ($^2H_2$) and/or a fully deuterated solvent under typical liquefaction conditions, 380-425°C and 1500-3000 psig (10-20 MPa). Solvent-fractionated products were analyzed to determine total deuterium content and deuterium incorporation by structural position.

In initial experiments, proton NMR deconvolution spectrometry was compared with deuteron NMR spectrometry to assess its potential for determining into which structural sites the deuterium was incorporated. In most cases, there was very good agreement between the values obtained by the two techniques. On the basis of the results, it was decided that deuteron NMR spectrometry would be used to assess the structural sites of deuterium incorporation in all subsequent experiments.

A hydrogenation index (HI) and an exchange index (EI) were formulated to measure the amount of deuterium incorporated due to hydrogenation and exchange reactions, respectively. The HI is the ratio of the net deuterium incorporated into the coal to the total deuterium incorporated into the coal. The EI is the ratio of the total deuterium incorporated minus the net deuterium incorporated, to the total deuterium incorporated into the coal.

When the coal-deuterium system was investigated, two significant results were obtained. Deuterium incorporation was found to vary both with product fraction and with structural position. Deuterium contents increased in the order oil < asphaltenes < preasphaltenes < residue. In each of the three soluble fractions, although some deuterium was incorporated into each of the structural positions, preferential incorporation occurred in the α-aliphatic positions. In addition, preferential incorporation was indicated in the γ-aliphatic positions of some fractions, as well as in the aromatic fractions of some oils.

The coal-deuterium-tetralin-d_{12} system gave fractions with uniform deuterium contents. Preferential deuterium incorporation was found in the α-aliphatic positions of all soluble products and in the aromatic positions of some oil and preasphaltene fractions. Important information with regard to the reaction mechanisms in the donor solvent system was obtained by analyzing the spent solvent mixture that was recovered from the reaction products. The results indicate that not only hydrogen donation but also hydrogen exchange involving the α-positions of tetralin can have a significant role in stabilizing the fragments that form during the thermal decomposition of the coal. In addition, evidence was obtained that there is also a direct route for deuterium incorporation into the coal from the gas phase without the participation of tetralin.

ACKNOWLEDGMENT

The authors wish to thank Dr. Ira B. Goldberg for helpful technical discussions and Mr. David K. Hadden and Mr. Thomas B. Johnson for their excellent technical assistance.

REFERENCES

1. B. S. Gudkov, N. E. Zlotina, L. A. Makhlis, and S. L. Kiperman, "Isotopic Exchange and Hydrogenation of Cyclohexene on Nickel," Inv. Akad. Nauk SSSR, Ser. Khim. 1970 (11) 2525-32; CA, 74:87184n (1971).
2. H. H. King and L. M. Stock, "Influence of Illinois No. 6 Coal and Coal-Related Compounds on the Exchange Reaction Between Diphenylmethane and Perdeuteriotetralin," Fuel 59:447 (1980).
3. Y. C. Fu and B. D. Blaustein, "Reaction of Coal and Graphite in a Microwave Discharge in H_2O and D_2O," Chem. Ind. 1257 (1967).
4. T. Kessler and A. G. Sharkey, Jr., "Use of High-Resolution Mass Spectrometry to Identify Products from Microwave Discharges in Coal-D_2O Mixtures," Spectrosc Lett., 1:1977 (1968).

5. R. P. Skowronski, J. J. Ratto, and L. A. Heredy, "Deuterium Tracer Method for Investigating the Chemistry of Coal Liquefaction," Technical Progress Report to U.S. ERDA, FE-2328-4 (1976).
6. F. K. Schweighardt, B. C. Bockrath, R. A. Friedel, and H. L. Retcofsky, "Deuterium Magnetic Resonance Spectrometry as a Tracer Tool in Coal Liquefaction Processes," Anal. Chem. 48:1254 (1976).
7. A. F. Gaines and Y. Yurum, "Pyrolysis of Deuterated and Reduced Coals," Fuel 55:129 (1976).
8. J. R. Kershaw and G. Barrass, "Deuterium Studies of Coal Hydrogenation," Fuel 56:455 (1977).
9. D. C. Cronauer, D. M. Jewell, Y. T. Shah, R. J. Modi, "Mechanism and Kinetics of Selected Hydrogen Transfer Reactions Typical of Coal Liquefaction," Ind. Eng. Chem. Fundam. 18:153 (1979).
10. J. A. Franz, "^{13}C, ^{2}H, ^{1}H NMR and GPC Study of Structural Evolution of a Subbituminous Coal During Treatment with Tetralin at $427°C$," Fuel 58:405 (1979).
11. R. P. Skowronski, J. J. Ratto, and L. A. Heredy, "Deuterium Tracer Method for Investigating the Chemistry of Coal Liquefaction," Technical Progress Report to U.S. ERDA, FE-2328-13 (1978).
12. R. P. Skowronski, J. J. Ratto, and L. A. Heredy, "Deuterium Tracer Method for Investigating the Chemistry of Coal Liquefaction," Technical Progress Report to U.S. DOE FE-2781-4 (1980).
13. I. B. Goldberg, H. R. Crowe, J. J. Ratto, R. P. Skowronski, and L. A. Heredy, "Study of the Products of Coal Hydrogenation and Deuteration by Electron Paramagnetic Resonance," Fuel 59:133 (1980).
14. R. P. Anderson, "Development of a Process for Producing an Ashless Low-Sulfur Fuel from Coal," The Pittsburgh and Midway Coal Mining Company, FE-486-T1 (1975).
15. B. K. Schmid and D. M. Jackson, "The SRC-II Process," Third Annual International Conference on Coal Gasification and Liquefaction," University of Pittsburgh (August 3-5, 1976).
16. L. E. Furlong, E. Efron, L. W. Vernon, and E. L. Wilson, "Coal Liquefaction by the Exxon Donor Solvent Process," presented at 1975 National AIChE Meeting, Los Angeles, California (November 18, 1975).
17. R. P. Skowronski, J. J. Ratto, and L. A. Heredy, "Deuterium Tracer Method for Investigating the Chemistry of Coal Liquefaction," Technical Progress Report to U.S. DOE, FE-2781-5 (1980).
18. R. P. Skowronski, J. J. Ratto, I. B. Goldberg, and L. A. Heredy, "Deuterium Tracer Method for Investigating the Chemistry of Coal Liquefaction," Technical Progress Report to U.S. DOE, FE-2781-6 (1980).
19. R. P. Skowronski, J. J. Ratto, I. B. Goldberg, and L. A. Heredy, "Deuterium Tracer Method for Investigating the Chemistry of Coal Liquefaction," Technical Progress Report to U.S. DOE, FE-11418-8 (1980).

THE REACTION OF CARBON-14-LABELED REAGENTS WITH COAL*

Clair J. Collins, Vernon F. Raaen, Cherie Hilborn,
W. Howard Roark, and Paul H. Maupin
Chemistry Division
Oak Ridge National Laboratory
Oak Ridge, Tennessee 37830

INTRODUCTION

The reasons for carrying out the research presented here were threefold: 1) to allow carbon-14-labeled reagents to react with coal and with coal-derived materials at low temperatures and thus to estimate a) the reagents' value in determining which functions are present in the original coal or b) have been added or removed during coal reactions; 2) to allow carbon-14-labeled compounds (containing structures similar to those known to be present in coal) to react with coal at or above 400°, then to determine the fates of these compounds using the carbon-14 as a tracer. In this way it was hoped conclusions could be drawn as to how the same kind of moiety behaves in the coal structure; and 3) to elucidate the mechanisms of the thermal reactions of the above-mentioned labeled compounds with coal.

RESULTS AND DISCUSSION

Reactions of Diazomethane-^{14}C and Dichlorocarbene-^{14}C with Coal and with Coal-Derived Materials

Diazomethane is a well-known reagent for the determination of acidic (phenolic OH or COOH) groups in coal.[1,2,3] Carbenes[4,5] and dihalocarbenes[5-7] are useful reagents for many organic reactions,

---

*Research sponsored by the Division of Chemical Sciences, Office of Basic Energy Sciences, U. S. Department of Energy, under contract W-7405-eng-26 with the Union Carbide Corporation.

but to our knowledge have not yet been applied to coal chemistry. We allowed coal and thermally treated coal-derived products to react with both of these reagents, compared results and attempted to draw conclusions concerning structural changes occurring during the thermal treatments. We report here the first preliminary results.

Diazomethane-^{14}C and dichlorocarbene-^{14}C were allowed to react (in a ball mill) with coal and with coal-derived materials. Carbon-14 analyses permitted us to determine the extents of reaction. Products of reaction with diazomethane-^{14}C were hydrolyzed with base, then acid, and each fraction was reassayed for carbon-14 content. The results are given in Tables 1 and 2. Diazomethane reacts with phenols and

Table 1. Action of Diazomethane-^{14}C Upon Illinois No. 6 Vitrain and Wyodak Coal

Type of Coal and Treatment	Uptake of *CH$_2$N$_2$ in Oxygen Equivalents	% Removed With HBr in O.E.[a]	% Removed With KOH in O.E.[a]
Illinois No. 6 Vitrain:			
1. None	7-9	3-4	1
2. Ball Mill 24 h, Aerosal & Air	5	3	0.5
3. Preheat 400°, 30 min	6	3	-
4. Tetralin 400°, 38 min, Extract Benzene	3	-	-
Wyodak:			
5. None	12	4	3.0
6. 400°, 30 min	12	3	2
7. 400°, 30 min HBr, 24 h	14	3	2
8. HBr, 24 h	18	6	2

[a] Oxygen equivalent = percent oxygen in sample which reacts. The percent CH$_2$ uptake = O.E. × 14/16. The percent CCl$_2$ uptake = O.E. × 82.9/16. The values are not corrected for the weights of CH$_2$ or CCl$_2$ in the samples.
[b] Basic hydrolysis was carried out first, then hydrolysis with HBr. When the procedure was reversed, the results were the same, as they also were when acid and base hydrolyses were done on separate aliquots.

Table 2. Action of Diazomethane-^{14}C and Dichlorocarbene-^{14}C on Heat-Treated Fractions (400°, 30 min) of Illinois No. 6 Vitrain

	Uptake of *CH$_2$N$_2$, O.E.[a]	Uptake of *CCl$_2$, O.E.[a]
Oil	2.8	
Asphaltene + Oil[b] Plus Air (24 h)	9.6	
Asphaltene	3–5	8.9[c]
Asphaltene Plus O$_2$		4.3
Preasphaltene	4.4	3.9
Preasphaltene Plus Air (24 h)	13	
Preasphaltene Plus O$_2$ (24 h)		2.2
Pyridine Insoluble		2.9

[a] Oxygen equivalent; see footnote, Table 1.
[b] Benzene extract of heat-treated vitrain
[c] The *CCl$_2$-treated fractions had been heated to 400° in the presence of tetralin.

with carboxylic acids, it inserts next to ketonic carbonyl, and also polymerizes. Base hydrolyzes carbomethoxy groups whereas HBr hydrolyzes both carbomethoxy esters and aryl alkyl ethers. The data (Table 1) are therefore indicative of minimum values for ArOH and -COOH groups. Air-exposed fractions (Table 2) exhibit enhanced reactivities with diazomethane-^{14}C, whereas oxygen-exposed asphaltenes and preasphaltenes exhibit reduced reactivities toward *CCl$_2$. We believe this means an increase in the ketonic and carboxyl content of the given fractions at the expense of phenolic hydroxyl, benzylic methylenes and other oxidizable moieties, e.g., alkoxyarenes, and (perhaps) pyrroles, and indoles. The data are in agreement with the reactions shown in Scheme 1, in which structures like A and A´ will react with CCl$_2$ but not with CH$_2$N$_2$, whereas B, B´, C, and C´ all will react with CH$_2$N$_2$ and not with CCl$_2$. From the last four entries of Table 1 it appears that HBr-treatment (24 h) of Wyodak coal must release phenolic hydroxyls (probably by cleavage of aryl alkyl ethers), since 6% O.E. of methyl groups are removed on acid hydrolysis compared with only 3% when the Wyodak is

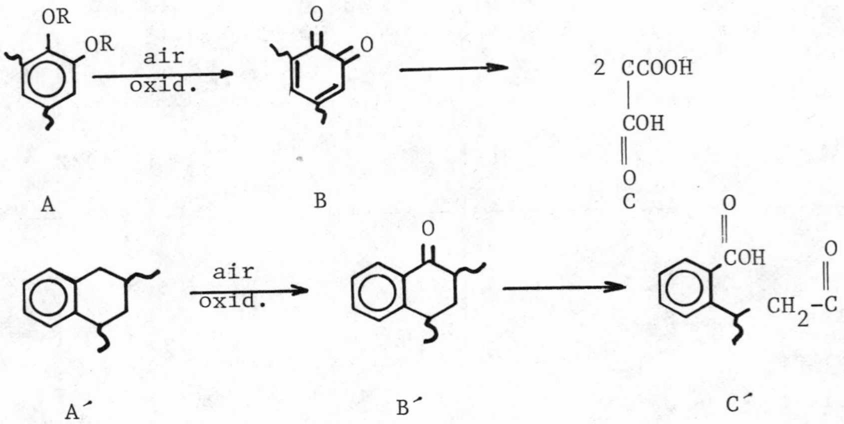

Scheme 1. Rationalization for increased uptake of $^{12}CH_2N_2$ and decreased uptake of $^{14}CCl_2$ on air oxidation of coal-heated fractions (see text for explanation).

first heat-treated and then allowed to react with *CH_2N_2. The heat-treatment must destroy aryl alkyl ethers to produce water, alkenes, and probably aromatic radicals which are then trapped by hydrogen-donor moieties in the coal.

It should be noted that the uptakes of diazomethane-^{14}C by coal and coal-derived samples [Tables 1 and 2] are not duplicable because of the varying degrees of polymerization exhibited by the diazomethane. This difficulty has already been commented upon by Blom, Edelhausen and van Krevelen.[2] The action of diazomethane upon coal has been classified by Given[3] as unsuccessful for this — and other — reasons. The acid and base hydrolyses of the diazomethane-^{14}C-treated samples, however, are nicely duplicable, since they are unaffected by the degree of polymerization. Although the diazomethane method[1,2,3] leads to estimates of phenolic content which are lower than those values estimated using other[3] techniques, the trends a) to increased uptake of diazomethane after air treatment of Illinois No. 6 asphaltenes and preasphaltenes, and b) to destruction of alkyl aryl ethers by heat treatment of Wyodak are unmistakable.

Reactions of Illinois No. 6 Coal with α-Naphthol-^{14}C and with β-Naphthol-^{14}C

It is well-known[8] that many coals contain phenolic hydroxyl groups, although the fate of these groups during thermolysis is not well understood. We therefore heated Illinois No. 6 with α-naphthol, and separately, with β-naphthol at or above 400°. Both naphthols

1-naphthol-*C₁,₈* + Illinois No. 6 + benzene →(400°, 20 h) naphthalene-*C* (31%)

+ methylnaphthalene-*C* (14%) + phenylnaphthalene-*C* (?) (38%) + biphenyl   (1)

0.500 g   0.500 g   2.5 ml

2-naphthol + Illinois No. 6 + V 131 Process solvent, Wilsonville →(400°, 20 h)

0.500 g   0.500 g   2.50 ml

1-methylindane-*C* (trace) + naphthalene-*C* (55%) + 2-methylnaphthalene-*C* (35%) + 1-methylnaphthalene-*C* (trace)   (2) were labeled with carbon-14 to facilitate the analysis. In this way we hoped to be able to draw some conclusions about what happens to the phenolic moieties present in the coal structure itself. In those cases in which tetralin or other hydrogen donor solvent was not present we used benzene as a medium to facilitate transfer in and out of the reaction vessels (20 cm × 12.7 mm o.d., 9 mm i.d., stainless steel tubing sealed at both ends with swagelok fittings). The results of two typical runs are shown in equations (1) and (2).

In the experiment illustrated in Eq. (1), 17% of the α-naphthol was present in the mixture of products, and the small amount of residual solid product, after washing and drying, contained radio-activity equivalent to 83% by weight of α-naphthol-^{14}C. In the experiment illustrated in Eq. (2), no β-naphthol was recovered, although the few mg remaining of solid residue were radioactive corresponding to 14% by weight of β-naphthol-^{14}C equivalent. From these experiments it is clear that both α- and β-naphthols under the reaction conditions lose their phenolic hydroxyl groups to yield water, naphthalene, and β-methylnaphthalene. The recovery of a trace of labeled methylindane, Eq. (2), may mean that β-naphthol is first reduced to tetralin which then rearranges to 1-methylindane through processes previously discussed.[9] The production of β-methyl-naphthalene (with traces of α-) then becomes intelligible on the basis of the reduction of the naphthols to dihydrotetralin — as shown, for example, in Eq. (3) — which then can react with alkyl

$$\text{naphthol-OH} \xrightarrow{2H} \text{ketone} \longrightarrow \text{OH-tetralin} \xrightarrow{-H_2O} \text{dihydronaphthalene} \quad (3)$$

radical from the coal (or H-donor solvent) to produce alkylated tetralin, Eq. (4), (5). The methyl tetralin then loses its hydrogen

$$\text{dihydronaphthalene} + RCH_2CH_2\cdot \longrightarrow \text{tetralin-CH}_2CH_2R \quad (4)$$

$$\text{tetralin-}\cdot CH_2CH_2R \xrightarrow{\text{H-donor cleavage}} \text{tetralin-CH}_3 + RCH_3 \quad (5)$$

(it can now act as a hydrogen donor) to yield 2-methylnaphthalene, Eq. (6). The biphenyl produced, Eq. (1) arises from the benzene used to facilitate material transfer to and from the steel tube bombs. (The biphenyl is unlabeled, which demonstrates that it

$$\text{1-methyltetralin} + \text{H-acceptor} \longrightarrow \text{2-methylnaphthalene} \quad (6)$$

cannot be formed from the labeled naphthols.) The 2-phenylnaphthalene, Eq. (1), tentatively identified through its nmr spectrum, could be produced by interaction of phenyl radicals with the postulated [Eqs. (3) and (4)] dihydronaphthalene, followed by dehydrogenation.

Reaction of Thiophenol-^{14}C with Illinois No. 6 Coal

Not much is known about the forms in which organic sulfur is present in coal. It is possible, however, that some of it could be present as thiophenolic groups. Consequently we prepared carbon-14-labeled thiophenol and heated it with Illinois No. 6 coal and benzene for 20 h at 400°. The composition of the liquid product, after removing the solid material, was as shown in Eq. (7). There

$$\text{PhSH}^* + \text{Illinois \#6} + \text{C}_6\text{H}_6 \xrightarrow[20\text{ h}]{400°} \text{C}_6\text{H}_6^* + \text{PhCH}_3^* +$$

1.00 ml   5.00 g   1.00 ml   31%   7%

$$\text{H}_2\text{S} + \text{PhCH}_2\text{CH}_3^* + \text{thianthrene}^{**} + \text{dibenzothiophene}^{**} \quad (7)$$

              4%   56%

was no residual thiophenol-^{14}C. There was some biphenyl produced, and this arises solely from the benzene present, since it contains no carbon-14. The solid residue after being washed with acetone and pyridine still contained carbon-14 equivalent to 2% by weight of thiophenol-^{14}C. We reported earlier[10] that thiophenol and tetralin at 400° for 18 h yield benzene, diphenyl sulfide and hydrogen sulfide. Thianthrene under similar conditions yields hydrogen sulfide and dibenzothiophene. The results for thermolysis of thiophenol with Illinois No. 6 coal, Eq. (7), are not easy to rationalize unless homolysis of both the S-H [Eq. (8)] and C-S [Eq. (9)] take place, in which case benzene, toluene and ethylbenzene could be produced as shown in Eqs. (10)-(13), and formation of thianthrene [Eqs. (14) and (15)] and of dibenzothiophene [Eqs. (16) and (17)] could proceed as shown.

$$\text{PhSH} \xrightarrow{R\cdot} \text{PhS}\cdot + RH \quad (8)$$

$$\text{PhSH} \longrightarrow \text{Ph}\cdot + \cdot SH \quad (9)$$

$$\text{Ph}\cdot + RH \longrightarrow \text{PhH} + R\cdot \quad (10)$$

$$\text{Ph}\cdot + RCH_2CH_2\cdot \longrightarrow \text{Ph-}CH_2CH_2R \quad (11)$$
(coal moiety radical)

$$\text{Ph-}CH_2CH_2R \xrightarrow{\text{H-Donor}} \text{Ph-}CH_3 + RCH_3 \quad (12)$$

$$\text{Ph-}CH_2CH_2R \xrightarrow{\text{H-Donor}} \text{Ph-}CH_2CH_3 + RH \quad (13)$$

$$\text{PhS}\cdot + \text{PhSH} \xrightarrow{-H\cdot} \text{Ph-S-C}_6H_4\text{-SH} \quad (14)$$

$$\text{Ph-S-C}_6H_4\text{-SH} \xrightarrow{-H\cdot} [\text{Ph-S-C}_6H_4\text{-S}\cdot] \xrightarrow{-H\cdot} \text{thianthrene} \quad (15)$$

$$\text{C}_6\text{H}_5\cdot + \text{C}_6\text{H}_5\text{SH} \xrightarrow{-\text{H}\cdot} \text{HS-C}_6\text{H}_4\text{-C}_6\text{H}_5 \quad (16)$$

$$\text{HS-C}_6\text{H}_4\text{-C}_6\text{H}_5 \xrightarrow{-\text{H}\cdot} \text{[dibenzothiophene radical]} \xrightarrow{-\text{H}\cdot} \text{dibenzothiophene} \quad (17)$$

CONCLUSIONS

Both asphaltenes and preasphaltenes from Illinois No. 6 vitrain exhibit enhanced uptake of $^{14}\text{CH}_2\text{N}_2$ after air exposure. Preasphaltenes however exhibit decreased $^{14}\text{CCl}_2$ uptake after oxygen treatment. These results are in agreement with a scheme in which aryl alkyl ethers and benzylic-type linkages are oxidized to ketones and then carboxylic acids.

Heat treatment of Wyodak coal appears to destroy most of the aryl alkyl ethers present, since HBr-treated Wyodak, after reacting with diazomethane-^{14}C, contains twice the acid-hydrolyzable methyl-^{14}C groups exhibited by heat-treated Wyodak.

Thermolysis of α-naphthol-^{14}C and of β-naphthol-^{14}C in the presence of Illinois No. 6 coal results in loss of hydroxyl to produce water as well as in alkylation of the naphthalene nucleus, processes which we presume also occur in the phenolic moieties in the coal itself. Thiophenol-^{14}C yields hydrogen sulfide, toluene, ethylbenzene, thianthrene, and dibenzothiophene. The results for the thermolyses of α- and β-naphthol-^{14}C and of thiophenol-^{14}C are explained through free radical mechanisms quite similar to those previously postulated[11] for the thermolyses of model compounds in the presence of tetralin.

EXPERIMENTAL

Diazomethane-^{14}C

The method of Heard, Jamieson, and Solomon[12] was used.

Chloroform-^{14}C for Generation of Dichlorocarbene-^{14}C

Chloroform-^{14}C was prepared from acetophenone-*methyl*-^{14}C by the method of Shuford, West, and Davis.[13] These authors also give the directions for preparing acetophenone-*methyl*-^{14}C.

α-Naphthol-1,4,5,8-^{14}C and β-Naphthol-1,4,5,8-^{14}C

Naphthalene-1-^{14}C was prepared by a method[14] previously described for the preparation of 2-methylnaphthalene-8-^{14}C. Naphthalene-1-^{14}C was sulfonated at 40° with conc. sulfuric acid to yield labeled α-naphthylsulfonic acid,[15] and at 160° to yield labeled β-naphthylsulfonic acid.[15] Treatment of the respective sulfonic acids with concentrated sodium hydroxide[16,17] yielded α-naphthol-1,4,5,8-^{14}C and β-naphthol-1,4,5,8-^{14}C$_1$.

Thiophenol-^{14}C

Benzene-^{14}C-sulfonyl chloride was prepared[18] from benzene-^{14}C,[19] and then reduced to thiophenol-^{14}C by the method of Adams and Marvel.[20]

Reactions of Coal and Coal Products with Diazomethane-^{14}C

The reactions took place in ball mills of .1, .3, and 1.5 liter volume. Steel balls of 1 cm diameter were used in the smallest mill, whereas alundum balls of 2 and 2.5 cm, respectively, were used in the two larger mills. In a typical run, 2.2 g of an Illinois No. 6 coal (<325 mesh) was further crushed under Argon in a (0.3 ℓ capacity) ball mill overnight. The diazomethane-^{14}C (in ether) generated from 6.1 g nitrosomethyl-^{14}C-urea was added, under argon, to the coal in the ball mill, and the mill was allowed to run from 1-3 days. The volume of dry ether was chosen so that the mill was never more than half filled. In earlier runs with THF as the liquid medium it was shown that the THF undergoes serious decomposition. After the appropriate time the mixture was filtered, washed with ether, dried, and assayed for carbon-14. For acid hydrolysis, 0.5 g of the sample (already treated with ^{14}CH$_2$N$_2$) was placed in 20 ml of 48% HBr and the mixture was heated under reflux for 24 h, filtered, washed with water, then absolute ethanol, then dried under high vacuum for 3 h and assayed for carbon-14 by the dry combustion method.[21] For basic hydrolysis, .5 g of a sample which had been treated with diazomethane-^{14}C was added to a mixture of 2.5 ml of 50% KOH and 22.5 ml of absolute methanol. The mixture was heated under reflux for 6 h, then cooled, filtered, the solid was washed with methanol, and then dried under high vacuum for several hours and assayed for carbon-14. In some cases the basic hydrolysis was performed on the sample which had been hydrolyzed with HBr. The results in both cases were the same within experimental error.

Reactions of Coal and Coal Products with Dichlorocarbene-^{14}C

These reactions also took place in ball mills (previously described). In a typical run, 0.5 g of Illinois No. 6 coal (<325 mesh), 4 ml benzene, 2.1 g 50% sodium hydroxide solution, and .11 g benzyltriethylammonium chloride (a phase-transfer catalyst) were cooled to -10° and placed in one of the small ball mills; 1.03 g chloroform-^{14}C was then added and the mill was allowed to rotate for 24 h. An argon atmosphere was employed. Another (1.02 g) portion of chloroform-^{14}C was added and the mill was allowed to rotate another 24 h. The mixture was washed out of the mill with benzene and absolute methanol and centrifuged. The solid was washed several times with water, dried i.v. and assayed for carbon-14.

Reactions with Carbon-14 Labeled α- and β-Naphthol and Thiophenol

These reactions were carried out as described previously[10] using stainless-steel tube-bombs 5 mm i.d. and 2 mm wall thickness with Swage-Lok fittings. In a typical run, 0.2 g of either α- or β-naphthol-^{14}C, 1.00 g Illinois No. 6 vitrain and 1.5 ml benzene were placed in a tube-bomb and heated at 400° for 20 h [A Techam Fluidized bed bath SBL-2 was used]. The contents of the bomb were then removed from the bomb, placed in a Soxhlet extractor and continuously extracted with a benzene-toluene mixture prior to g.c. analysis of the soluble products. [Barber-Coleman Series 5000 and a Packard Model 894 both with radioactivity monitors were employed.] The reactions with thiophenol and thiophenol-^{14}C were carried out in the same way except that the wash solvents employed were acetone and ether.

ACKNOWLEDGEMENTS

The authors thank Drs. B. M. Benjamin, E. W. Hagaman, and Mr. L. L. Brown for carrying out the proton and ^{13}C-nmr analyses by which some of the products were identified.

REFERENCES

1. W. Fuchs and W. Stengel, Hydroxyl and Carboxyl Groups in Humic Acids, Brennst.-Chem. 10:303 (1929); W. Fuchs and A. G. Sandhoff, Investigations Concerning Hydroxycarboxylic Acids from Bituminous Coals, FUEL 19:45 (1940).
2. L. Blom, L. Edelhausen, and D. W. van Krevelen, Chemical Structure and Properties of Coal XVII. Oxygen Groups in Coal and Related Products, FUEL 36:135 (1957).
3. P. H. Given, The Reactivity of Coal in Organic Chemical Reactions, Brennst.-Chem. S. Bd 39:14 (1958).
4. T. L. Gilchrist and C. W. Rees, "Carbenes, Nitrenes and Arynes," Meredeth Corporation, New York (1969).
5. W. Kirmse, "Carbene Chemistry," Academic Press, New York (1964), 2d Edition (1971).
6. J. Hine, "Divalent Carbon," Ronald Press, New York (1964).
7. D. Seyferth, Phenyl(trihalomethyl)mercury Compounds: Exceptionally Versatile Dihalocarbene Precursors, Acc. Chem. Res. 5:65 (1972).

8. D. W. van Krevelen, Coal as Plant Debris and Functional Group Analysis in "Coal," Elsevier, Amsterdam (1972).
9. B. M. Benjamin, E. W. Hagaman, V. F. Raaen, and C. J. Collins, Pyrolysis of Tetralin, FUEL 58:386 (1979).
10. B. M. Benjamin, V. F. Raaen, P. H. Maupin, L. L. Brown, and C. J. Collins, Thermal Cleavage of Chemical Bonds in Selected Coal-Related Structures, FUEL 57:269 (1978).
11. C. J. Collins, V. F. Raaen, B. M. Benjamin, P. H. Maupin, Coal Chemistry. Reactions of Tetralin with Coal and with Some Carbon-14-Containing Compounds, J. Amer. Chem. Soc. 101:5009 (1979).
12. R. D. H. Heard, J. R. Jamieson, and S. Salomon, The Synthesis of Methylamine-^{14}C and Diazomethane-^{14}C, J. Amer. Chem. Soc. 73:4985 (1951).
13. C. H. Shuford, Jr., D. L. West, and H. W. Davis, Allylic Rearrangement of Hexachloropropene-1-^{14}C, J. Amer. Chem. Soc. 76:5803 (1954).
14. C. J. Collins, Carbon-14 Synthetic Studies. 2-Methyl-1,4-Naphthoquinone-8-^{14}C, J. Amer. Chem. Soc. 73:1038 (1951).
15. D. F. Othmer, J. J. Jacobs, Jr., and W. J. Bushmann, Sulfonation of Naphthalene, Ind. Eng. Chem. 35:326 (1943).
16. C. E. May, Beta Naphthol, J. Amer. Chem. Soc. 44:650 (1922).
17. F. Wilson and K. H. Meyer, Die Überführung aromatischer Sulfosäuren in Phenole, F. Wilson, K. H. Meyer, Ber der deutsch. Chem. Ges. 47:3160 (1914).
18. H. T. Clark, G. C. Babcock and T. F. Murray, Benzene Sulfonyl Chloride, in "Organic Syntheses," Collective Volume I, p. 85, H. Gilman, Ed. J. Wiley & Sons, Inc., New York (1941).
19. H. S. Turner and R. J. Warne, Benzene-^{14}C in "Organic Syntheses with Isotopes," p. 820, Part I, A. Murray III and D. L. Williams, Eds., Interscience Publishers, Inc., New York (1958).
20. R. Adams and C. S. Marvel, Thiophenol, in "Organic Syntheses," Collective Volume I, p. 504, H. Gilman, Ed., J. Wiley & Sons, New York (1958).
21. V. F. Raaen, G. A. Ropp, and H. P. Raaen, Carbon-14, McGraw-Hill Book Co., New York, page 240 (1968).

NATURAL PERMEABILITY REDUCTION IN POROUS MEDIA DUE TO THE PRESENCE OF KAOLINITE

Witold Kubacki

Instituto de Investigaciones Petroleras
Universidad del Zulia
Maracaibo, Venezuela

INTRODUCTION

The permeability to water of a clay medium is always a topic of controversy. The majority of engineers consider that different clays show different behaviour under fluid invasion being especially affected by the acidity of the fluid. Among the naturally occuring clays however, Kaolinite is considered to be very stable chemically and hence not altered by fluid invasion[1].
Its weakness as a mineral in a petroleum bearing formation is due to its low binding force to the Quartz grains which leads to ready migration in any fluid flow through the formation.

This work shows that in fact a certain amount of chemical reactivity is observed with Kaolinite. This conclusion was reached as a result of displacement tests using different fluids with varying pH. The Kaolinite in these samples occurred as a cementing material of the sand grains in a petroleum formation saturated with a medium viscosity crude. It is also found that a suitable chemical treatment can be used to reduce this reactivity considerably.
These results to some extent are contrary to generally accepted beliefs.

EXPERIMENT:

The quantitative and qualitative determinations of the minerals as well as their chemical elemental composition were carried out out by means of X-ray diffraction and fluorescence analysis. X-ray diffraction studies showed the presence of Quartz grains (SiO_2), Kaolinite $Al_2(OH)_4Si_2O_5$ and traces of Illite $1.3(K,Na)_2O.0.6(Mg,Fe)O.3.3(Fe,Al)_2O_3.16(Si,Al)O_2.5H_2O$ as well as organic constituents.

After the X-ray diffraction studies, the samples were pulverised and melted at 1050°C to convert them to a homogeneous form. They were then triturated and deposited by applying pressure on a boric acid pellet base for analysis by X-ray fluorescence. This analysis makes use of the characteristic emitted energies resulting from electronic transitions induced by the X-rays for distinguishing the different chemical elements. By this technique it is possible to detect the elements Na, K, Mg, Fe, Si and Al, which make up the minerals Quartz, Kaolinite and Illite. By establishing a mass balance between the experimental values of the chemical elements of these three minerals with the theoretically calculated values it is then possible to find the concentration of each of the minerals.[2] The results of this calculation are given in Table 1.

The samples used for the displacement tests were selected on the basis of porosity and absolute permeability data. Cylindrical samples with a base diameter of 1" and a length of 1.5" were dry cut from the original cores. The cylindrical samples were washed in toluene and then in distilled water at room temperature. They were then dried at 110°C for 24 hours and finally vacuum saturated with a particular displacement liquid. The cylindrical area of the samples was then painted with an impermeable lead oxide (Pb_3O_4) based paint. They were then placed in the sample holders and hermetically sealed with the alloy "Cerrobent."

The displacement liquid is contained in a calibrated recipient from which it can be pumped out at different rates which correspond to different displacement speeds of the piston of the Ruska pump. The rates are variable between 2.3ml/min. up to 23ml/min.
As the liquid is displaced into the cylindrical sample core, the entrance pressure produced is recorded by a manometer placed at the entrance to the sample container. The outlet pressure from the sample container is the same as the atmosphere. The displaced liquid is recovered and measured after it leaves the sample container. The permeability of the sample is calculated by means of Darcy's equation:

$$K = \frac{Q \cdot \mu \cdot L}{\Delta P \cdot A} \quad \ldots \ldots \ldots \ldots \ldots \ldots \ldots \ldots (1)$$

where: Q - rate of displacement of the liquid,
        μ - viscosity of the fluid,
        L - length of cylindrical sample
        ΔP - pressure difference between the entrance and exit of the cylindrical sample core,
        A - cross sectional area of the cylindrical sample core.

The liquid used to saturate the samples prior to the first series of displacement experiments is saline solution (1.6% NaCl), which corresponds closely to the formation water. The liquid first used in the displacement experiments was distilled water, considered

Table 1

Mineralogical composition of the sample cores

Sample core N°	Porosity %	Mayor components %	
		Kaolinite	Quartz
1	12.24	34.4	61.4
2	10.26	34.7	60.5
3	10.17	30.0	61.3
4	7.77	37.5	58.0
5	7.76	31.5	67.0
6	8.05	28.0	71.0
7	7.98	44.5	55.0
8	8.13	46.0	49.0

to be the least compatible with the mineralogical composition of the formation. The observed effect of distilled water on the permeability of the sample cores leads us to conclude that it damages the formation. The permeability is seen to decrease with respect to the displaced porous volume as shown in Figure 1. The damage permeability is also seen to reach a stable level. The core samples at this point are then used in displacement experiments with a 2% $AlK(SO_4)_2.12H_2O$ solution, which is known to stabilize dispersed Kaolinite[3]. Such a solution partially reestablishes the permeability that has been damaged by the distilled water. The dotted curves in Figure 1 show this restoration of the permeability in five core samples for when this study was done.

Once the displacement experiments have been completed, three cuts parallel to the flat surfaces were made in each sample core. The first cut is of the invasion front of the liquid as it enters the core. The second is of the middle of the core and the third at the exit. These cuts are shown in Figure 2. Each cut was pulverised and then analysed both for their mineralogical and chemical content, so as to determine the change in Kaolinite concentration along the core as a result of dispersion and migration during the displacement of the liquids. The results of these analyses are shown in Table 2. The concentration of Kaolinite is seen to be diminished at the entrance and to increase towards the middle and exit region of the core samples. Similar cuts were made of core samples subjected solely to distilled water displacement and then submitted to thermogravimetric analysis (TGA). None of the five samples analysed indicated that there were hydrates present.

Tests were performed on core sample N$\underline{o}$ 8 to restore the damaged permeability by treatment with 2% and 5% $AlK(SO_4)_2.12H_2O$ solutions. These results given in Figure 3 showed that the 2% solution was preferable.

Core sample N$\underline{o}$ 6 was vacuum saturated with 2% $AlK(SO_4)_2.12H_2O$ for 24 hours. Displacement experiments were then carried out with distilled water. The results of this test are shown in Figure 4. During this test, the pH at the exit of the nucleus was continuously monitored. A certain degree of stabilization of the permeability was observed after 800 porous volumes were displaced. The distilled water at this point was replaced by a cement filtrate of pH = 12. This produced an immediate decrease in the permeability with almost a complete sealing of the sample pores.

Figure 4 showed that distilled water damages the permeability of the cores even when they are first saturated with a 2%$AlK(SO_4)_2.12H_2O$ solution which is known to restore damaged permeability.

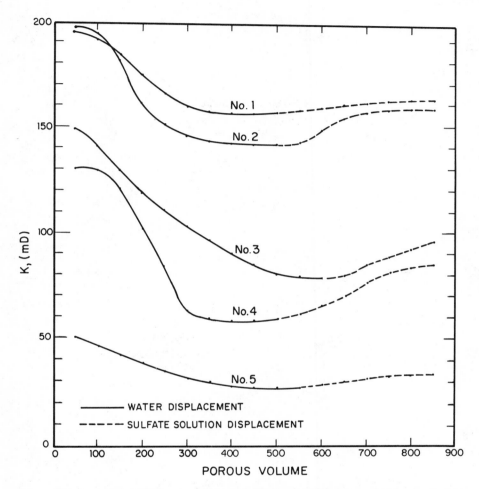

Fig. 1. Permeability variation after water and sulfate solution displacement.

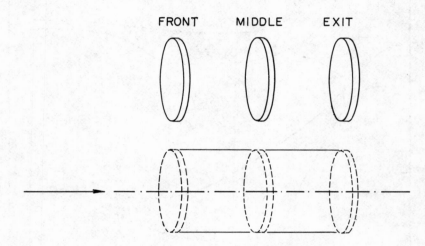

Fig. 2. Sample cuts used in the analysis of the Kaolinite concentration changes of the samples after water displacement.

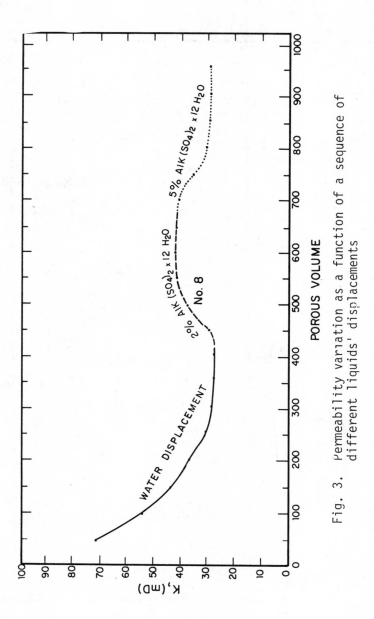

Fig. 3. Permeability variation as a function of a sequence of different liquids' displacements

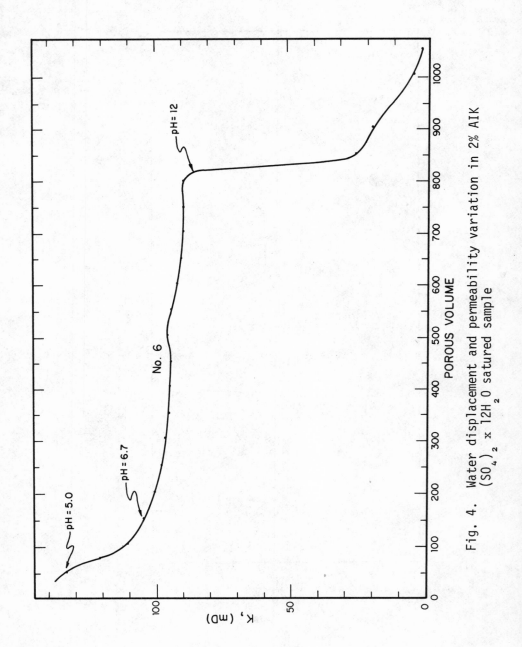

Fig. 4. Water displacement and permeability variation in 2% AlK$(SO_4)_2 \times 12H_2O$ saturated sample

To further study this problem, two core samples N° 7, previously vacuum saturated with the 2% $AlK(SO_4)_2 \cdot 12H_2O$ solution were subjected to displacement tests with the same solution. For the first sample a flow rate of 2.3 ml/min. was used, while a rate of 5.1 ml/min. was used for the second. The permeability variation in these tests is shown in Figure 5. A flow rate of 2.3 ml/min. was found to produce a smaller initial permeability reduction than one of 5.1 ml/min.

A final series of tests was done on two sample cores N° 8; the first of which was previously vacuum saturated with 2% $AlK(SO_4)_2 \cdot 12H_2O$. The first of these core samples was submitted to a displacement test with the same solution, while for the second, distilled water was the displacement liquid. The results are shown in Figure 6. A greater permeability reduction is observed with distilled water displacement.

DISCUSSION OF RESULTS:

Permeability Reduction by Water Displacement:

Distilled water displacement has been seen to cause a reduction in the original permeability of the core samples analysed. Thermo-gravimetric analysis did not indicate that hydrates were formed in the cores as a result of water displacement. This type of chemical reaction is therefore excluded. Yet Kaolinite was found to be displaced in the liquid flow as evidenced by its reduction at the entrance and increased concentration towards the center and outlet region of the core samples (see Table 2). In addition, it is noted that no colloidal emulsion was found in the liquid collected at the exit of the core sample.

The displacement of Kaolinite along the core sample is generally thought to be due to its dispersion in water. This dispersion is thought to be caused by the breaking of the hydrogen bonds which thereby separates the structural units of the Kaolinite in the dispersing aqueous medium[3,4]. Some of the dispersed particles are of colloidal size and can migrate in the liquid flow. The dispersed Kaolinite collects in the central and posterior regions of the core samples, thereby reducing the diameter of the capillary pores and hence their natural porosity and permeability.

Partial Restoration of the Damaged Permeability:

In all the displacement tests with 2% $AlK(SO_4)_2 \cdot 12H_2O$ solution, a partial restoration of the permeability after water displacement damage was observed. Figure 1 shows that this restoration is generally about 35%, calculated as a percentage of the amount of damage that has occured.

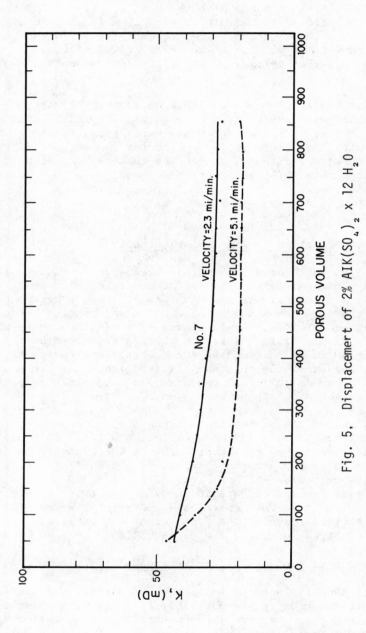

Fig. 5. Displacement of 2% $AlK(SO_4)_2 \times 12 H_2O$

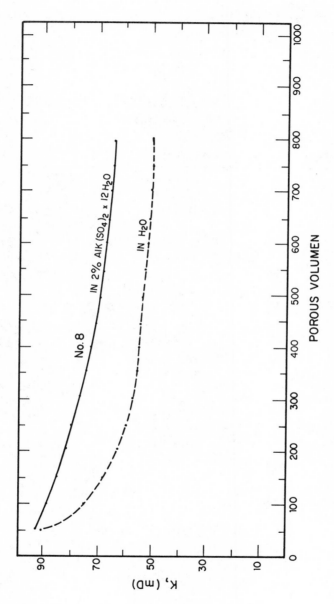

Fig. 6.- Permeability reduction due to Kaolinite dispersion.

Table 2

Change in Kaolinite Composition Along the Sample Cores After Water Damage

Sample core N°	Position with Respect to Flow	Kaolinite %
1	Front	28.5
	Middle	32.3
	Exit	36.1
2	Front	30.2
	Middle	31.8
	Exit	35.6
3	Front	25.7
	Middle	31.2
	Exit	33.4
4	Front	30.0
	Middle	34.6
	Exit	41.2
5	Front	29.0
	Middle	32.0
	Exit	31.5

A possible mechanism for understanding this permeability restoration can be found by considering certain properties of colloids. Kaolinite particles dispersed in water are generally thought to consist of micelles with negative electrostatic charge[5,6]. Introducing a positive charge in the dispersing medium attracts the micelles, thereby causing coagulation. This grouping of the previously dispersed particles causes precipitation. At a pH = 5, the introduction of 2% $AlK(SO_4)_2 \cdot 12H_2O$ solution provokes hydration of the aluminium as follows:

$$2/AlK(SO_4)_2 \cdot 12H_2O/ \xrightarrow{\text{in water}} 2/Al(H_2O_6)^{3+}/ + 3\ SO_4^{2-} + K_2SO_4 + 12H_2O$$
$$\dots\dots\dots\dots(2)$$

giving rise to a positively charged complex ion $/Al(H_2O_6)/^{3+}$. This positive ion is responsible for attracting the colloidal Kaolinite particles, thereby grouping them into a more compact form.

Figure 7a shows schematically the suspension of colloidal particles of Kaolinite in a transversal capillary section. Due to their random motion the particles increase the resistence to the flow of the displacing liquid in the capillaries. When the particles are precipitated, increased free space is produced which allows easier displacement of the liquid through the capillaries. This effect representing the action of the 2%$AlK(SO_4)_2 \cdot 12H_2O$ solution is shown in Figure 7b. This removal of the freely moving suspended particles which resist liquid flow is thought to be responsible for the partial restoration of the permeability.

Damage Prevention by the Use of $AlK(SO_4)_2 \cdot 12H_2O$ Solution:

In the previous paragraph the mechanism for a partial restoration of the permeability of the damaged core samples was discussed. The same 2% $AlK(SO_4)_2 \cdot 12H_2O$ can also be used to prevent this damage by saturating the sample cores with this solution prior to distilled water displacement. Figure 4 shows an experiment of this nature in which sample Nº 6 is previously saturated with the solution and then subjected to distilled water displacement. At the beginning of the displacement a drop in the core permeability was observed. Later the permeability continued to decrease more slowly, due to the change in the pH of the displacing solution, but eventually reached a virtually stable value. After 800 porous volumes were displaced, the pH was increased to pH = 12, by means of a cement filtrate. A considerable decrease in the permeability was observed supposedly due to the precipitation of $Al(OH)_3$[7,8].

As shown in Figure 4, distilled water displacement produced a permeability reduction of some 20% of the original value. The mechanism in this case is thought to be entirely due to mechanical

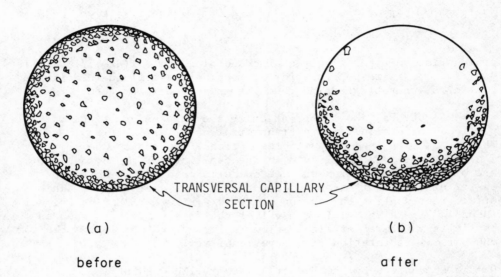

Fig. 7.- Coalescence of Kaolinite particles due to the effect of Al^{3+} ions

entrainment of the Kaolinite particles and their migration in the porous medium which leads to plugging in regions, where the porous capillaries are narrow. The mechanism was tested in experiments with sample core N⁰ 7. Two samples with the same physical and chemical properties were submitted to displacement rates with 2% $AlK(SO_4)_2 \cdot 12H_2O$ solution, but at different displacement rates. Curve N⁰ 1 in Figure 5 is for a displacement rate of 2.3ml/min., whereas curve N⁰ 2 is for a rate of 5.1ml/min. The greater displacement rate produced a greater permeability reduction which is in agreement with the proposed mechanical entrainment theory.

Effect of Chemical Alteration and Mechanical Entrainment on the Permeability Reduction:

In these tests the sample cores N⁰ 8 were chemically and physically identical. One sample was subjected to displacement with the 2% $AlK(SO_4)_2 \cdot 12H_2O$ solution after being saturated with the same solution. The permeability change graphed against the displaced porous volume is shown in curve N⁰ 1 of Figure 5. The second sample was submitted to distilled water displacement without previous saturation. The permeability reduction is greater in this second case as shown in curve N⁰ 2 of Figure 5.

Indeed the reduction in curve N⁰ 1 reaches an approximate value of 20%, whereas in curve N⁰ 2 the reduction is about 45%. Comparing these values, one can arrive at the following conclusions about the causes for the observed difference:

1) the permeability reduction due to mechanical entrainment of Kaolinite in the 2% $AlK(SO_4)_2 \cdot 12H_2O$ solution is 20%,
2) the permeability reduction due to mechanical entrainment plus colloidal dispersion and migration of Kaolinite in distilled water is 45%.

CONCLUSIONS:

1) The permeability reduction of the core samples studied was found to be due to Kaolinite dispersion followed by migration of the dispersed particles through the porous medium which eventually become clogged, thereby obstructing liquid flow.
2) The quantity of Kaolinite particles that migrate depend on:
    a. the amount of colloidal dispersion that occurs as a result of the rupture of the crystal lattice bonds in the liquid,
    b. the amount of mechanical entrainment which is proportional to the liquid flow rate.
3) A 2% $AlK(SO_4)_2 \cdot 12H_2O$ solution precipitates the Kaolinite particles by causing them to coagulate and thereby prevents their dispersion when the pH is less than 5.0.
4) By saturation with the 2% $AlK(SO_4)_2 \cdot 12H_2O$ solution to precipitate the Kaolinite particles prior to distilled water displace-

ment, the permeability reduction is made to depend solely on mechanical entrainment of the particles.

ACKNOWLEDGEMENT:

The author wishes to thank Dr. Clarence A. Gall for his assistance in preparing this article and for useful discussion.

REFERENCES:

1. W. R. Almon and D. K. Davies, Clay Technology and Well Stimulation. Transaction-Gulf Coast Association of Geological Societies, vol. 28, 1978.
2. R. Jenkins and J. L. de Vries, Practical X-Ray Spectrometry, Philips Technical Library. Hazell Watson & Viney Ltd., Aylesbury, Bucks, 1975.
3. W. Kubacki, Stabilization of Dispersed Clays by Ionic Exchange Concentration. Symposium on Stabilization and Floculation, Division of Colloid Chemistry, American Chemical Soc., Honolulu (Hawaii), 1979.
4. L. V. Azaroff, Introduction to Solids. N. York-Toronto-London, 1960.
5. A. Bielanski and J. Haber, Physical Chemistry. Edit. PWN, Warsow-Cracow, 1970.
6. E. Gorlich, Silicates Chemistry. Edit. W/G Warsow, 1957.
7. C. D. Veley, How Hydrolyzable Metal Ions React with Clays to Control Formation Water Sensitivity. J. of Petroleum Technology, Sept. 1969, p. 1111.
8. K. H. Stern, Chemical Reviews. $\underline{54}$, 1954, p. 54.

RESEARCH METHODOLOGY IN USED OIL RECYCLING

Donald A. Becker

Recycled Oil Program
National Bureau of Standards
Washington, DC 20234

ABSTRACT

Legislation and activities in the United States on the subject of used oil recycling have increased dramatically in the past several years. However, a substantial portion of both industry and government have some concerns about the lack of scientific and technical research and data on certain aspects of the quality and consistence of recycled petroleum oils, particularly re-refined engine oils. Further, there are some significant environmental concerns about pollution aspects of used oils and their recycling by-products and wastes. Since 1976, the (U.S.) National Bureau of Standards (NBS) has had a legislatively mandated program to ". . . develop test procedures for the determination of substantial equivalency of re-refined or otherwise processed used oil . . . with new oil for a particular end use" (42 U.S. Code 6363c). The NBS research includes identification of problem areas in the characterization of used and recycled oils, research into new measurement methods for determination of novel constituents in these materials, and the development and evaluation of appropriate test procedures and standards for recycled oil products. Aspects of this research discussed in this paper include analysis of total elemental content and speciation studies on lead and on the halogens (chlorine and bromine) and hydrocarbon type characterization studies on lubricating oil fractions.

INTRODUCTION

The National Bureau of Standards' (NBS) Recycled Oil Program was initiated in 1976 to respond to a legislative mandate to ". . . develop test procedures for the determination of substantial equivalency of re-refined or otherwise processed used oil . . . with new oil for a particular end use." (Energy Policy and Conservation Act; 42 U.S.C. 6363c, EPCA). The end uses considered for recycled petroleum oils included both energy recovery as a fuel and reuse as a lubricant. In reflecting back on these past four years and on the topic for this conference, I believe it will be useful to briefly review the program plan developed to address this problem, and our utilization of atomic and nuclear methods for several specific research and development projects.

When EPCA was enacted, NBS had no substantial existing effort in petroleum oil testing and evaluation. In retrospect, this lack may have been advantageous, since it allowed us to design the program plan without a strong preconceived bias. During this planning period, two important needs were identified: 1) the test procedures recommended for any particular end use needed to be highly performance oriented, rather than property - or compositionally-related (i.e., "will it do the job?" rather than "is it the same as substance X or Y?"); and 2) significant information related to potential environmental effects needed to be communicated, even though these were, strictly speaking, properly the domain of EPA rather than NBS. The primary emphasis was on strong, unbiased technical evaluation of existing test procedures, particularly in the highly visible and rapidly changing situation currently facing oil recycling. Effective cooperation needed to be established with industry and with other government agencies involved in oil recycling.

The detailed program plan that was developed is shown in Figure 1. This plan comprises six steps, beginning with identification of the specific characteristics of that particular product which needs to be evaluated and monitored for acceptable performance in use. A useful first draft of these characteristics was obtained from existing specifications for similar products. As an example, with recycled oil used as burner fuel, the existing virgin fuel oil specification contained eleven test procedures. Upon completion of our work on this product, we recommended that three new test procedures be added for recycled oil used as burner fuel, plus recommended substantial modifications in three other existing test procedures for virgin fuel oils.[1] Also, a number of test procedures for contaminants of potential environmental concern were evaluated and, shold EPA find them to be necessary, are recommended for adoption.

In some cases, we found that only one test procedure among two or three well established tests for a specific characteristic was actually able to adequately monitor such a characteristic of the recycled oil. This failure of even well established and accepted test procedures resulted from the different chemical and physical characteristics of recycled oil as compared with the virgin material and was due primarily to the increased contaminant levels found.

As a result of the above, a very important part of our effort has been the use of a wide variety of analytical techniques to extensively characterize the various types of used and recycled oils. The atomic and nuclear techniques utilized in this continuing effort include neutron activation analysis, atomic absorption spectroscopy, optical emission spectroscopy, x-ray fluorescence analysis, organic and inorganic mass spectrometry, nuclear magnetic resonance spectroscopy, and infrared spectroscopy. The concentrations of typical used lubricating oil contaminants (including used additive components) often are very high, ranging up to 20,000 ppm for lead, over 3000 ppm for barium, and close to 3000 ppm for both chlorine and bromine. While the concentrations of lead and the halogens are decreasing due to required reductions in the use of leaded gasolines, substantial amounts are likely to be used for some time. These concentration values provide some insight into the magnitude of the contamination levels in many used lubricating oils.

In addition, we utilize many of these same analytical techniques to determine very low levels of contaminants and impurities in some of the recycled oils, particularly re-refined engine oil basestocks. In this paper, evaluated analytical techniques for lead and the

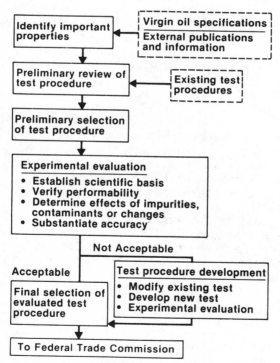

Figure 1. NBS Test Procedure Evaluation Process.

halogens will be reported, along with on-going speciation studies on the chemical forms of these elements in used or recycled oils. Further, efforts in progress on hydrocarbon type characterization of lubricating oil fractions will also be reported.

ANALYSIS FOR LEAD

The lead content of a recycled oil is important both in relation to the control of ash and deposits in burner systems, and to the reduction of emissions which could adversely affect the environment. While the lead content is negligible in virgin oils, (<0.01 weight percent), it is found in concentrations of up to 2 percent by weight in used crankcase drain oils. Lead and lead compounds found as impurities in recycled oil used as fuel result primarily from the use of leaded gasolines. At the present levels of consumption of leaded gasoline, significant concentrations of lead in used automotive crankcase oils (and therefore in oil recycling feed stocks) may be anticipated for some time.

The accurate determination of lead in recycled oils is particularly difficult, due in part to the tendency of particulates containing lead to settle to the bottom of the container. To assure homogeneous distribution of the various types of lead species present, a 20 minute shaking of the sample container on a very vigorous mechanical shaker (i.e., a paint shaker or equivalent) was employed prior to sub-sampling for analysis. In addition, it was found that during the atomic absorption analyses, the diluted oil samples had to be agitated immediately prior to analysis (e.g., 30 inversions of a 100 ml volume of 1 percent oil in methyl isobutyl ketone was sufficient to provide a homogeneous solution for 2 minutes).

Because of the observed discrepancies of the method even with the above modifications, the atomic absorption method used, ASTM Test Method D 2788-72 was evaluated to reduce this uncertainty. Additional investigation into the method established that (1) the lead calibration curve developed as part of the test procedure is apparently non-linear over the 0 to 30 ppm (by weight) range indicated as acceptable in the test procedure, and (2) the presence of zinc and calcium compounds in the oil samples (present as additive components) along with the particulate nature of the lead appeared to enhance the lead signal for certain oils, even using the method of additions as described in D 2788-72.

The problems described above appear to be due primarily to the particulate nature of the lead and lead compounds found in used automotive crankcase oils (Figures 2 and 3). Additional evaluation of the test procedure indicated the problems could be substantially reduced by the introduction of a dispersant along with ultrasonic agitation of the diluted sample just prior to final

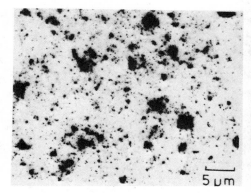

Figure 2. Electron Photomicrograph of Particles in Used Automotive Oil Extract

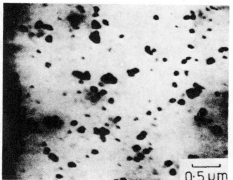

Figure 3. Electron Photomicrograph of Particles in Used Automotive Oil Extract

dilution and analysis. The modified D 2788-72 procedure has been described in detail elsewhere[1].

Using this modified atomic absorption method, analytical results for lead in four used or recycled oils were compared to results using a reference method, in this case isotope dilution mass spectrometry (IDMS). This IDMS method, although too costly and impractical for routine use, has been established at NBS as a highly precise absolute method for the determination of lead in a wide variety of sample materials, including petroleum products[2].

Results obtained from these analyses are found in Table 1, and these data indicate that the precision of the test procedure is acceptable (although a full-scale precision study has not been made). Results from the modified D 2788-72 procedure agreed reasonably well with the Reference Method for three of the four oils analyzed. The results from the fourth oil (W023) agreed reasonably well with the value found by the Reference Method when the relatively large sampling variability (as indicated by the large relative standard deviation of the Reference Method for that sample) is considered.

Lead Speciation Study

A sample of used automotive crankcase drainings having a total lead content of 1.4 percent by weight was further treated to evaluate the chemical and physical status of the lead present in the oil. As anticipated from commercial re-refining data[3], the original lead content of 1.4 percent was reduced to 0.7 percent total

Table 1. Comparison of lead values from the modified D 2788-72 method and the reference method.

Type	Number	REFERENCE METHOD [a]				MODIFIED D 2788-72 [b]			
		Percent Lead [c]	Standard deviation	Relative SD (%)	n	Percent Lead [c]	Standard Deviation	Relative SD (%)	n
Composite used crankcase oil (Source A)	W022	1.311	0.009	0.7	8	1.33	0.054	4.1	12
Composite used crankcase oil (Source B)	W011	0.657	0.006	0.9	3	0.662	0.013	2.0	3
Re-refiners feed-stock used oil	W023	0.477	0.030	6.3	2	0.526	0.027	5.2	12
Recycled fuel oil	F134	0.533	0.001	0.3	2	0.491[d]	0.042	8.5	3

[a] The Reference Method used was isotope dilution-mass spectrometry (see text).
[b] See reference 1 for modifications of the D 2788-72 procedure.
[c] Results are in percent by weight.
[d] The sampling sequence for these three samples was 0.534, 0.488, and 0.451 percent lead, indicating rapid settling of lead-containing particulates in the original oil sample. Additional evidence for the particulate problem with sample F 134 were obtained during this study and can be found in reference 1.

lead following ultracentrifugation in 1:1 benzene solution. Further microscopic examination of the supernatant oil revealed a stable suspension of relatively uniform, spherical opaque particles of less than 0.3 micron size (Figures 2 and 3).

Samples of the centrifuged oil were further diluted and injected into the high performance liquid chromatographic-graphite furnace atomic absorption (HPLC-GFAA) system for lead specific analysis. Details of the experimental protocol employed and the specific analytical procedure can be found in reference 4.

A number of batches of supernatant oil was examined, employing both isochratic and solvent gradient flow programs designed to establish optimal solvent strengths (polarities) consistent with efficient column performance. Earlier, we found that solvent pairs of low strength (hexane: methylene chloride = 95:5) served to provide useful retention volumes (V_R) on a 10 micron silica gel column for a high molecular weight, nonpolar organolead, hexaphenyldilead. In the present survey, solvent pairs of much higher strength, such as methylene chloride-acetonitrile, were applied in order to insure elution of both higher molecular weight and polar lead-containing components.

Preliminary results show that the dissolved lead species can be eluted having two or more distinct peaks, depending upon the solvent program used and the column loading incurred by the amount of sample. Figure 4 depicts a HPLC-GFAA chromatogram for one methylene chloride-acetonitrile step-gradient program run. Note that in this experiment, as with all the others, no column blockage arose as a result of the presence of the submicron suspension previously noted. Neither did frequent lead "spikes," nor high lead background, appear between separate peaks of lead-containing eluants. This indicates that the fine particle suspension moves through the HPLC column with little difficulty, and it may be that lead content associated with the very small particles still remaining in the centrifuged oil is low.

The two lead-containing peaks shown in Figure 4 and those obtained similarly in other runs cannot yet be identified but are estimated to comprise about 65 percent of the residual dissolved lead not removed by ultracentrifugation. Inasmuch as no collection of crankcase oil samples has yet been speciated for lead, we cannot yet infer that this result is typical.

Clearly, several molecular lead species, possibly organoleads, are present in the centrifuged samples obtained from this sample of crankcase oil. Compared with authentic organolead compounds [(e.g., hexaphenyldilead or triphenyllead (chloride)], the chromatographic peak shapes found support this view. However, the much larger retention volumes (>20 ml) obtained here with stronger solvents suggest that these eluants may be of high molecular weight and/or polar character as oil-soluble complexes.

The concurrent use of an ultraviolet detector [(UV) operating at 254 nm] with the AA detector (Figure 4) indicates that significant chromophores may be associated to one of the lead-containing peaks (V_p = 27 mL). Since most automotive lubricating oils are treated with performance "additive packages" containing a variety of metal solubilizing ligands, apparently these ligands are available for complexing the various metals in lubricating oils. Any such groups may survive engine service temperatures to yield discrete organometallics or metallo-organic complexes. Many of these, though highly polar, would form hydrocarbon-soluble molecules possessing strongly UV-absorbing moieties. Obviously, future work in this area would be desirable to further characterize these lead species.

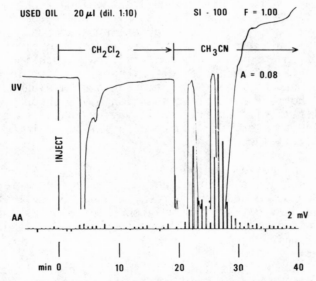

Figure 4. HPLC-GFAA chromatogram of used oil sample indicates elution of two major lead-containing components (AA detector) and several strong UV absorbing eluants, one of which may contain lead.

ANALYSIS FOR HALOGENS (CHLORINE AND BROMINE)

While chlorine and bromine are not normally found at significant levels nor measured in most petroleum products, they are found at high concentrations in many used oils available for recycling that have been derived from used automobile crankcase oil. For example, used oils from vehicles utilizing leaded gasolines have been analyzed and found to contain 0.1 to 0.3 percent by weight of both chlorine and bromine, owing primarily to the industries' use of halogenated hydrocarbons as "scavengers" for lead. Also, high levels of chlorine are found in certain types of industrial cutting oils, which may be recycled. While there appear to be no specific studies on either the effects of halogens on the performance of recycled lubricating oils or on the environmental effects of halogens in burner fuel oils, the toxicity of some halogenated hydrocarbons is well established.

Analysis of these halogens by neutron activation analysis (NAA) was found to be both repeatable and consistent with results from other analytical techniques.[1] All of the other techniques investigated (x-ray fluorescence, oxygen bomb combustion with gravimetric or titriametric determination, (ASTM Method D 808-63), or sodium alcoholate dissolution with ion chromatographic determination (ASTM Method D 1317-64, modified) had significant problems in terms of sensitivity, accuracy, or precision when compared to NAA.[1]

The NAA procedure employed for the determination of high levels of chlorine and bromine as found in used oil is described in reference 1. An important part of that procedure required vigorous shaking in a paint shaker to homogenize the highly particulate laden oil, followed by immediate pouring into pre-weighed and pre-cleaned polyethylene vials. This technique resulted in chlorine blank levels of 14 + 2 µg/g of oil, due to the polyethylene vial (bromine < 1 µg/g). These vials are thus inappropriate to use with oils having low concentrations of chlorine.

For measurement of the very low levels of chlorine found in re-refined petroleum oils, this procedure was modified to use considerably larger sample sizes (2.5 g vs 100 mg) and post-irradiation transfer of the oil sample into a non-irradiated container for counting and quantitation. With this revised method, the limit of detection (blank level) for the procedure used was 0.05 µg/g for chlorine and 0.01 µg/g for bromine. While this method does allow possible loss of volatile Cl or Br species during transfer, this problem was not observed.

The results obtained by these two NAA methods were acceptably precise, with reproducibilities of ±3-5%. The estimated accuracy was ±10% (standard deviation of a single determination).[5] Actual values obtained for used automotive crankcase oils ranged from a few µg/g (diesel oil) to 3300 µg/g for chlorine and < 1 µg/g (diesel oil) to 2900 µg/g for bromine. For re-refined lubricating oil basestocks, chlorine concentrations measured to date range from 0.49 µg/g to 371 µg/g, while for bromine the range is 0.05 µg/g to 13.0 µg/g.

Chlorine Speciation Study

The concern over halogen levels in used and recycled petroleum oils results from 1) the potential effects of these halogens on the performance of recycled lubricating oils in the field, and 2) the environmental implications of these halogens for used oil recycling and disposal. At present, our immediate need is to evaluate whether the presence of these halogens has a significant effect on the performance of lubricating oils, Therefore, our initial research efforts on speciation have been in this direction. Since chlorine has been found to be present in re-refined lubricating oils at substantially higher concentrations than bromine,[6] we have initially focused our efforts on this element. In addition, there is some evidence to suggest that the presence of chlorine in a lubricating oil may cause a decrease in the performance of that oil.[7]

The chlorine present in used oils can be expected to originate from two major sources. The first source is from the ethylene dichloride and ethylene dibromide found in leaded gasolines. These compounds are known to substantially decompose during the combustion process, with formation of different molecular species.[8] Expected products from this source include inorganic compounds containing the halogens combined with lead, phosphorus, oxygen, and ammonium groups.[8] No references have been located which indicate that organic halogen compounds result from this source, although it would seem reasonable that they might be formed and migrate to the crankcase oil.

The second major source of chlorine in used oil recycling is from lubricating oil additives, particularly additives for industrial cutting oils. Chlorinated paraffin cutting oils containing up to several percent chlorine by weight are often used in commerce. This source provides at least one mechanism for organochlorine compounds to get into the used oil system. Again, the effects of low to moderate concentrations of such compounds on lubricating oil performance or on the environment is not well established.

The first question on chlorine speciation concerns possible inorganic chlorine compounds in the re-refined oil, particularly chlorides. An agueous extraction of the oil (5-20 grams oil, diluted with about 25 grams heptane to facilitate surface contact between organic and aqueous phases) followed by ion chromatographic determination indicated that less than one percent of the chlorine in the oil was present as aqueous extractable chloride. For one sample of re-refined oil containing almost 200 μg/g total chlorine, less than 0.02 percent was present as aqueous extractable chloride.

Further effort was then directed towards the characterization of organochlorine compounds or hydrocarbon-soluble complexes in these oils. Since this effort is closely connected to the following research on hydrocarbon type characterization, this chlorine speciation study will be continued in that section.

HYDROCARBON TYPE CHARACTERIZATION

In our hydrocarbon type characterization effort, a number of different methods of column chromatography utilizing clay, silica gel, and alumina are used to separate an oil by gradient elution into its various fractions (saturates, aromatics, and polar fractions). These fractions are then characterized using a variety of techniques including nuclear magnetic resonance, mass spectrometry and chemical analysis.

Thin-layer chromatographic methods using an ultraviolet detector (254 nm) were applied to the fractions in order to evaluate how well the various column fractionation methods actually separated the hydrocarbon groups. We found that ASTM Method D 2007-75 [9] provided the best separation of the fractions, but the limited sample size capacity (10 grams of starting oil) severely limited evaluation of the resulting product fractions. Analysis of the fractions for total chlorine by neutron activation from two column materials (alumina and clay-silica gel) supported this conclusion that the clay-gel column fractionation (ASTM D 2007-75) provided the most complete separation of differing fractions (Table 2). These results also indicated that the chlorine species in re-refined oils are highly concentrated in the polar fraction. The nature of these chlorine species is not known at this time, and further work is being carried out.

CONCLUSIONS

Since the studies reported here are far from complete, final conclusions cannot be made. However, there are two conclusions which can be stated at this time. First, the high concentrations of lead, chlorine and bromine found in some used lubricating oils indicate that suitable precautions should be taken to reduce indiscriminate or uncontrolled burning or disposal. Second, while there are measurable differences between re-refined lubricating oils and virgin lube oils, it is still not clear how significant these differences are or the effect on the performance of that oil.

Table 2. Analysis of hydrocarbon fractions for halogens by neutron activation.

Starting Oil	Column Material	Halogen Determined	Halogen Concentration (µg/g)			
			Starting Oil	Saturate Fraction	Aromatic Fraction	Polar Fraction
Re-refined Oil A	Alumina	Chlorine	13.2	3.3	11.0	180
Re-refined Oil A	Alumina	Bromine	7.3	2.6	7.2	47
Re-refined Oil A	Clay-Gel[a]	Chlorine	13.2	<1	ND	ND
Re-refined Oil A	Clay-Gel[a]	Bromine	7.3	<1	ND	ND
Re-refined Oil B	Clay-Gel[a]	Chlorine	189	2	ND	ND

[a] Clay-Gel = ASTM Method D2007-75 (see text)
[b] ND = Not Determined

ACKNOWLEDGMENTS

The author gratefully acknowledges the substantial contributions of the various NBS scientists working on projects for the Recycled Oil Program, particularly T. Rains, F. Brinkman, R. Fleming, R. Koch and S. Hsu. Thanks are also due to Virginia Davis, Alberta Epstein and Anna Marinoff for their untiring efforts on behalf of the Program.

REFERENCES

1. D. A. Becker and J. J. Comeford, "Recycled Oil Program: Phase I-Test Procedures for Recycled Oil Used as Burner Fuel," NBS Technical Note 1130, U.S. Department of Commerce, Washington, DC (August 1980).
2. I. L. Barnes, T. J. Murphy, J. W. Gramlich, and J. R. Shields, Lead separation by anodic deposition and isotope ratio mass spectrometry of microgram and smaller samples, Anal. Chem. 45, 1881-1884 (1973).
3. T. D. Coyle and A. R. Siedle, A survey of metals in oil: occurence and significance for reuse of spent automotive lubricating oils, in "Proceedings, Conference on Measurements and Standards for Recycled Oil-II" NBS Special Publication 556, U.S. Department of Commerce, Washington, DC (September 1979), pp. 189-216.
4. F. E. Brinkman and W. R. Blair, Speciation of metals in used oils: recent progress and environmental implications of molecular lead compounds in used crankcase oils, in "Proceedings, Conference on Measurements and Standards for Recycled Oil," ibid, pp. 25-38.
5. R. Fleming, National Bureau of Standards unpublished data available by calling (301) 921-2166.
6. E. A. Frame and T. C. Bowen, Jr., "U.S. Army/Environmental Protection Agency Re-refined Engine Oil Program," AFLRL Report No. 98, U.S. Army MERADCOM, Ft. Belvoir, VA (May 1978), p. 7.
7. F. Sam, Chevron Research's experience with re-refined oils, in "Proceedings, Conference on Measurements and Standards for Recycled Oil," NBS Special Publication 488, U.S. Department of Commerce, Washington, DC (August 1977) pp. 69-71.
8. D. A. Hirschler, et al., Particulate lead compounds in automobile exhaust gas, Ind. and Eng. Chem. 49, pp. 1131-1142 (July 1957).
9. ASTM Method D2007-75, Characteristic groups in rubber extender and processing oils by the clay-gel adsorption chromatographic method, "1978 Annual Book of ASTM Standards, Part 24" (ASTM Philadelphia, PA, 1978).

THE APPLICATION OF PHOTON INDUCED X-RAY FLUORESCENCE FOR THE SIMULTANEOUS DETERMINATIONS OF COBALT, NICKEL AND MOLYBDENUM IN HYDRODESULFURIZATION CATALYSTS

J.J. LaBrecque, C.A. Peña, E. Marcano, P. Rosales and W.C. Parker.

IVIC, Apartado 1827
Caracas 1010-A, Venezuela

INTRODUCTION

Hydrodesulfurization catalysts are one of the most important types of catalyst employed in the petroleum industry for adjusting crude oils to meet the requirements of the market. Catalysts containing Co-Mo and Ni-Mo have been used extensively, and more recently a Co-Ni-Mo catalyst has been employed as an alternative.

A great interest exists in he preparation of the optimum catalyst (for economic reasons); not only does the chemical composition determine the activity but also the physical properties affect the activity too (1). Some of the precipitation techniques are precipitation, gel formation, impregnation or simple mixing of the components.

Usually the optimization of selectivity, activity, stability and ultimately the commercial success of a catalyst is controlled partly by the selection of the chemical composition of the primary catalyst (2). Thus, accurate and precise analytical techniques are necessary to determine the chemical composition of the native catalysts.

Recently, both atomic absorption spectroscopy (3) and neutron activation techniques (4) have been employed with success but both of these instrumental methods have inherent shortcomings. In atomic absorption, as in spectrophotometric methods (5,6), the sample is required in a complete dissolved form. The normal support (carrier) of hydrodesulfurization catalysts is γ - alumina with small amounts of silica. Therefore to ensure a complete decomposition it is necessary to use concentrated acids (H_2SO_4, HNO_3, HF etc) and sometimes under pressure in a bomb. This procedure is long and tedious. The major disadvantage of neutron activation techniques is that a high flux of neutrons is required, which is very costly in most cases and not readily available to most laboratories.

Conventional X-ray fluorescence (wavelength-dispersive) has been applied by Sato et al. (7) but this method suffers from the fact that two different X-ray tubes have to be employed for the determination of Co and Mo. Finally, Co and Mo have been determined using energy dispersive X-ray fluorescence (8) by the same technique presented in this paper, but Ni was not contained in the catalysts studied. The reasons why Ni was not considered in the original paper were: 1) the problem of the resolution of the peaks of the adjacent elements, Co and Ni; 2) the interference of the Co_{K_β} peak with the Ni_{K_α} peak and; 3) at this time only one catalyst with all these elements was available.

In this paper a rapid radioisotope induced energy-dispersive X-ray fluorescence procedure for the simultaneous determinations of cobalt, nickel and molybdenum in hydrodesulfurization catalysts is presented. The sample preparation is simple and rapid, it utilizes both an internal standard as well as a thin film technique. Synthetic standards are prepared simply by mixing thoroughly the appropriate amounts of the elements of interest as their respective oxides with the matrix (usually Al_2O_3 or graphite) and the internal standards. Some of the advantages of this technique are: 1) the weighing of a small amount of standard or sample (10-20 mg) and ensuring that all of it is quantitatively transferred to the sample hold can now be neglected. Only the original amounts in the mixture are used in the calculations with this method. 2) Geometrical errors are eliminated by the use of the relative fluorescence ratio factor, that is, the sample and standard don't have to be placed in the same exact arrangement with the detector and source. 3) In most cases the absorption enhancement effects are negligible because of the thin sample. If there are interelement effects they can be seen by the differences in the F_{JL} factor for a series of standards. Finally a comparison of this sample preparation-analysis technique with the more conventional methods of analysis for X-ray fluorescence for the effectiveness for the compensation of errors is given in Table I.

Table I.- The Compensation of Errors by Analytical Methods for X-Ray Spectrometric Analysis.

Source of Error	This Work (Thin Film and Internal Standard)	Standard addition dilution	Calibration Standardization
Instrumental (Short Term)	Complete or Substantial Correction	---	---
Instrumental (Long Term)	Complete or Substantial Correction	---	---
Operational	Partial	---	---
Sample Absorption and Enhancement	Complete or Substantial Correction	Complete or Substantial Correction	Only if the Sample and Standard are Very Similar
Sample Homogeneity	Able to Check with Different Portions	---	---
Sample Position	Complete or Substantial Correction	Complete or Substantial Correction	---
Sample Surface	Complete or Substantial Correction	---	---
Sample Particle Size	Complete or Substantial Correction	---	Only if the Sample and Standard are Very Similar

EXPERIMENTAL

Sample preparation for X-ray fluorescence

Three to five grams of sample are ground to completely pass a 150 mesh sieve (106 µm) before a 800 mg portion is mixed with 200 mg of $KBrO_3$ (the internal standard) with steel balls in a minivial employing a SPEX MILL for 30 minutes. The internal standard was previously ground to less than 150 mesh and dried at 102°C. Another portion of the sample (about 2-3 grams) is used to determine the moisture separately. Three portions of about 20 mg each of the mixture of sample and internal standard are placed in separate sample holders for analysis. A simplified diagram of the sample preparation is given in figure 1.

Radioisotope - excited X-ray fluorescence system

The X-ray fluorescence system employed is based on a PDP-11/05 processor which is used as a multichannel analyser as well as to control the complete system. This system has been described in detail elsewhere (9). The detector is a high resolution Si(Li) semi-conductor manufactured by Nuclear Semi-conductors with a measured resolution of less than 155 eV (FWHM) at 5.9 KeV. While the excitation system consists of a ^{109}Cd (7mCi) annular radioisotope source and housing supplied by New England Nuclear. A comparison of excitation spectra for the ^{109}Cd source and a Cr X-ray tube is given in Figures 2 and 3. Note the flat background in the region of analysis for the ^{109}Cd source compared to the conventional Cr X-ray tube. The complete system is shown in figure 4. The whole system has been automated for the simultaneous determination of Co-Ni-Mo by means of a pre-written computer program in FLEXTREN, a language developed by Tracor Northern listed in Table II. The commands incorporated in this program have been previously described elsewhere (10).

PHOTON INDUCED X-RAY FLUORESCENCE

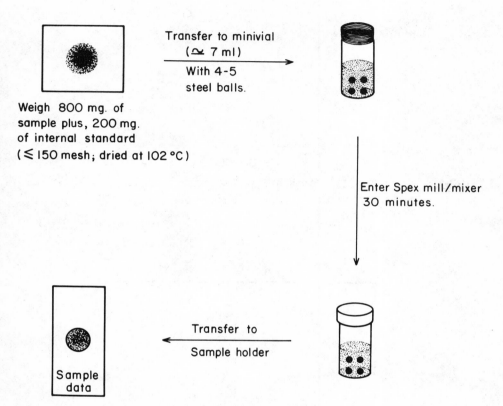

Figure 1. Sample Preparation for Radioisotope Induced X-Ray Fluorescence Analysis

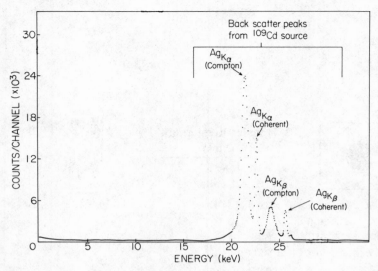

Figure 2. A Typical Excitation Spectrum from the ^{109}Cd Source

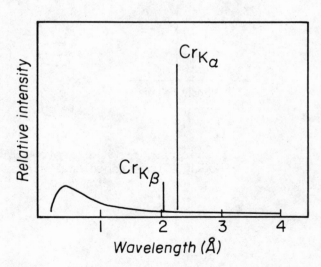

Figure 3. A Typical Excitation Spectrum from a Conventional Chromium X-Ray Tube

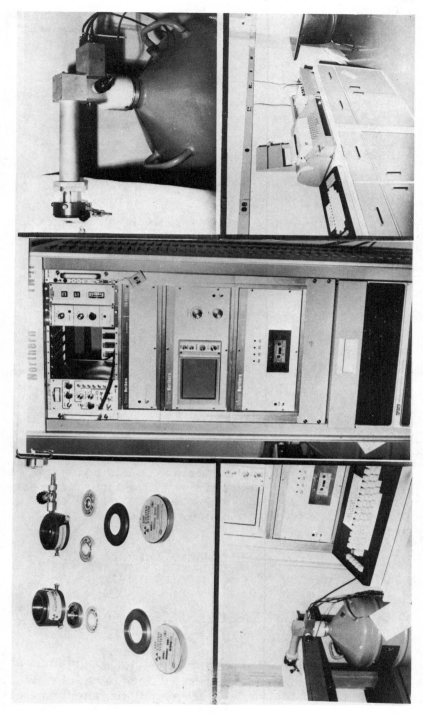

Figure 4. Components of the X-Ray Fluorescence System

Table II. The computer program for the simultaneous determinations of Co-Ni-Mo in catalytic materials.

```
FL
+EC 10    3
AA (11)   1   40   300
DS (11)   "         "
```

DB1	6520	6560	Co K_α	
DB2	7120	7160		
DB3	7080	7120	Ni Y_α	
DB4	7800	7840		
DB5	7880	7920	Ni K_β	Background regions
DB6	8520	8560		
DB7	11640	11680	Br K_α	
DB8	12440	12480		
DB9	17280	17320	Mo K_β	
DB10	18280	18320		
DB11	6600	7080	Co K_α	
DB12	7160	7760	Ni K_α	
DB13	7960	8480	Ni K_β	Peak areas
DB14	11720	12400	Br K_α	
DB15	17360	18240	Mo K_α	

```
IB(11)    11    5    10
```

Calculations of Relative fluorescence ratio-factors

The relative fluorescence ratio-factors are determined experimentally by integrating the peaks of interest (elements to be determined) and the internal standard in a series of prepared standard using the following general equation:

$$FJ_L = \frac{\text{intensity of internal standard peak}}{\text{intensity of peak of interest}} \times \frac{\text{wt. of elements of interest}}{\text{wt. of internal standard}} \times A$$

that is, F_{JL} is equal to the integral of counts above background for the internal standard peak divided by the integral of counts above background of the peak of interest times the weight of the element of interest (mg) divided by the weight of the internal standard (mg) times A; where A is conversion factor, e.g. to convert to percentage, the pure element to oxide or vice versa.

The standards are prepared simply by weighing the appropriate amounts of the elements to be determined and the internal standard as oxides in the matrix for a given concentration range. All the reagents are pure fine powders (> 150 mesh particle size) and dried at 102°C before being weighed into a minivial for mixing with steel balls in a SPEX mill for 30 minutes. About 20 mg of the standard mixture are transferred to a separate sample holder for analysis. The sample holder is a piece of 2 x 5 cm IBM computer card with a 1 cm aperture in which the sample is distributed between two pieces of Scotch magic tape.

RESULTS

A statistical evaluation of the errors in this technique has been studied and the results are shown in Table III. The homogeneity of the sample and internal standard can be checked by determining the values of Co-Ni-Mo in different portions of the sample mixture. In our experiments the sample preparation error (homogeneity) has been about the same as the standard counting, instrumental and operational error. Thus, the sample preparation error is small since the method used to determine it also includes the standard counting, instrumental and operational errors.

A typical spectrum of a hydrodesulfurization catalyst with the internal standard ($KBrO_3$) is given in figure 5. The calculations of the quantities of Co-Ni-Mo in the sample were performed by the following general equation:

$$\%X = \frac{\text{intensity of element of interest}}{\text{intensity of internal standard}} \times \frac{\text{weight of internal standard}}{\text{sample weight}} \times F_{JL}$$

where F_{JL} is the relative fluorescence-ratio factor defined previously. These results are summarized in Table IV and compared with values obtained from atomic absorption

Table III. Different types of deviations for a typical Co-Ni-Mo catalyst.

Type of Deviations	Method of calculations	$\bar{X} \pm$ SD		
		CoO%	NiO%	MoO_3%
Total	Five independent prepared samples	1.40±0.07	2.70±0.1	11.3±0.5
Sample preparation standard counting instrumental and operational	Five independent portions of the same sample preparation	1.37±0.05	2.70±0.1	11.1±0.3
Standard counting instrumental and operational	the same portion of the sample preparation measured 5 times	1.40±0.07	2.70±0.1	11.5±0.1

conventional neutron activation, and prompt-gamma spectroscopy.

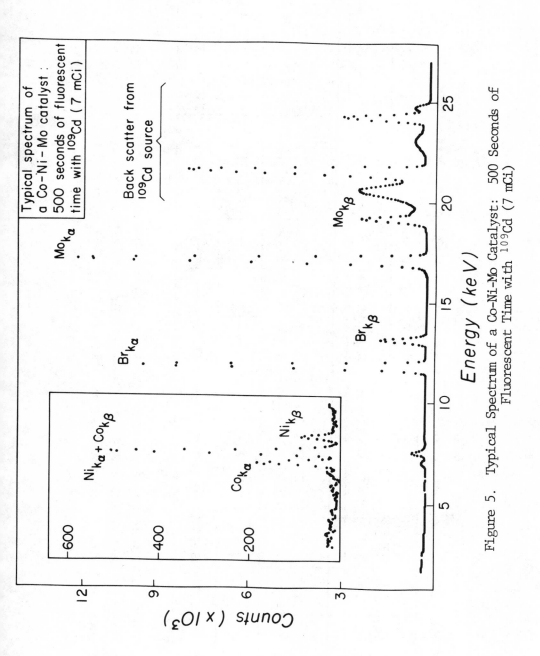

Figure 5. Typical Spectrum of a Co-Ni-Mo Catalyst: 500 Seconds of Fluorescent Time with ^{109}Cd (7 mCi)

Table IV. The comparison of results for Co-Ni-Mo by different methods

Sample Code	CoO% *			NiO% *				MoO₃% *			
	This work	AAS[a]	NAA[b]	p-γ[b]	This work	AAS[a]	NAA[b]	p-γ[b]	This work	AAS[a]	NAA[b]	p-γ[b]
A	1.43	1.25	1.54	1.23	2.75	3.09	3.03	2.68	15.17	14.76	15.70	13.92
B	2.01	2.34	1.87	1.74	4.04	4.14	3.82	4.21	9.32	10.80	8.97	9.04
C	3.67	3.97	2.59	2.89	—	—	—	—	16.08	15.92	15.29	16.14
D	3.71	3.57	—	3.19	—	—	—	—	13.07	13.01	14.61	14.49

*The values are the mean of three or more determinations
a These values are from Applied Spectroscopy 30 (1976) 625
b These values are from Analytical Chemistry 48 (1976) 1969

REFERENCES

1. J.M. Smith, "Chemical Engineering Kinetics", 2nd. Ed., Mc. Graw-Hill Book Company, New York, New York, 1970 p. 320.
2. National Science Foundation Workshop on "Research Needs and Instrumental Requirements in Catalysis", University of Maryland, College Park, Maryland, June 1978, p. 24.
3. J.J. LaBrecque, Applied Spectroscopy 30 (1976) 625.
4. M. Heurtebise, H. Buenafama and J.A. Lubkowitz, Anal. Chem. 48 (1976) 1969
5. A.G. Sokolov, Yw. Yw. Yanpoliskii, Nefteperelab Neftehim (Moscow), 12 (1972) 34.
6. G.G. Dubinina, G.A. Berg, Nefteperelab Neftehim (Moscow) 11 (1971) 38.
7. T. Sato, T. Sekine, N. Todo, Tokyo Kogyo Strikensho Hokoku, 66 (1971) 450.
8. J.J. LaBrecque, Journal of Radioanal. Chem. 41 (1977) 127.
9. J.J. LaBrecque, Proc. of 3rd. Intern. Conf. of Computers in Chemical Research, Education and Technology, Caracas, Venezuela, 1976, p. 66-81.

QUANTITATIVE ELECTRON PROBE MICROANALYSIS OF FLY ASH PARTICLES

R. L. Myklebust, J. A. Small, D. E. Newbury

Center for Analytical Chemistry
National Bureau of Standards
Washington, DC 20234

ABSTRACT

Fly ash particles or other similar particles may be quantitatively analyzed with a flat-sample, matrix-correction method that has been modified to include the peak-to-background ratio for each element as a normalizing factor. The effects of the different matrix corrections on particles are discussed. Examples of analyses of standard reference material glass particles by both a standard matrix correction program (FRAME C) and a modified correction program (FRAME P) are presented as well as analyses of fly ash (SRM-1633).

A quantitative procedure for analyzing individual fly ash particles by electron probe microanalysis must take into account several difficulties not encountered in the analysis of conventional specimens. The matrix correction techniques for quantitative microanalysis were all developed for the analysis of bulk samples that have been polished flat. Therefore, it should not be assumed that any matrix-correction method will be able to provide acceptable analyses of small particles. With care, particles larger than 20μm can be analyzed as bulk specimens and particles smaller than 0.2μm can be treated as thin films[1]. Many fly ash particles fall in the size range between these and must be treated as a separate case.

For the Monte Carlo computations we will assume that the fly ash particles are spherical in shape, which, as shown in Fig. 1, is a reasonable assumption. The normal matrix corrections employed in electron probe microanalysis have been developed for flat, "infinitely thick" specimens; therefore, these corrections give

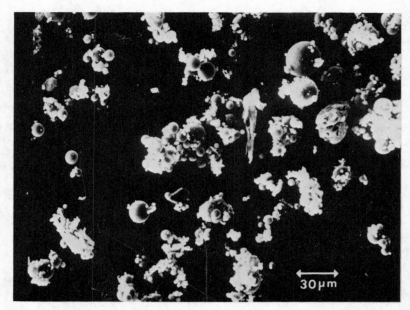

Fig. 1. Scanning electron micrograph of fly ash particles (SRM-1633).

different results when applied to small particles depending on the size and shape of the particles. Each of the matrix corrections will be examined to see what effect it has on the calculation of x-rays generated and emitted by small particles when a quantitative analysis is to be attempted with conventional microprobe standards.

The first effect is concerned with the volume of primary x-ray generation (see Fig. 2). As long as the particle is larger than this volume, no correction is necessary; however, when the particle is smaller than this interaction volume for a bulk specimen, electrons capable of exciting x-rays will be lost by scattering out the sides and bottom of the particle. The resulting x-ray intensity generated in the particle will then be less than the x-ray intensity generated in a bulk specimen of the same composition as the particle. The steep drop in x-ray intensity with decreasing particle size shown in Fig. 3 is the result of the particle becoming smaller than the interaction volume in the bulk material.

The absorption correction for the generated x-rays is the second effect that is different for particles and bulk specimens. The x-ray absorption is a function of the path length from the point of generation to the surface of the specimen. For a flat, bulk specimen this path length can be satisfactorly computed and the

ELECTRON PROBE MICROANALYSIS

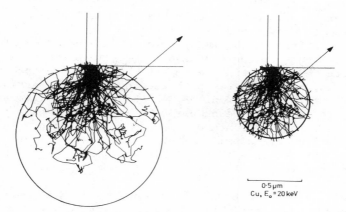

Fig. 2. Monte Carlo electron trajectory plots in two spherical particles of copper. The arrows indicate the x-ray paths out of the particle. The horizontal line represents the position of the surface of a flat specimen.

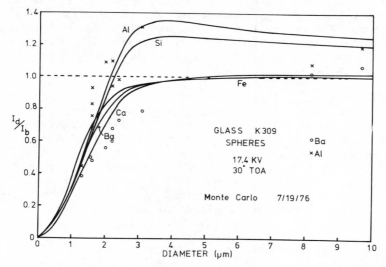

Fig. 3. Ratio of intensities from spherical particles to bulk intensity as a function of particle diameter. The curves are computed by Monte Carlo calculations and the points for barium and aluminum are experimental measurements.

absorption expression can be described mathematically as a function of the depth of x-ray generation within the sample[2]. In order to derive a similar expression for particles, the equation of the surface of the particle must be known as well as the point on the surface where the electron beam impacts the particle (see Fig. 2). In general, the average x-ray path lengths out of a particle are less than the path lengths out of a bulk specimen. This effect, then, tends to increase the number of x-rays that escape a particle as opposed to a bulk specimen of the same composition. The lower energy x-rays from low atomic number elements are affected by this to a greater extent since these x-rays have much larger absorption corrections than x-rays of higher energy (see Fig. 3).

The third effect to be considered is the loss of x-ray intensity due to secondary excitation by both characteristic x-rays and continuum x-rays generated within the particle. Since the range of both types of x-rays is much greater than the diameter of small particles, the x-rays escape the particle before they can excite a lower energy characteristic x-ray. Therefore, this correction has little effect on the x-rays generated within the particle and causes an error of less than 1% for particles below 5μm. However, it cannot be ignored when calculating the bulk standards for the analysis since both secondary excitations do have an effect in the bulk material. The loss of excitation by the continuum can be minimized by selecting lower energy x-ray lines for the analysis (e.g. the L-line instead of a K-line for the heavier elements). Reducing the beam energy also helps to minimize this problem by decreasing the high energy continuum available for exciting the high energy characteristic x-ray lines.

Methods to generalize the matrix corrections to include particles have been attempted[3]. These methods have produced models that require extensive operator input that includes estimates of particle size and shape as well as the normal input required for the matrix corrections. Individual particles may also be analyzed with the aid of Monte Carlo calculations to predict electron trajectories, x-ray generations and x-ray absorptions within the particles[4]. This method, however, is quite lengthy (and expensive) since it requires computing enough particles of different size and composition to set up calibration curves for particle analysis.

Recently, a method for compensating for the geometrical effects in particles has been suggested which is based on the observation that the ratio of a characteristic x-ray peak to the continuum intensity at the same energy is a constant regardless of the shape or size of the particle[5,6]. In this model if P is the background-corrected peak and B is the continuum intensity for the same energy as the peak, then:

$$(P/B)_{particle} = (P/B)_{bulk} \tag{1}$$

Table 1. Line-Background and Characteristic Intensity Ratios for K-309 Microspheres.[a]

Element	Al		Si		Ca		Ba		Fe	
Sphere Diameter in μm	A[b]	B[c]	A	B	A	B	A	B	A	B
1.5	1.2	0.76	0.89	0.77	0.89	0.74	0.95	0.72	0.93	0.73
1.8	1.2	0.44	0.94	0.35	0.79	0.36	0.66	0.32	0.93	0.44
1.8	0.95	0.40	0.96	0.43	0.74	0.41	0.66	0.39	0.80	0.47
3.4	0.92	0.70	0.94	0.73	0.90	0.81	0.76	0.71	0.90	0.80
6.2	0.87	0.77	0.91	0.84	0.93	0.95	0.81	0.86	0.91	0.90
8.0	1.0	0.64	0.98	0.67	0.93	0.75	0.96	0.80	0.92	0.77
Average	1.0	0.62	0.94	0.63	0.86	0.67	0.80	0.63	0.90	0.69

[a] Spheres were analyzed by scanning the beam over an area slightly less than the cross-sectional area of the sphere.
[b] Line-to-background ratios normalized to bulk values.
[c] Characteristic intensity ratios.

We have used this assumption to predict a modified peak intensity P^* that would be the intensity observed in the absence of the particle geometric effects.

$$P^* = P_{bulk} = P_{particle} \times B_{bulk}/B_{particle} \qquad (2)$$

All of the terms in equation 2 can be directly measured except B_{bulk} which is estimated within the matrix correction loop of the analysis program from:

$$B_{bulk} = C_i B_{i,E} \qquad (3)$$

where C_i is the concentration of element i and $B_{i,E}$ is the continuum from pure element i at energy E. The x-ray continuum (I_i) from each element i may either be measured from the pure element or computed for each element from a measurement of a single reference element I_x by:

$$I_i = I_x \frac{R_i a_i Z_i^{n_i}}{R_x a_x Z_x^{n_x}} \qquad (4)$$

where R is the fraction of the total primary x-ray continuum generation that is not lost due to backscattered electrons, Z is the atomic number and a and n are a function of the beam voltage, and the excitation energy of the continuum x-rays. Equation 4 was determined from a large number of pure elements at several different beam energies (see Fig. 4).

Actual experimental measurements of $(P/B)_{particle}/(P/B)_{bulk}$ and $P_{particle}/P_{bulk}$ made on glass spheres of known composition are listed in Table 1. In general, the peak-to-background ratio is not as strong a function of particle size as the peak intensity, although some deviations in (P/B) are noted. These deviations may arise from a failure in the background fitting model.

Analyses for individual spherical particles of a standard glass (K961), which closely resembles fly ash in composition, are presented in Table 2. These analyses are compared with results obtained from the same data but analyzed with the normal matrix correction procedure (FRAME C). In all cases, oxygen was determined by stoichiometry. The FRAME C results were normalized to 100%. Both methods seem to produce similar results, however, in normalizing the data, there is a danger that some elements may be missed entirely. Table 3 contains average totals for several analyses before normalization. Note the wide range in totals obtained with FRAME C as opposed to the much narrower range for the particle program (FRAME P).

Table 2. Analyses of Standard Glass K961 Spheres (2-7μm in diameter).

Mass Fraction

		Normalized FRAME C		FRAME P	
	True	Average	Range	Average	Range
Na	0.0225	0.023	0.020-0.030	0.026	0.021-0.033
Mg	0.0289	0.029	0.028-0.030	0.031	0.029-0.033
Al	0.0602	0.061	0.060-0.064	0.061	0.058-0.066
Si	0.2924	0.303	0.300-0.305	0.293	0.286-0.296
K	0.0241	0.025	0.024-0.027	0.022	0.021-0.024
Ca	0.0374	0.040	0.037-0.043	0.036	0.034-0.038
Ti	0.0107	0.012	0.011-0.013	0.010	0.009-0.011
Fe	0.0350	0.040	0.037-0.044	0.034	0.033-0.038

Table 3. Comparison of Unnormalized Analysis Totals from FRAME C and FRAME P. (Totals are in weight percent)

	FRAME C	FRAME P
Area Scans		
Average:	54.0%	97.0%
Range:	43.1-64.8%	94.6-101.0%
Point Beam		
Average:	98.6%	98.9%
Range:	70.5-116.0%	96.3-101.0%

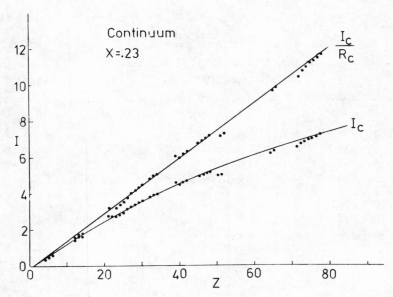

Fig. 4. Continuum intensity generated at 4.6 keV by a 20 keV electron beam (X=4.6/20=0.23) vs. atomic number. Both the generated intensity (I_c) and the generated intensity corrected for backscatter loss (I_c/R_c) are shown.

The analyses for individual fly ash particles (NBS SRM-1633) are presented in Table 4. Since the compositions of the fly ash particles vary, it is not advisable to directly compare the results with the certified composition. It is interesting to note that, in the case of iron, most of the particles measured have less iron than the certified value. This effect may arise from the heterogeneous distribution of iron as a function of particle size, and the preferential sampling of particles with a diameter less than 20μm in the analyses listed in Table 4.

Table 4. FRAME P Analyses of Fly Ash Particles (SRM-1633).

	Na	Mg	Al	Si	S	K	Ca	Ti	Fe	O	Total
					Mass Fraction						
				Top Fraction							
1	0.009	0.019	0.019	0.389	0.001	0.003	0.011	0.002	0.006	0.478	0.929
2	0.012	0.016	0.174	0.187	0.004	0.006	0.045	0.006	0.009	0.414	0.876
3	0.011	0.016	0.140	0.241	0.001	0.030	0.014	0.005	0.050	0.445	0.958
4	0.016	0.021	0.171	0.204	0.005	0.010	0.047	0.004	0.013	0.439	0.933
5	0.010	0.037	0.062	0.247	<0.001	0.003	0.121	0.005	0.011	0.421	0.919
6	0.007	0.006	0.016	0.389	<0.001	0.004	0.006	0.001	0.006	0.471	0.908
7	0.009	0.018	0.128	0.244	0.003	0.030	0.014	0.004	0.033	0.434	0.918
8	0.008	0.060	0.084	0.078	0.014	0.002	0.234	0.008	0.010	0.329	0.829
9	0.011	0.030	0.121	0.204	0.006	0.021	0.036	0.010	0.064	0.417	0.924
10	0.009	0.013	0.182	0.203	0.001	0.017	0.024	0.012	0.033	0.437	0.936
				Bottom Fraction							
11	0.007	0.011	0.110	0.243	0	0.048	0.011	0.006	0.083	0.426	0.949
12	0.003	0.003	0.145	0.167	0	0.026	0.028	0.032	0.062	0.378	0.849
13	0.010	0.034	0.190	0.189	<0.001	0.003	0.096	0.003	0.012	0.456	0.992
14	0.012	0.008	0.156	0.247	<0.001	0.029	0.006	0.016	0.019	0.454	0.950
15	0.017	0.011	0.091	0.253	<0.001	0.032	0.004	0.002	0.109	0.423	0.947
16	0.013	0.012	0.130	0.232	<0.001	0.018	0.039	0.014	0.046	0.435	0.943
17	0.009	0.012	0.134	0.236	<0.001	0.027	0.018	0.011	0.033	0.429	0.912
18	0.008	0.011	0.115	0.307	<0.001	0.019	0.005	0.004	0.013	0.476	0.961
19	0.011	0.131	0.085	0.046	0.006	<0.001	0.236	0.006	0.013	0.329	0.862
20	0.012	0.012	0.129	0.232	<0.001	0.018	0.040	0.013	0.048	0.434	0.943

SRM-1633 Certified values:
0.0017 0.00455 (0.14) 0.228   ND   0.0188   0.0111   (0.008)   0.094
Values in ( ) are not certified.

REFERENCES

1. J. A. Small, K. F. J. Heinrich, D. E. Newbury, and R. L. Myklebust, "Scanning Electron Microscopy, II", SEM, Inc., AMF O'Hare, Il., 807 (1979).
2. K. F. J. Heinrich, and H. Yakowitz, Anal. Chem., 47:2408 (1975).
3. J. A. Small, Int'l. J. Scanning Electron Microscopy, (1981) in press.
4. R. L. Myklebust, D. E. Newbury, K. F. J. Heinrich, J. A. Small, and C. E. Fiori, Proc. 13th MAS Conf., Ann Arbor, Mich., 61 (1978).
5. J. A. Small, K. F. J. Heinrich, C. E. Fiori, D. E. Newbury, and R. L. Myklebust, Proc. 13th MAS Conf., Ann Arbor, Mich., 56 (1978).
6. P. Statham, and J. Pawley, "Scanning Electron Microscopy, I", SEM, Inc., AMF O'Hare, Il., 445 (1978).

SPECTROSCOPY AND ASPHALTENE STRUCTURE

James G. Speight and Robert B. Long

Exxon Research and Engineering Company
P. O. Box 45
Linden, NJ 07036

ABSTRACT

The spectroscopic (and chemical) methods used to determine the "structure" of petroleum asphaltenes, coal and coal asphaltenes have added valuable information about the character of these materials. Nevertheless, attempts to be too literal in the interpretation of the data and the continued insistence that these complex materials have a well-defined molecular structure is of questionable value to fossil fuel technology and is certainly beyond the scope of the available methods to derive such a formula. In fact, petroleum, coal and coal asphaltenes would best be described in terms of several structural types rather than definite molecular structures.

INTRODUCTION

Projected shortages of liquid fuels have led refineries to "look deeper into the barrel" for further sources of hydrocarbon liquids. Thus, the heavy ends of petroleum are assuming a "popularity" never before imagined. It is, however, the misfortune of these heavier fractions to be rich in the asphaltene portion of petroleum.

In short, the asphaltenes are not very amenable to refinery processes and usually are responsible for catalyst destruction and coke lay-down. Thus, a considerable effort has been applied to defining asphaltenes in terms of structural and functional moieties on the presumption that knowledge of the structure will assist in the design of suitable conversion sequences in the refinery.

The conventional definition of asphaltenes is based on the solution properties of petroleum residua in various solvents (1,2,3). This generalized concept has been extended to fractions derived from other carbonaceous sources, such as coal and oil shale. Thus, there are "petroleum asphaltenes", "coal liquid asphaltenes", "coal tar asphaltenes", "shale oil asphaltenes", "tar sand bitumen asphaltenes" and the like. With this extension, there has followed considerable scientific effort to further define asphaltenes in terms of molecular structures (4,5,6).

Nevertheless, it must always be recognized that asphaltenes (from whatever the source) are, in fact, a solubility class (Figure 1) and that the definition is, in fact, an operational one; that is, asphaltenes are soluble in benzene and insoluble in pentane. Usually, for virgin petroleum samples, the residuum is completely soluble in benzene. However, with heat-soaked samples or coal derived liquids, the benzene insolubles can be appreciable.

In addition to the further classification of benzene insolubles, there also has been a growing tendency to classify asphaltenes by the particular paraffin used to precipitate them from the benzene-soluble portion of the feed. Thus, there are pentane-asphaltenes, hexane-asphaltenes, heptane-asphaltenes, and so on with the yield of paraffin-insolubles decreasing with increasing carbon number of the paraffin (9,10,11). For example, whereas liquid propane will precipitate approximately 50% of a natural bitumen as "asphaltenes" the yields of asphaltenes using n-pentane, n-heptane and n-decane are 17%, 11% and 9%, respectively (10), with very little difference in the amount precipitated for higher molecular weight n-paraffins (10,11). However, it must be stressed that the addition of a minimum of 40 volumes of the liquid hydrocarbon is required for complete precipitation of the asphaltene fraction; use of much lower proportions of the precipitating medium may lead to errors not only in the determination of the amount of asphaltenes in the crude oil but also in the determination of the compound type. For example, when insufficient proportions of the precipitating medium are employed, resins are adsorbed on to the asphaltenes from the supernatant liquid and can be released later by reprecipitation in the correct manner. Thus, questionable isolation techniques throw serious doubt on any conclusions drawn from any subsequent work done on the isolated material.

THE NATURE OF PETROLEUM ASPHALTENES

The molecular nature of the asphaltene fractions of petroleum and bitumens has been the subject of numerous investigations (4,5,6) but determining the actual structures of the constituents of the asphaltene fraction has proved to be difficult. It is, no doubt, the great complexity of the asphaltene fraction which has hindered the

SPECTROSCOPY AND ASPHALTENE STRUCTURE

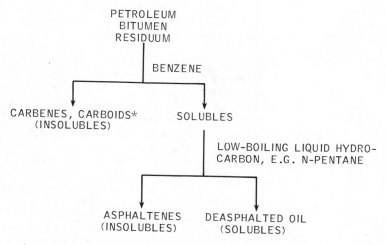

Figure 1. Schematic representation of Asphaltene separation.

*In the case of coal liquids, these "insolubles" are often referred to as "preasphaltenes", "asphaltols", etc. based on solubility in solvents such as pyridine, quinoline, etc. (7,8).

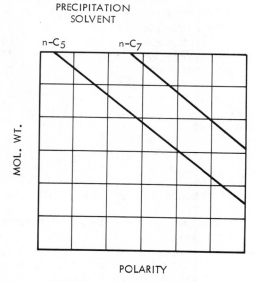

Figure 2. Petroleum Asphaltene precipitation.

formulation of the individual molecular structures. Nevertheless, the various investigations have brought to light some significant facts about asphaltene composition and structure.

The elemental compositions of asphaltenes isolated by use of, as near as can be determined, excess (greater than 40) volumes of n-pentane as the precipitating medium (Table 1) shows that the amounts of carbon and hydrogen usually vary over only a narrow range: 82+3 percent carbon; 8.1+0.7 percent hydrogen (13). These values correspond to H/C atomic ratios of 1.15+0.05, although values outside of this range are sometimes found. The near constancy of the atomic H/C ratio is itself surprising when the number of possible molecular permutations involving the heteroelements (nitrogen, oxygen, and sulfur) are considered. In fact, this property, more than any other, is the cause for the general belief that unaltered asphaltenes from virgin petroleum have a definite composition and that asphaltenes are precipitated by hydrocarbon solvents because of this composition but not because of solubility properties. However, notable variations do occur in the proportions of the heteroelements, in particular in the proportions of oxygen and sulphur. Oxygen contents vary from 0.3 to 4.9 percent, so the O/C atomic ratios vary from 0.003 to 0.045; sulfur contents vary from 0.3 to 10.3 percent so atomic S/C ratios range from 0.001 to 0.049.

In contrast, the nitrogen content of the asphaltenes appears to remain relatively constant; the amount present varies from 0.6 to 3.3 at the extremes and atomic N/C ratios are usually about 0.015 $\pm$ 0.008. It should be noted, however, that exposing asphaltenes to atmospheric oxygen can substantially alter the oxygen content and exposing a crude oil to elemental sulfur, or even to sulfur-containing minerals, can result in excessive sulfur uptake. Perhaps oxygen and sulphur contents vary more markedly than nitrogen contents because of these conditions.

In addition, it is worth noting that the use of heptane as the precipitating medium yields a product which is substantially different from the pentane-insoluble material (Table 2). For example, atomic H/C ratios of the heptane-precipitate are markedly lower than those of the pentane-precipitate, indicating a higher degree of aromaticity in the heptane precipitate; atomic N/C, O/C, and S/C ratios are usually higher in the heptane-precipitate, indicating higher proportions of the heteroelements in this material (13).

With regard to their molecular nature, asphaltenes may be regarded as consisting primarily of polar aromatics with an appreciable content of naphthene-aromatics and traces of saturates (11). There are also indications that with increasing molecular weight of the asphaltene fraction (from one particular source), both aromaticity and the proportion of the heteroelements increase(14,15,16) i.e. there is an increase in the proportion of polar aromatics.

Table 1. ELEMENTAL COMPOSITIONS OF PETROLEUM ASPHALTENES

Source (Country)	Composition (weight percent)					Atomic Ratios			
	C	H	N	O	S	H/C	N/C	O/C	S/C
Canada	79.0	8.0	1.0	3.9	8.1	1.21	0.011	0.037	0.038
	79.5	8.0	1.2	3.8	7.5	1.21	0.013	0.036	0.035
	88.5	8.2	1.6	1.4	0.3	1.11	0.015	0.012	0.001
	86.8	10.2	1.3	1.1	0.6	1.41	0.013	0.010	0.003
	85.1	11.1	0.7	2.5	0.6	1.56	0.007	0.022	0.003
	81.9	8.1	1.2	1.0	7.8	1.19	0.012	0.009	0.036
	82.2	8.2	1.6	0.4	7.6	1.19	0.017	0.004	0.035
	80.4	7.8	2.6	2.0	7.2	1.17	0.028	0.019	0.034
	82.7	7.8	2.8	1.0	5.8	1.12	0.029	0.009	0.026
	88.7	8.5	0.7	0.5	1.7	1.15	0.007	0.004	0.007
	82.8	7.5	2.1	1.6	6.0	1.09	0.022	0.015	0.025
	83.3	7.8	1.5	1.9	5.6	1.12	0.015	0.017	0.025
	82.3	7.7	2.3	1.3	6.4	1.13	0.024	0.012	0.029
	82.6	8.2	1.4	1.6	6.2	1.19	0.014	0.014	0.028
	82.9	8.2	1.5	0.5	6.8	1.19	0.015	0.005	0.031
	84.4	7.9	1.6	1.1	5.0	1.12	0.017	0.009	0.022
	84.8	6.9	1.8	0.9	5.5	0.98	0.018	0.008	0.024
	85.3	7.4	1.9	0.8	4.6	1.05	0.019	0.007	0.020
	87.1	9.5	1.0	1.7	0.8	1.30	0.010	0.015	0.003
	84.9	8.7	1.4	3.7	1.3	1.23	0.014	0.033	0.006
	85.7	7.8	1.4	1.3	5.8	1.09	0.014	0.011	0.025
	84.5	7.5	2.4	1.2	4.5	1.06	0.024	0.011	0.020
	87.9	7.6	2.2	1.8	0.5	1.04	0.022	0.015	0.002
	86.5	7.7	1.3	3.1	1.4	1.07	0.013	0.026	0.006
	86.4	9.0	1.7	2.1	0.8	1.25	0.016	0.018	0.003
Iran	83.7	7.8	1.7	1.0	5.8	1.19	0.017	0.009	0.026
	83.8	7.5	1.4	2.3	5.0	1.07	0.014	0.021	0.022
Iraq	80.6	7.7	0.8	0.3	9.7	1.15	0.009	0.003	0.045
	80.9	7.5	2.6	2.6	9.0	1.11	--	--	0.042
	78.3	7.9	0.7	4.5	8.6	1.21	0.008	0.043	0.041
	81.7	7.9	0.8	1.1	8.5	1.16	0.008	0.010	0.039
Kuwait	82.2	8.0	1.7	0.6	7.6	1.17	0.017	0.005	0.035
	81.8	8.1	0.9	1.7	7.5	1.18	0.009	0.016	0.034
	81.6	8.0	0.8	1.8	7.8	1.18	0.008	0.017	0.036
	82.1	8.1	0.6	1.3	8.0	1.18	0.006	0.012	0.037
	81.6	8.1	1.0	1.5	7.8	1.19	0.011	0.014	0.036
	82.1	8.0	1.7	0.6	7.6	1.17	0.017	0.005	0.035
	82.4	7.8	0.9	1.5	7.4	1.14	0.009	0.014	0.034
Mexico	81.4	8.0	0.6	1.7	8.3	1.18	0.006	0.016	0.038
Sicily	81.7	8.8	1.5	1.8	6.3	1.29	0.016	0.017	0.029
	81.3	8.5	trace	4.9	5.2	1.25	--	0.045	0.024
	78.0	8.8	trace	3.0	10.2	1.35	--	0.033	0.049
	78.9	7.8	trace	3.1	10.3	1.19	--	0.029	0.049
U.S.A.	84.5	7.4	0.8	1.7	5.6	1.05	0.008	0.015	0.025
	88.2	8.1	1.7	1.3	0.6	1.10	0.017	0.011	0.003
	82.9	8.9	2.3	-	6.5	1.29	0.024	--	0.029
	88.6	7.4	0.8	2.7	0.5	1.00	0.008	0.023	0.002
	84.2	7.6	0.8	1.6	5.8	1.08	0.008	0.014	0.026
	84.0	7.9	1.9	4.1	2.1	1.13	0.019	0.037	0.009
	83.2	8.3	2.3	4.8	1.4	1.20	0.024	0.043	0.006
	85.5	8.1	3.3	1.8	1.3	1.14	0.033	0.016	0.006
Venezuela	84.2	7.9	2.0	1.6	4.5	1.13	0.020	0.014	0.020
	83.5	8.3	1.0	1.5	2.7	1.19	0.010	0.013	0.012
	84.7	8.0	0.9	1.0	5.5	1.13	0.009	0.009	0.024
	84.0	7.9	2.0	1.6	4.5	1.13	0.020	0.014	0.020
	81.1	7.8	0.2	4.2	6.7	1.15	0.002	0.039	0.031
	81.2	7.9	2.0	2.0	6.9	1.17	0.021	0.018	0.032

Table 2. ELEMENTAL COMPOSITIONS OF "ASPHALTENE" FRACTIONS PRECIPITATED BY DIFFERENT SOLVENTS

Source	Precipitating Medium	Composition (weight percent)					Atomic Ratios			
		C	H	N	O	S	H/C	N/C	O/C	S/C
Canada	n-pentane	79.5	8.0	1.2	3.8	7.5	1.21	0.013	0.036	0.035
	n-heptane	78.4	7.6	1.4	4.6	8.0	1.16	0.015	0.044	0.038
Iran	n-pentane	83.8	7.5	1.4	2.3	5.0	1.07	0.014	0.021	0.022
	n-heptane	84.2	7.0	1.6	1.4	5.8	1.00	0.016	0.012	0.026
Iraq	n-pentane	81.7	7.9	0.8	1.1	8.5	1.16	0.008	0.010	0.039
	n-heptane	80.7	7.1	0.9	1.5	9.8	1.06	0.010	0.014	0.046
Kuwait	n-pentane	82.4	7.9	0.9	1.4	7.4	1.14	0.009	0.014	0.034
	n-heptane	82.0	7.3	1.0	1.9	7.8	1.07	0.010	0.017	0.036

More recently, a new method for describing asphaltenes has emerged (17) and involves, initially, a consideration of molecular weight vs. polarity for the molecular types found in petroleum residua and/or heavy oils (Figure 2). The polarity scale may be in definable arbitrary units, such as relative adsorptive strength on a solid, e.g. Attapulgus clay and/or silica gel (18), or by solubility in a variety of solvents of increasing polarity as practiced on separation of coal liquid fractions (8); this scale is thus a molecular weight-independent portion of the solubility parameter. For any particular precipitating medium, say n-heptane, the precipitated asphaltenes will, therefore, consist of less polar materials of higher molecular weight and more polar materials of lower molecular weight; with n-pentane as the precipitating agent, both less polar and lower molecular weight materials are included in the precipitate and the total amount of precipitate increases (10,11). This concept can also be extended to asphaltenes from coal liquids (Figure 3) which are essentially of much lower molecular weight, but higher polarity, than petroleum asphaltenes (Table 3).

An additional subtlety in choice of solvent is shown by cyclohexane and methylcyclohexane. In these cases, petroleum asphaltenes are completely soluble, while coal derived asphaltenes are only sparingly soluble. This would imply that the precipitation line for cyclohexanes run completely outside the petroleum range but does pass through the coal liquids range (Figure 3). The concept also allows for a polarity range within the various asphaltenes where varying heteroatom (nitrogen, oxygen, and sulfur) contents of the petroleum asphaltenes (Table 1) or variations in the source (and method of preparation) of the coal asphaltenes would be expected to influence polarity.

THE STRUCTURE OF PETROLEUM ASPHALTENES

(a) The Carbon Skeleton

The issue of the structural nature of petroleum asphaltenes is not a recent issue. For example, an early theory (19) invoked the concept of polymeric analogues of aromatic and naphthenic ring systems which also allowed the possibility of including nitrogen, oxygen and sulfur into these ring systems (Figure 4a). Another theory (20) also invoked the concept of kata-condensed ring systems with the nitrogen, oxygen, and sulfur, again, scattered throughout the rings (Figure 4b).

Much of the information available on the carbon skeleton of asphaltenes has been derived from spectroscopic studies of asphaltenes isolated from various petroleums and natural asphalts (4,5,6). The data from these studies "support" the hypothesis that asphaltenes are permutations of condensed polynuclear aromatic ring systems

TABLE 3
COMPARISON OF COAL AND PETROLEUM ASPHALTENES

Inspection	Coal Asphaltenes	Petroleum Asphaltenes
Carbon, wt.%	86.93	81.7
Hydrogen, wt.%	6.83	7.60
Nitrogen, wt.%	1.36	1.23
Sulfur, wt.%	1.09	7.72
Oxygen, wt.%	3.8	1.7
Vanadium, ppm	9	1200
Nickel, ppm	3	390
VPO, Mol. Wt.	726	5400
H/C, Atom Ratio	0.94	1.12

TABLE 4
AROMATIC NUCLEI OF THE TYPE PRESUMED TO BE PRESENT IN ASPHALTENES

HYDROCARBON

NAME	STRUCTURE	NAME	STRUCTURE
		OVALENE	
DIBENZO(de,hi)NAPTHACENE			
ANTHRO(defghi)NAPTHACENE		DINAPHTHO[2,7-hijk-2,7-stuv]-OVALENE (Circopyrene)	
CORONENE		CIRCOCORONENE	

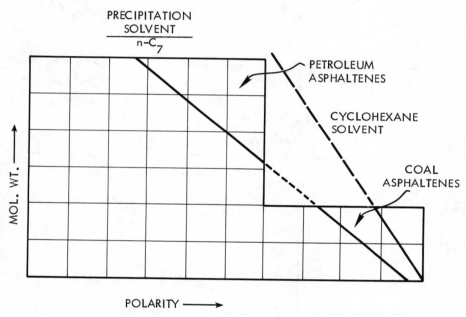

Figure 3. Comparison of petroleum and coal asphaltenes

(A)

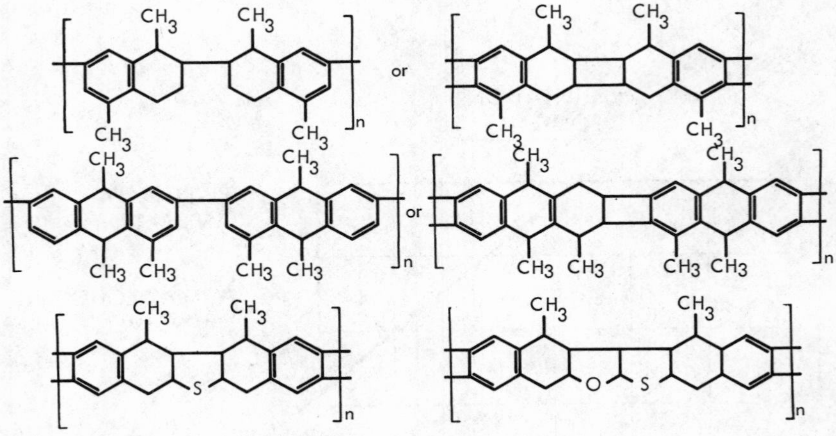

WHERE n > 1;

(B)

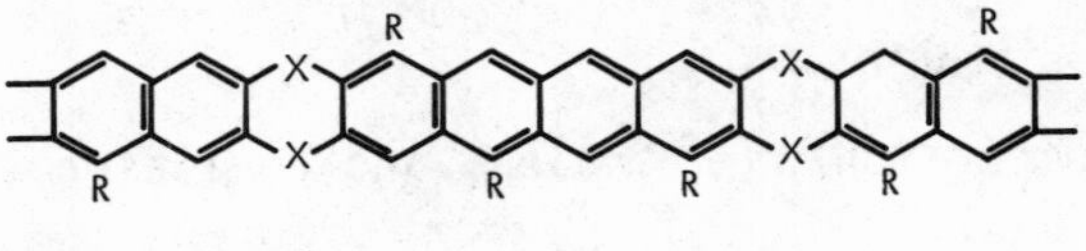

WHERE X = NH, O or S

Figure 4. Representations of early structures suggested for petroleum asphaltenes.

(Table 4) and bear alkyl side chains. The number of rings apparently varies from as low as six in smaller systems to 15 to 20 in more massive systems (4,5).

Attempts have also been made to describe the total structures of asphaltenes (Figure 5) in accordance with nuclear magnetic resonance (nmr) data and results of spectroscopic and analytical techniques although it may be difficult to visualize such well-defined structures as representing the asphaltene molecule. The fact is that all methods employed for structural analysis involve, at some stage or another, assumptions which, although based on data concerning the more volatile fractions of petroleum, are of questionable validity when applied to asphaltenes.

Asphaltenes have also been subjected to X-ray analyses in order to gain an insight into their macromolecular structure (21); the method is reputed to yield information about the dimension of the unit cell such as interlamellar distance (c/2), layer diameter (L_a), height of the unit cell (L_c), and number of lamellae (N_c) contributing to the micelle (Figure 6).

As with any particular spectroscopic techniques, it is possible to note various "trends" which occur with various asphaltenes. For example, fractionation of an asphaltene by stepwise precipitation with a series of hydrocarbon liquids (n-heptane to n-decane) allows separation of the asphaltene by molecular weight. The structural parameters determined using the X-ray method show a relationship to the molecular weight (22). For the particular asphaltene in question (Athabasca), the layer diameters, L_a, increase with molecular weight to a "limiting" value (Figure 7); similar relationships also appear to exist for the interlamellar distance (c/2), micelle height (L_c), and even the number of lamellae (N_c) in the micelle (Table 5). However, the occurrences of "coincidences" of this type have not been thoroughly explained nor are they fully understood. The progression from simpler molecules to the more complex molecules as the carbon number (and solvent power) of the precipitating medium is increased is, obviously, one explanation. However, the method does not indicate the molecular nature of the material nor does it suggest how any errors might be compounded by use of these procedures.

Application of Diamonds's X-ray diffraction matrix method to the problem of determining asphaltene structure produced some novel results (23). For example, asphaltenes (precipitated by n-pentane from Athabasca bitumen and conventional Alberta crude oils) which are soluble in decane gave histograms completely different from those obtained with the decane-insoluble material (Figure 8), suggesting the existence of two different molecular types in the asphaltene fraction. The molecular types predominant in the decane-soluble material appear to be based on simple

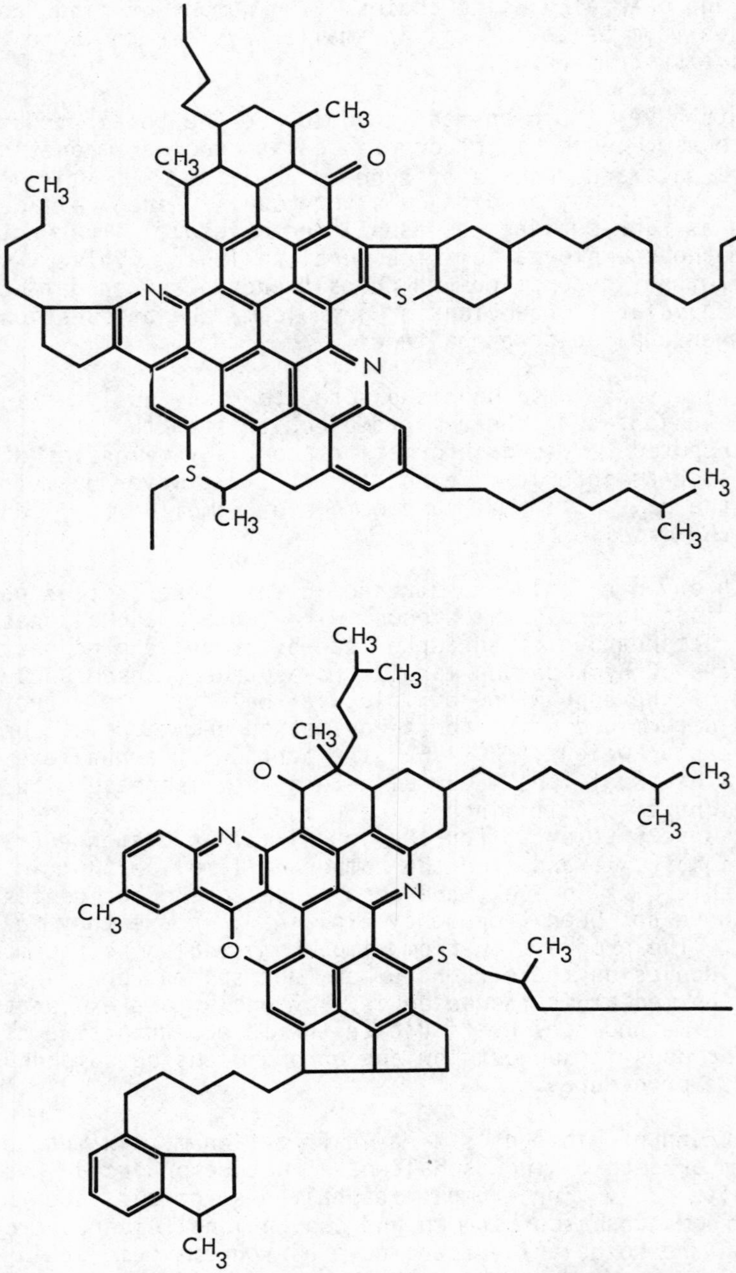

Figure 5. Hypothetical Structures for Asphaltenes from
(A) Venezuelan Crude Oil
(B) Californian Crude Oil

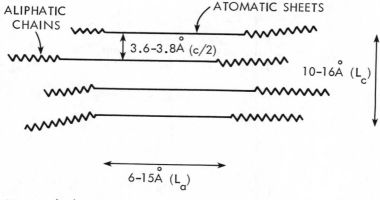

Figure 6. Diagrammatic Representation of an Asphaltene from X-ray Analysis.

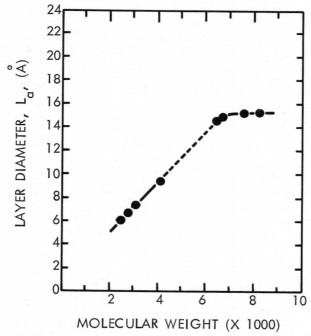

Figure 7. Relationship of Layer Diameter, L_a, to Molecular Weight.

Table 5
Structural Parameters of Asphaltene Fractions Derived by the X-Ray Method

Molecular Weight	Structural Parameters			
	$L_a(\text{Å})$	$c/2(\text{Å})$	$L_c(\text{Å})$	N_c
2694	6.0	3.81	10.7	2.8
2704	6.3	3.79	10.2	2.7
3185	7.3	3.79	11.0	2.9
4338	9.1	3.74	11.6	3.1
6427	13.9	3.65	12.5	3.4
6530	14.5	3.65	13.8	3.8
7603	14.7	3.64	13.8	3.8
8158	14.8	3.64	14.0	3.9

Table 6
Elemental Compositions of Coal Asphaltenes

Asphaltene Source

Process-Type Coal	Synthoil Pittsburgh	Synthoil Pittsburgh	Synthoil Pittsburgh	SRC West Kentucky	SRC West Kentucky
C, wt%	87.85	87.9	85.15	85.2	86.3
H, wt%	6.1	6.4	6.4	6.07	5.33
N, wt%	1.95	2.00	1.80	2.0	2.32
O, wt%	3.4	3.05	5.45	6.3	5.2
S, wt%	0.70	0.69	1.21	-	0.8
Molecular Weight	525	505	590	630	-
Reference	51	51	51	54	54

condensed aromatic units with only about six aromatic rings per unit. On the other hand, the decane-insoluble fraction contained aromatic systems of much greater complexity, but these systems appeared to be "collections" of simpler units which were actually similar to those in the lower molecular weight material but linked head-to-tail in a cyclic system.

In the case of asphaltenes of lower molecular weight, the absence of any evidence in the histograms to support "hole" structures clearly suggests that any heteroatoms located in the cyclic systems are found at peripheral sites. However, there is some evidence of "hole" structures within the carbon lamellae of the higher molecular weight asphaltenes. To accommodate heteroatoms in such "holes" would require a significant modification of the nmr model to a much larger, more ponderous structure. In contrast, insertion of the heteroelements into the X-ray model (Figure 8) could be achieved quite conveniently without any significant structural alterations. However, "structural acceptability" is not the only criterion for "proof of structure" when deducing structural formulae by any of the physico-chemical techniques. In fact, the method itself may require proof of acceptability.

Investigations of the X-ray diffraction patterns of various low temperature (450-750°C) synthetic carbons (24), carbon black blended with polyethylene (21), condensed aromatics of known structure where the maximum diameter of the sheets is approximately 14 A (21) as well as mixtures of condensed aromatics and porphyrins (25) may be cited as evidence supporting the concept of condensed aromatic sheets (having a tendency to stack) as the structure of asphaltenes. However, it is perhaps this "ease" with which the X-ray diffraction of the asphaltenes can be reproduced which dictates that caution is necessary in the interpretation of the data - an empty polyethylene sample holder will exhibit a similar X-ray diffraction pattern to that of the asphaltenes. It, therefore, appears debatable whether the data derived by this method or, for that matter, by any other method are absolute or whether their sole value is limited to comparison between the various parameters.

Indeed, an excellent example of misrepresentation by a spectroscopic method (i.e., nuclear magnetic resonance) is provided by an nmr examination of the asphaltenes from Athabasca bitumen where alkyl side chains are deduced to contain approximately four carbon atoms (26,27). The pyrolysis (350-800°C) of this asphaltene produces substantial amounts of alkanes (C_{34}) in the distillate (28,29,30). The presence of these alkanes in the pyrolysates is thought to reflect the presence of such chains in the original asphaltene (30) but this is difficult to rationalize on the basis of an "average" structure derived from the nmr data. Obviously, recognition of the inconsistencies of the spectroscopic methods with respect to the paraffinic moieties must lead to recognition of

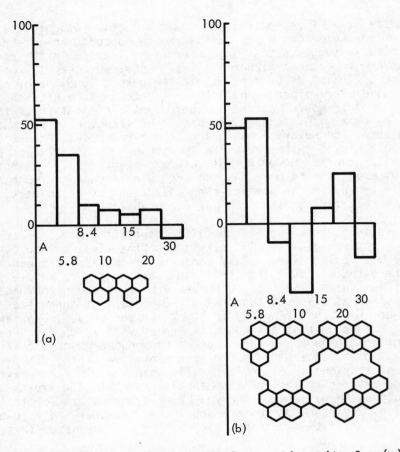

Figure 8. Histograms and hypothetical aromatic units for (a) decane-soluble asphaltenes and (b) decane-insoluble asphaltenes from Alberta crude oils.

similar inconsistencies when considering the aromatic nucleus of asphaltenes.

(b) Nitrogen, Oxygen, and Sulfur in Petroleum Asphaltenes

Unfortunately, in all these studies, too little emphasis has been put on determining the nature and location of the nitrogen, oxygen, and sulfur atoms in the asphaltene structure. However, mass spectroscopic investigations (31) of a petroleum asphaltene have allowed the identification of fragment peaks which indicate that at least some of the heteroatoms exist in the ring systems (Figure 9). A study of the thermal decomposition of the asphaltenes from a natural bitumen (28,29) indicated that only 1 percent of the nitrogen was lost during the thermal treatment while substantially more sulfur (23 percent) and almost all of the oxygen (81 percent) were lost as a result of this treatment. The tendency for nitrogen and sulfur to remain during thermal decomposition, as opposed to the easy elimination of oxygen, supports the concept that nitrogen and sulfur have stability due to their location in ring systems (30,32,33). On the other hand the concept of asphaltenes being a sulfur polymer of the type:

or even a regular hydrocarbon polymer of the type:

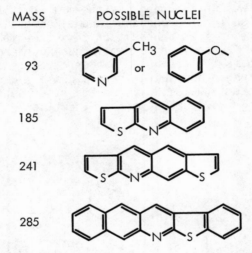

Figure 9. Suggested heteroatom locations from mass spectroscopic fragmentation peaks.

has arisen because of the nature of the products obtained by reacting an asphaltene with potassium naphthalide (34). However, interpretation of the experimental evidence leading to this type of sulfur-containing structure has been questioned (6,35) and the behavior of the sulfur-structure under thermal conditions (28,29,30) as well as to reagents such as Raney nickel (6) is not consistent with this type of structure. It is evident, therefore, that experimental organic chemistry suffers from as many inconsistencies as spectroscopy when acquired data are interpreted in terms of asphaltene structure.

Of the heteroatoms in petroleum there is more data relevant to the locations of the oxygen atoms than to sulfur and nitrogen. However, the majority of the data relates to oxygen functions in blown (oxidized) asphalts and residua (36,37) which may actually be of little relevance to the oxygen functions in the native materials. Of the limited data available, there are indications that oxygen in asphaltenes can exist as non-hydrogen-bonded phenolic hydroxyl groups (38,39). Other information on the presence and nature of oxygen in asphaltenes has been derived from infrared spectroscopic examination of the products after interaction of the asphaltenes with acetic anhydride 8,40) and produced evidence for the presence of ketones and/or quinones as well as indications that a considerable portion of the hydroxyl groups present in asphaltenes may exist as collections of two, or more, hydroxyl functions on the same aromatic ring, or on adjacent peripheral sites on a condensed ring system:

or even on sites adjacent to a carbonyl function in a condensed ring system:

In the context of polyhydroxy aromatic nuclei existing in Athabasca asphaltenes it is of interest to note that pyrolysis at 800°C results in the formation of resorcinols (30).

The form in which nitrogen exists in asphaltenes is even less well understood than oxygen and sulfur. Attempts have been reported to define organic nitrogen in terms of various structural and basic/non-basic types as a result of the titration of various asphaltenes with perchloric acid (41):

On the other hand, spectroscopic investigations (42) suggest that carbazoles might be a predominant nitrogen-type in the asphaltenes from Athabasca bitumen thereby supporting earlier mass spectroscopic evidence (31) for nitrogen-types in asphaltenes.

THE NATURE OF COAL AND COAL ASPHALTENES

Coal has been used as a combustible fuel throughout recorded (and unrecorded) history but underwent a decline in popularity as a fuel throughout the middle decades of the present century because of the availability of supposedly plentiful supplies of cheap petroleum.

Coal itself has been the subject of numerous spectroscopic investigations (43 and references cited therein) which led to the postulation of chemical structures for coal based on, for example, the ten-ring condensed aromatic compound ovalene to more complex three-dimensional structures. However, spectroscopic techniques do offer valuable primary information about coal structure (43,44) but mathematical manipulations to reach beyond this primary information must be handled with extreme care.

The recent emergence of coal as a medium to long-term source of liquid fuels has required an understanding not only of the conversion processes but, more particularly, an understanding of the products of these processes (45,46). Thus, structural investigations of the products of coal refining processes have paralleled, and even overtaken, the popularity of the structural investigations of coal(s).

As with petroleum constituents and products, identification of the lower molecular weight products from coal refining processes is often facilitated by their volatility and, hence, suitability to many of the analytical procedures. The converse is true for the higher molecular weight (or nonvolatile) products of coal refining processes and particular attention has been paid to these more complex products of coal refining, the so-called asphaltenes (45,46).

In accordance with the nomenclature employed in petroleum technology, coal asphaltenes are also a solubility class; these materials are soluble in benzene but insoluble in a low-boiling liquid hydrocarbon such as n-pentane (46,47,48; see also Figure 1). The origin of coal asphaltenes has been the subject of speculation (49,50,51) insofar as they have been considered to be not only the initial products of coal liquefaction:

$$coal \longrightarrow asphaltenes \longrightarrow oil$$

but also the secondary products of coal liquefaction:

$$coal \longrightarrow oil \longrightarrow asphaltenes.$$

Although current concepts tend to favor the former hypothesis (52), the matter is still not completely resolved. Nevertheless, structural studies of the asphaltenes from coal liquefaction processes have progressed to the point where various molecular types have been identified.

Coal asphaltenes are believed to be quite different in nature from petroleum asphaltenes (Table 6; see, for example, Tables 1 and 3). The molecular weight of coal asphaltenes may be some eight to ten times lower than the observed molecular weight of petroleum asphaltenes although this latter can be revised to lower values for a variety of reasons (39,53).

Coal asphaltenes have frequently been defined in terms of acid-base complexes (52) with the asphaltene existing as a composite of the two systems (52):

Although a more recent report (51) indicates that the pentane-insolubility of the bulk of the coal asphaltenes cannot be ascribed to hydrogen bonding effects between the acidic and basic components.

Examinations of coal asphaltenes by various spectroscopic techniques (46,55,56) has resulted in the identification of many compound types. Thus, from mass spectroscopic investigations major oxygen compound types include phenols, dihydroxybenzenes, and the hydroxy and dihydroxyderivatives of phenanthrenes, pyrenes, benzopyrenes, coronenes, and of other condensed aromatics. Nitrogen compounds present include carbazoles, benzocarbazoles, and hydroxycarbazoles. Aromatic and aromatic sulfur compounds found include naphthenonaphthalenes, phenanthrenes, pyrenes, benzopyrenes, benzperylenes, coronenes, and dibenzothiophenes.

Nuclear magnetic resonance spectroscopy indicates that the asphaltenes consist predominantly of aromatic molecules carrying short alkyl groups that may be either open chains or saturated rings condensed to the aromatic rings. In addition, data from ultraviolet spectroscopy has provided indications (46) that the aromatic portion of certain coal asphaltenes may contain condensed structures composed of more than five aromatic rings, but the uv data excludes the presence of anthracene and tetracene homologues. Finally, there are also indications from the various techniques that high proportions (up to 85%) of the total carbon exists in aromatic locations.

Again, it is necessary to proceed with caution in deriving molecular structures by mathematical methods when the spectroscopic techniques offer valuable primary information relating to structural types.

EPILOGUE

It is very questionable whether the intensive investigation of the so-called average structures in the heavier fractions of petroleum (and the analogous fractions of coal) is worthy of any effort. Indeed, the continued insistence that these particular materials have a well-defined (but "average") structure is of questionable value to petroleum and coal technology. The well-meaning scientist who postulates an average molecular structure for these materials but then falls into the trap of continually trying to prove that such a structure actually exists may only add confusion to a very complex issue.

In spite of these negative comments, some meaningful data have been derived by the various spectroscopic techniques but these data relate to structural types. Thus, a key structural feature in the current concept of petroleum, coal and coal asphaltenes is the occurrence of condensed polynuclear aromatic systems as well as various heterocyclic (and non-heterocyclic) locations for the nitrogen, oxygen, and sulfur. Such findings allow more realistic projections of the behavior of these materials with respect to, say, processability and interactions with catalysts as well as affording valuable geochemical information. Indeed, the evolution of the spectroscopic techniques appears to be advancing to the stage where subtle differences in the paraffin chains of petroleum asphaltenes and coal asphaltenes can be identified (57). In fact, it is becoming increasingly obvious that the occurence of long-chain n-paraffins in the volatile products from the thermal decomposition of petroleum asphaltenes is not an isolated occurrence (58,59). Hypothetical petroleum asphaltene formulae, although in some instances still based on the highly condensed aromatic nuclei (Table 4), are now beginning to appear with the longer alkyl side chains (60).

Obviously, many of the assumptions required to derive the earlier structural formula for these complex materials may have to be re-evaluated in order to account for observed behavioral characteristics of petroleum asphaltenes, coal and coal asphaltenes. Thus, it is perhaps more advantageous to describe these carbonaceous materials in terms of several structural types rather than definite (and, most probably incorrect or inadequate) molecular structures.

REFERENCES

1. "Annual Book of ASTM Standards; American Society for Testing and Materials, Philadelphia, Part 24, Standard No. D-2006", 1975, withdrawn 1976.

2. "Annual Book of ASTM Standards; American Society for Testing and Materials, Part 15, Proposed Standard for Asphalt Composition".

3. "Standards for Petroleum and Its Products, Standard No. IP 143/57", Institute of Petroleum, London.

4. Speight, J. G., Applied Spectroscopy Reviews, 1972, $\underline{5}$, 211, and references cited therein.

5. Yen, T. F., PREPRINTS, American Chemical Society, Division of Petroleum Chemistry, 1972, $\underline{17}$(4), F102, and references cited therein.

6. Speight, J. G., Preprints, Am. Chem. Soc., Div. Petrol. Chem., 1979, $\underline{24}$(4), 910.

7. Brooks, J. D., and Taylor, G. H., Chem. & Phys. of Carbon, Vol. 4, 243 (1968) Marcel Decker, New York.

8. Farcasiu, M., Mitchell, T. O., Whitehurst, D. D., A.C.S. Div. Fuel Chemistry Preprints, Vol. 21, No. 7, 11, (1976).

9. Bland, W. F.; Davidson, R. L. (Editors), "Petroleum Processing Handbook", McGraw-Hill, New York, 1967, p. 3 et seq.

10. Mitchell, D. L.; Speight, J. G., Fuel, 1973, $\underline{52}$, 149.

11. Corbett, L. W., and Petrossi, U., Ind. Eng. Chem. Prod. Res. Dev. $\underline{17}$, 342 (1978).

12. Speight, J. G., unpublished data, 1969.

13. Speight, J. G., "The Structure of Petroleum Asphaltenes", Information Series No. 81, Alberta Research Council, 1978; and Speight, J. G., "Geochemical Influences on Petroleum Constitution and Asphaltene Structure" presented at the Div. of Geochem., American Chemical Society, Washington, D. C., September, 1979.

14. Yen, T. F., Fuel, 1970, $\underline{49}$, 134.

15. Yen, T. F., Fuel, American Chemical Society, Division of Fuel Chemistry, 1971, $\underline{15}$(1), 57.

16. Koots, J. A.; Speight, J. G., Fuel, 1975, 54, 179.

17. Long, R. B., Preprints, Am. Chem. Soc., Div. Petrol. Chem., 1979, 24(4), 891.

18. Corbett, L. W., Anal. Chem. 41, 576 (1969).

19. Hillman, E., and Barnett, B., Proc. 4th Ann. Mtg., A.S.T.M., 1937, 37(2), 558.

20. Murphy, B., J. Inst. Petrol., 1945, 31, 475.

21. Yen, T. F.; Erdman, J. G.; Pollack, S. S., Anal. Chem., 1961, 33, 1587.

22. Speight, J. G., paper presented at United Bureau of Mines Symposium on Fossil Chemistry and Energy, Laramie, Wyoming, July 23-27, 1974.

23. Speight, J. G., Proc. National Science Foundation Symposium on the Fundamental Organic Chemistry of Coal, Knoxville, Tennessee, 1975, 125.

24. Speight, J. G., unpublished data, 1971.

25. Speight, J. G., unpublished data, 1977-1978.

26. Speight, J. G., Fuel, 1970, 49, 76.

27. Speight, J. G., Fuel, 1971, 50, 102.

28. Speight, J. G., Fuel, 1970, 49, 134.

29. Speight, J. G., , Am., Chem., Soc., Div. Fuel Chem., 1971, 15(1), 57.

30. Ritchie, R. G. S.; Roche, R. S.; Steedman, W., Fuel, 1979, 58, 523.

31. Clerc, R. J.; O'Neal, M. J., Anal. Chem., 1961, 33, 380.

32. Moschopedis, S. E.; Parkash, S.; Speight, J. G., Fuel, 1978, 57, 431.

33. Speight, J. G.; Penzes, S., Chem. and Ind., 1978, 729.

34. Ignasiak, T.; Kemp-Jones, A. V.; Strauz, O. P., Org. Chem., 1977, 42, 312.

35. Speight, J. G., and Moschopedis, S. E., Fuel, 1980, 59, 440.

36. Knotnerus, J., PREPRINTS, Am. Chem. Soc., Div. Petrol. Chem., 1971, 16(1), D37.

37. Petersen, J. C.; Barbour, F. A.; Dorrence, S. M., Proc. Assoc. Asphalt Paving Technol, 1974, 43, 162.

38. Moschopedis, S. E.; Speight, J. G., PREPRINTS, American Chemical Society, Division of Fuel Chemistry, 1976, 21(6), 198.

39. Moschopedis, S. E.; Speight, J. G., Fuel, 1976, 55, 187.

40. Moschopedis, S. E.' Speight, J. G., Fuel, 1976, 55, 334.

41. Nicksic, S. W.; Jeffries-Harris, M. J., J. Inst. Petroleum, 1968, 54, 107.

42. Moschopedis, S. E.; Speight, J. G., PREPRINTS, Am. Chem. Soc., Div. Petrol. Chem., 1979, 24(4), 1007.

43. Speight, J. G., "Assessment of Structures of Coal by Spectroscopic Techniques" in "Analytical Methods for Coal and Coal Products", C. A. Karr, Jr. (Editor), Academic Press Inc., New York, 1978, Chapter 22.

44. Retcofsky, H. L., "Applications of Spectrometry in Studies of Coal Structure" in "Scientific Problems in Coal Utilization", B. R. Cooper (Editor), U. S. Dept. of Energy, 1978, 79.

45. Dooley, J. E., Thompson, C. J., and Scheppele, S. E., "Characterizing Syncrudes from Coal" in "Analytical Methods for Coal and Coal Products", C. A. Karr, Jr., (Editor), Academic Press Inc., New York, 18, Chapter 16.

46. Aczel, T., Williams, R. B., Brown, R. A., and Pancirov, R. J., "Chemical Characterization of Synthoil Feeds and Products" in "Analytical Methods for Coal and Coal Products", C. A. Karr, Jr. (Editor), Academic Press Inc., New York, 1978, Chapter 17.

47. Mima, M. J., Schultz, H., and McKinstry, W. E., "Method for the Determination of Benzene Insolubles, Asphaltenes and Oils in Coal Derived Liquids" in "Analytical Methods for Coal and Coal Products", C. A. Karr, Jr. (Editor), Academic Press Inc., New York, 1978, Chapter 19.

48. Steffgen, F. W., Schroeder, K. T., and Bockrath, B. C., Anal. Chem., 1979, 51, 1164.

49. Schweighardt, F. K., Retcofsky, H. L., and Raymond, R., Preprints, Am. Chem. Soc., Div. Fuel Chem., 1976, 21(7), 27.

50. Farcasiu, M., Chem. Tech., 1977, 7, 680.

51. Bockrath, B. C., Delle Donne, C. L., and Schweighardt, F. K., Fuel, 1978, 57, 4.

52. Sternberg, H. W., Raymond, R., and Schweighardt, F. W., Preprints, Am. Chem. Soc., Div. Petrol. Chem., 1976, 21(1), 198; and S. Yokoyama, S. Ueda, Y. Maekawa and M. Shibaoka, Preprints, Am. Chem. Soc., Div. Fuel Chem., 1979, 24(2), 289.

53. Moschopedis, S. E., Fryer, J. F., and Speight, J. G., Fuel, 1976, 55, 227.

54. Farcasiu, M., Mitchell, T. O., and Whitehurst, D. D., Preprints, Am. Chem. Soc., Div. Fuel Chem., 1976, 21(7), 11.

55. Bartle, K. D., Ladner, W. R., Martin, T. G., Snape, C. E., and Williams, D. F., Fuel, 1979, 58, 413.

56. Curtis, C. W., Hathaway, C. D., Guin, J. A., and Tarrer, A. R., Fuel, 1980, 59, 575.

57. Melchior, M. T., Exxon Research and Engineering Co., private communication.

58. Schucker, R. C., and Keweshan, C. F., Preprints, Am. Chem. Soc., Div. Fuel Chem., 1980, 25(3), 155.

59. Steedman, W., and Simm, I., Fuel, 1980, 59, 669.

60. Bukowski, A., and Gurdzinska, E., Preprints, Am. Chem. Soc., Div. Petrol. Chem., 1980, 25(2), 293.

GEOLOGIC FACTORS THAT CONTROL MINERAL MATTER IN COAL

C. B. Cecil[1], R. W. Stanton[1], F. T. Dulong[1], and J. J. Renton[2]
[1]U.S. Geological Survey, National Center, Reston, VA 22092
[2]West Virginia University, Morgantown, WV 26506

INTRODUCTION

Elements other than organically derived and bound C, H, N, O, and S constitute mineral matter in coal. Mineral matter may consist of discrete minerals such as calcite, quartz, clays, and pyrite, and/or organic compounds that contain organically bonded elements such as Ca and Cl. Processes that may affect the association and content of mineral matter are operable from the initial peat-forming stage to the time of utilization of the coal. However, the objective of this paper is to discuss some of the various geologic processes that may influence content and associations in unmined coal.

MINERAL MATTER CONTENT

Of the continuum of processes controlling mineral matter content, those which are operable during peat formation are perhaps the most important. Sources of mineral matter during peat formation and accumulation may include (1) elements incorporated in plants that ultimately become peat, (2) elements incorporated in the peat by ion exchange from interstitial waters, (3) detrital minerals that are washed or blown into the peat-forming environment, and (4) chemically precipitated minerals such as sulfides and carbonates.

In coals currently being mined in the United States, our contention is that vegetal matter is the dominant source of mineral matter exclusive of Fe, S, Ca, and carbonate ($CO_3^=$).

The concentrations of Fe, S, Ca, and $CO_3^=$ are controlled mainly by the interrelated factors of the pH of the environment, availability of ions and the microbial activity that produces reduced sulfur species (H_2S, $S^=$, etc.) and CO_2. Leaching can remove mineral matter during the peat-forming stage. Leaching takes place especially when the peat-forming environment is highly acidic (pH 3-4.5); in contrast, when the pH is nearly neutral, conditions are favorable for the retention of mineral matter. Loss of organic matter by degradation (either biological or chemical) also may concentrate mineral matter.

The various geochemical and microbial factors affecting mineral-matter content at the time of peat formation are interrelated; however, the importance of pH cannot be overemphasized. Low pH (3-4.5) implies low ionic strength of interstitial waters in the peat. If the ionic strength and composition of interstitial water were the same as average river water, then a nearly neutral pH would result from hydrolysis.

The conditions of low pH and low ionic strength are optimum for leaching most metal ions. For example, iron is reduced from the ferric to the ferrous state at approximately +600 mev and at a pH of 3 (Garrels and Christ, 1965). Thus, under acid conditions, iron is reduced at a relatively high redox potential to the ferrous state which is readily soluble in the absence of high concentrations of sulfide ions. The solubility of other metal ions such as those of the alkaline and alkaline-earth series are not affected by changes in Eh conditions. However, the solubility of ions such as Ca^{++} is strongly pH dependent. At nearly neutral pH, calcium salts of organic acids are relatively stable, whereas under acid conditions Ca will be soluble.

Under low pH conditions, the activity of $SO_4^=$ reducing bacteria is minimal (Baas Becking and others, 1960). Limited $SO_4^=$ reduction by bacteria may be related to a low $SO_4^=$ content of interstitial waters, and/or to low pH. Fermentation reactions that produce CO_2 and methane may also be inhibited by low pH. Although few studies have been conducted and data are limited, it appears that methane production is minimal in acid environments such as the Okefenokee Swamp of Georgia (Swain, 1973). A further indication of the antiseptic nature of acid peat waters is supported by the historical fact that fresh water from the St. Mary's River, which drains the southeastern corner of the Okefenokee Swamp, did not "turn sour" in the barrels of ocean-going vessels even after months at sea (Spackman and others, 1976).

The importance of pH during peat formation on the mineral matter content of coal has been suggested by Cecil and others (1978, 1979a, 1979b, 1980a, 1980b). Recent independent work by

A. D. Cohen (oral commun.) and the writers have demonstrated that low-ash peat (2-4 wt.%, dry basis) is associated with low pH (3-4) in modern peat-forming environments of the Atlantic Coastal Plain. The ash (mineral matter) content of the modern peats studied increases as the pH increases. The ash content of modern peats containing nearly neutral interstitial waters is commonly high (>50 wt.%, dry basis); thus, such material would probably yield a carbonaceous shale upon diagenesis/ coalification.

Studies of the Upper Freeport coal bed (Cecil and others, 1980b) in west-central Pennsylvania have shown that the mineral matter and macerals in this coal can be divided statistically into two groups (table 1). The statistical groupings probably have genetic implications as the positive correlations within each group do not necessarily indicate element-mineral associations in the coal. For example, Zn, Pb, Cu, and Cd occur mainly as sulfide minerals in the coal although they are correlated with SiO_2 and Al_2O_3 and not with sulfur. These correlations suggest that the source of the Zn, Pd, Cu and Cd was the same as for SiO_2 and Al_2O_3. The elements and minerals that correlate positively (95% confidence level) with vitrinite (dominantly telocollinite), fusinite, semifusinite, and macrinite (group I of table 1) probably accumulated contemporaneously with the peat. Possible sources include 1) mineral-matter incorporated by plants, 2) detrital minerals, and/or 3) dissolved elements incorporated in the peat by ion exchange or complexing. Element concentrations resulting from mixed sources such as deposition of detrital minerals, sorption of dissolved ionic species, and mineral matter of plant origin probably would not lead to a strongly interrelated assemblage; therefore, the statistical correlations imply a single dominant source. Most vitrinite is generally considered to be derived from plant cell walls. Most fusinite and semifusinite are generally believed to be coalified products of partially carbonized and/or oxidized cell-wall material. The carbonization process produces a loss of organic matter and concentrates residual mineral matter. Therefore, plant inorganic matter in conjunction with ion exchange is considered by the writers to be the dominant source of elements in group I (table 1), and detrital mineral contributions are considered negligible. This interpretation is consistent with 1) the moderate ash content (8-15 wt.%), 2) the positive correlations of those elements of group I (table 1) with coal macerals derived from plant cell walls (vitrinite, fusinite, semifusinite), and 3) positive correlation of the group I elements and macerals with ash content.

The statistical relationship of the elements, minerals, and macerals of group II may also be related to genetic processes which were controlled by pH, ionic strength of interstitial waters, and microbial activity. Irwin and Curtis (1977) summarized CO_2 generating processes in marine sediments including aerobic oxidation, sulfate reduction, fermentation, and thermal generation (fig. 1). Presumably, the reactions in marine sediments can be extrapolated to peat and coal if the factors of pH and ionic strength are considered. Preliminary carbon isotope data on calcite fillings in cleats of the Upper Freeport coal bed indicate that the isotopically heavy carbonate in the cleat was derived from fermentation reactions. Gould and Smith (1979) suggest that siderite in some Permian coal beds in Australia was fixed by fermentation reactions, whereas the isotopically heavy calcite in fractures near the top of the beds was the result of isotope exchange between CO_2 and CH_4.

Table 1. Statistical groupings of coal macerals, elements and minerals (based on as many as 75 samples, Upper Freeport coal bed, Homer City, Pennsylvania study area).

	Group I	Group II
Macerals	Vitrinite Fusinite Semifusinite Macrinite	Sporinite Micrinite
Elements oxides and sulfur forms	SiO_2 Eu Sb Al_2O_3 F Sc MgO Ga Se Na_2O Gd Sm K_2O Hf Sn TiO_2 La Tb B Li U Ba Lu V Be Nb Y Cd Nd Yb Ce Ni Zn Cr Pb Zr Cs Rb Cu	CaO Pyritic sulfur Total sulfur Organic sulfur Fe_2O_3 As Ge Hg
Minerals	Quartz Illite Kaolinite	Pyrite Calcite

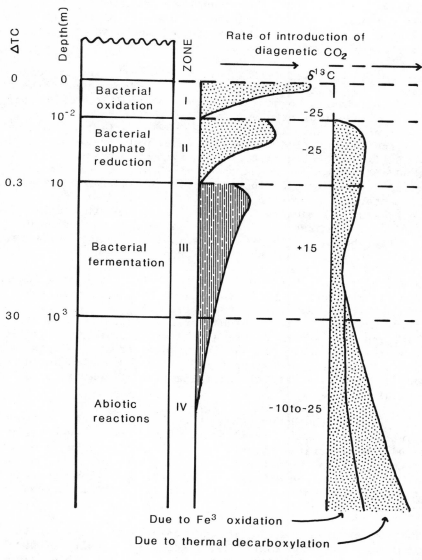

Figure 1. Production of diagenetic CO_2 within different diagenetic zones. (after Irwin and others 1977)

The pyrite and marcasite in the Upper Freeport coal bed apparently were fixed by sulfate-reducing bacteria with the mineral phase being controlled by the pH of the system. Arsenic was determined to be associated with pyrite on the basis of electron microprobe analyses (Minkin and others, 1979).

Calcium can be fixed during peat formation by ion exchange and/or as calcium salts of organic acids. The subsequent liberation of CO_2 and organically bound calcium by microbial reactions and coalification reactions may have resulted in the formation of calcite in macerals and cleats. The buffering effect of calcium-carbonate species supplied to the ancestral peat partly controlled the pH. Sulfate-reducing bacterial activity is greatest at neutral pH and minimal or nonexistent at pH 4.0 or less (Bass Becking and others, 1960). Therefore, pH values near neutral favor 1) bacterial generation of sulfide species if sulfate ions are present, 2) bacterial degradation of peat that could result in the concentration of mineral matter, 3) retention of mineral matter because of decreasing solubility as pH increases, and 4) fixation of iron and sulfur as iron sulfides.

In contrast, low pH (<4.0) would favor formation of a low-ash (<8 wt.%), low-sulfur (<1 wt.%) coal. Partial neutralization of waters of the ancestral peat of the Upper Freeport coal bed is indirectly indicated by the presence of 1) calcite in coal macerals and cleats and 2) mixed carbonate and clastic sediments directly underlying the coal bed. The samples studied from the Upper Freeport coal bed are medium ash (8-15 wt.%) and medium sulfur (1-3 wt.%), indicating partial neutralization of peat and coal interstitial waters.

The pyrite in the Upper Freeport coal bed commonly is concentrated in the upper and/or basal zones of the bed. Genetically, pyrite and the associated trace elements As, Ge, and Hg, are statistically unrelated to the amount of the coal ash. Although pH conditions may have been suitable for pyrite formation, pyrite concentrations were controlled by the availability of ferrous iron and appropriate sulfur species.

The more massive forms of pyrite commonly associated with fusain may have been controlled by 1) the high porosity of fusain and 2) microgeochemical environments which were favorable for bacterial action and sulfide generation. Most fusain generally is considered to be the product of fires; thus, alkali and alkaline earth metal ions should be enriched in charcoal layers. Hydrolysis of the enriched alkalies could create neutral to slightly alkaline microgeochemical environments favorable for pyrite formation.

COMPARISON OF THE UPPER FREEPORT TO OTHER COAL BEDS

A study by Williams and Keith (1963) has shown that the sulfur content of the Upper Freeport coal bed is less than the sulfur content in the Lower Kittanning coal bed where the lower Kittanning is overlain by marine sediments (table 2). The Upper Freeport coal bed was interpreted by Williams and Keith to have formed in a fresh-water environment, which is consistent with our study, and the difference in sulfur content was ascribed to marine versus nonmarine influences. However, a comparison of the sulfur content of these two coals with two other coals that are considered to be of fresh water origin (i.e. Pocahontas No. 3, $\bar{X}= 0.6\%$, and the Vermillion Creek coal, $S = 6\%$) readily shows a greater range in the sulfur content of coals of fresh-water origin than between the Upper Freeport coal and the marine-influenced Lower Kittanning coal.

A striking relationship exists between sulfur in coals and calcium-carbonate content of associated rocks. The low-ash, low-sulfur coals of the Pocahontas basin of West Virginia are associated with noncalcareous rocks. In general, the rocks of the Allegheny, Conemaugh, and Monongahela Formations of northern West Virginia, eastern Ohio, and southwestern Pennsylvania tend to be calcareous, and the ash and sulfur content of the associated coal beds are relatively high. The high-sulfur Vermillion Creek coal bed of Wyoming contains interbedded marls. The absence of dissolved calcium carbonate must have caused the ancestral peat swamps of the Pocahontas basin coals to have been highly acidic (pH<4.5). Highly acidic conditions would have 1) enhanced leaching of mineral matter causing ash content to be low, 2) inhibited the activity of sulfate-reducing bacteria, and 3) enhanced the reduction of ferric iron to ferrous iron.

As availability of dissolved calcium carbonate increases as in the Upper Freeport or Vermillion Creek coal beds: 1) leaching of mineral matter would be less intense; 2) sulfur-reducing and organic-decomposing bacterial activity would increase; and 3) fixing of iron in sulfide or carbonate phases would increase. Thus, the ash and sulfur contents of the coal would increase.

The macerals of the exinite group (sporinite, cutinite, and resinite) generally are considered to be the most resistant to decay. Therefore, an enrichment of these macerals relative to vitrinite should take place when microbial degradation of organic matter is increased, and as a result, the exinite macerals should correlate with sulfur and calcium. These relationships have been observed in the current study of the Upper Freeport coal bed.

Table 2. Comparison of total sulfur content for three coals of nonmarine origin (i.e. Upper Freeport, Pocahontas No. 3 and Vermillion Creek) and one having marine influence (Lower Kittanning).

SAMPLING UNITS		N	$\overline{X}$	s
* Lower Kittanning coal	- total area	255	2.64	1.16
	- marine area	71	3.15	1.17
	- transitional area	48	2.45	1.03
	- continental area	125	1.73	0.81
* Upper Freeport coal	- total area	183	2.09	0.85
	- area III	96	2.21	0.88
	- area II	32	1.89	0.85
	- area I	45	1.90	0.66
** Pocahontas No. 3 (no. sample = 43) coal			0.67	0.31
*** Vermillion Creek (Wyoming) coal			4-8% sulfur	

N = number of mines sampled (average 4 sulfur analyses per mine)

$\overline{X}$ = arithmetic mean value of sulfur in %

s = standard deviation

area III   Lower Kittanning marine area

area II   -   -   transitional area

area I   -   -   continental area

* from Williams and Keith (1963)
** National Coal Resources Data System (U.S. Geological Survey)
*** Hank Roehler, personal commun.

SOME EFFECTS OF COALIFICATION ON MINERAL MATTER CONCENTRATIONS

In the transformations of plant material to peat and of peat to coal, all the processes that operate result in a net loss of organic matter. Therefore, at any stage of coal formation, loss of organic matter has the potential effect of concentrating the mineral matter. Furthermore, from experimental work and theoretical considerations, it can be shown (Cecil and others, unpub. data) that the transformation of well-preserved peat (from the Okefenokee Swamp of Georgia) into a medium volatile bituminous coal equivalent in rank to the Upper Freeport coal involves a minimum compaction ratio of 10:1 and a 50% weight loss of the original peat organic matter. Therefore, a 10 cm thickness of peat having 10 wt. % ash (dry basis) will become 1 cm of bituminous coal with an ash content of 18 wt.%, assuming that mineral matter is not added or lost during coalification. A peat having 25 wt.% ash would be transformed into a bituminous coal with 40 wt.% ash. Changes of these magnitudes have profound implications on the interpretations of the sedimentary history and depositional environments of coal beds. For example, a 6 inch (15 cm) thick parting or bone coal layer containing 50 wt. % ash (20% by volume) may represent the coalification product of 60 inches (150 cm) of peat whose ash content (dry basis) would have been approximately 33 wt. % (13% by volume). This hypothetical parting or bone-coal layer would not be visibly recognizable as such in the peat stage of coal formation. This probably explains why laterally extensive parting precursors are seldom, if ever, recognized in modern peat-forming environments. Such parting precursors have the physical appearance of "normal" peat in that they are non-banded and mineral matter is dispersed. However, during compaction, coalification, and diagenesis, fissility and/or banding develops (often confused with sedimentary layering) and the mineral matter will appear in discrete bands. Banding is generally not pronounced even at the lignite stage of coalification. Thus, the commonly held opinion that all mineral-rich bands are detrital in origin does not have a sound physical or theoretical basis.

SUMMARY

From the study of samples from the Upper Freeport coal bed and comparisons made with other coals, a hypothetical model (fig. 2) has been developed that can be used to help in understanding ash and sulfur variations in coal. During peat formation, specific chemical conditions are requisite to ash concentration and sulfur fixation.

Coals containing small amounts of ash and sulfur probably are derived from peats which formed in highly acidic anaerobic environments. Highly acidic conditions would 1) leach those elements

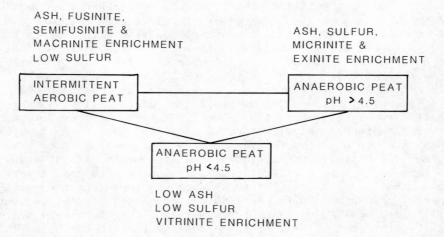

Figure 2. Hypothetical model illustrating possible controls on mineral matter and maceral variation in coal

which tend to be soluble under acid conditions and 2) inhibit sulfate-reducing and fermentation microbial activity and peat degradation.

Carbonization and/or oxidation taking place in an acid environment with intermittent aerobic conditions would tend to concentrate mineral matter and correspondingly increase certain inertinite macerals. Sulfur fixation by anaerobic bacterial activity would be inhibited. Therefore, acid-aerobic conditions would result in a coal that would have a low sulfur content; the ash and inertinite content would be relatively enriched.

Anaerobic conditions at or near neutral pH are optimum for sulfur fixation and mineral-matter retention. As pH increases (4.5-7.5) the following would probably occur: 1) minimal leaching of elements that tend to be less soluble at neutral pH; 2) increased sulfate-reducing bacterial activity; and 3) increased degradation of organic matter resulting in a relative increase in concentration of the more resistant exinite and micrinite pre-macerals.

Maximum bacterial activity takes place where pH conditions are neutral, or nearly so; such conditions favor sulfate reduction and peat degradation. This pH model is consistent with Schopf's (1952) suggestion "that the sulfur content of a coal may give an indirect indication of the extent of anaerobic decay". The commonly observed relationship between marine roof rock and high sulfur coal is a special case of the pH model. Transgression of marine or brackish waters over a fresh-water peat-forming environment would cause neutralization of interstitial waters at least near the top of the peat body.

Peat which may become a commercial quality coal cannot form under continuous marine or brackish-water influence because of ash and sulfur enrichment. Therefore, mineable coal beds in the United States were derived from peat that formed in fresh-water environments where there was little or no marine influence during the time of peat formation.

Major factors to be considered in understanding the variations of ash and sulfur in coal currently being mined are: 1) the availability of dissolved calcium carbonate that controlled pH in the ancestral peat environment; 2) ionic strength of interstitial waters including the availability of iron and sulfate ions; and 3) plant paleoecology.

References

Baas Becking, L. G. M., Kaplan, I. R., and Moore, D., 1960, Limits of the natural environment in terms of pH and oxidation-reduction potentials: Journal of Geology, v. 68, 3, p. 243-284.

Cecil, C. B., Stanton, R. W., Allshouse, S. D., and Finkelman, R. B., 1978, Geologic controls on mineral matter in the Upper Freeport coal bed: Proceedings: Symposium on coal cleaning to achieve energy and environmental goals, p. 110-125.

Cecil, C. B., Stanton, R. W., Allshouse, S. D., Finkelman, R. B., and Greenland, L. P., 1979a, Geologic controls on elements concentration in the Upper Freeport coal (abs): American Chemical Society, Preprints of papers, v. 24, no. 1, p. 230-235.

Cecil, C. B., Renton, J. J., Stanton, R. W., and Finkelman, R. B., 1979b, Mineral matter in coals of the central Appalachian Basin (abs): IX-International Carboniferous Congress Abstracts of Papers, p. 32.

Cecil, C. B., Stanton, R. W, Dulong, F. T., and Renton J. J., 1980a, Geologic controls on mineral-matter content of Coal in the central Appalachian Basin: American Association of Petroleum Geologist Annual Meeting, Abstracts of papers, p. 38-39.

Cecil, C. B., Stanton, R. W., and Dulong, F. T., 1980b, Geology of contaminants in coal: Phase I Report of Investigations (submitted to the Environmental Protection Agency for publication, October, 1980).

Garrels, R. M., and Christ, C. L., 1965, Solutions, minerals, and equilibria: New York, N.Y., Harper and Row, 450 p.

Gould, K. W., and Smith, J. W., 1979, The Genesis and isotopic composition of carbonates associated with some Permian Australian coals: Chemical Geology, v. 24, p. 137-150.

Irwin, H. and Curtis, C., 1977, Isotopic evidence for source of diagenetic carbonates formed during burial of organic-rich sediments: Nature, v. 269, p.209-213.

Minkin, J. A., Finkelman, R. B., Thompson, C. L., Cecil, C. B., Stanton, R. W., and Chao, E. C. T., 1979, Arsenic-bearing pyrite in the Upper Freeport coal, Indiana Co., PA (abs.): IX-International Carboniferous Congress, Abstracts of papers, p. 140-141.

Schopf, J. M., 1952, Was decay important in origin of coal: Journal of Sedimentary Petrology, v. 22, no. 2, p. 61-69.

Spackman, W. H., Cohen, A. D., Given, P. H., and Casagrande, D. J., 1976, The comparative study of the Okefenokee Swamp and the EvergladesMangrove swamp-marsh complex of southern Florida: Coal Research Section, The Pennsylvania State University, p. 22.

Swain, F. M., 1973, Marsh gas from the Atlantic Coastal Plain, United States: Advances in Organic Geochemistry, p. 673-687.

Williams, E. G. and Keith, M. L., 1963, Relationship between sulfur in coals and the occurrance of marine roof beds: Economic Geology, v. 58, p. 720-729.

STRUCTURE ANALYSIS OF COALS BY RESOLUTION ENHANCED SOLID STATE ^{13}C NMR SPECTROSCOPY[1]

Edward W. Hagaman and M. C. Woody

Chemistry Division
Oak Ridge National Laboratory
Oak Ridge, Tennessee 37830

Solid state ^{13}C nmr spectroscopy employing the experimental techniques of dipolar decoupling,[2,3] magic angle spinning[4,5] and ^{1}H-^{13}C cross polarization[3] (CP/MAS-^{13}C nmr spectroscopy) yields spectra which approach solution ^{13}C-FT-nmr spectra in resolution and sensitivity. Resonance linewidths of 10-50 Hz in discrete organic substances are typical. Those in amorphous solids and glassy polymers generally are degraded to 50-150 Hz (2-6 ppm at 2.35T) by chemical shift dispersion and/or residual dipolar broadening.[5-10] In homogeneous systems, this level of resolution permits the measurement of isotropic chemical shift and relaxation time parameters in solids.

The potential of this spectroscopy in applications to amorphous, heterogeneous materials, e.g., coals[11-15] and oil shales,[16,17] has eluded full realization since the extreme chemical complexity of these substances results in spectra which consist of a continuum of resonances within the separate sp^2- and sp^3-carbon resonance manifolds. CP/MAS-^{13}C nmr spectroscopy uniquely provides a direct measure of the "aromatic" and "aliphatic" carbon content of these materials and constitutes the major application of this method in the study of these materials. Zilm, et al. have reported significant changes in the broadened sp^3-carbon resonance envelope of heavy coal liquids in comparison with the CP/MAS-^{13}C-spectra of the coals from which they were derived, and interpreted the differences in context of the chemical modification of the coals upon liquefaction.[14]

Resonance bandshapes of coal spectra are rarely symmetrical envelopes devoid of structure. We have been interested in the decomposition of these broad envelopes into components which reveal a more detailed structural description of the organic matter in coal. Using a combination of nmr and chemical resolution enhancement techniques, the sp^2- and sp^3-carbon resonance manifolds each can be partitioned into fractions representing protonated and non-protonated carbons, and these in turn can be divided again using chemical shift criteria into subgroups reflecting the presence or absence of directly bonded oxygen substituents. Thus, within the approximation that the organic matter of coal can be considered to be composed of carbon, hydrogen and oxygen, the area under each hybridization resonance envelope may be divided into at least three fractions reflecting the immediate bonding environment of carbon. The techniques which prove of value in this spectral decomposition, i.e., convolution difference methods, partial polarization spectroscopy, spectral subtraction and ^{13}C-chemical enrichment, are discussed in the following paragraphs and employed in the analysis of the products from a reductive alkylation of a typical bituminous coal.

CONVOLUTION DIFFERENCE SPECTROSCOPY

Spectrum A and A' (Figure 1) are typical CP/MAS-^{13}C-nmr spectra of a bituminous (Illinois No. 6 vitrinite) and subbituminous (Wyodak) coal, respectively. Processing the free induction decay (FID) of spectrum A(A') by standard convolution difference (CD) techniques that discriminate against components of the FID with fast T_2^* results in spectrum B(B').[18-20] The CD spectrum reveals that the carbon magnetization of these coals possesses an inherent relaxation time dispersion which can be used to enhance the asymmetric lineshape in both the sp^2- and sp^3-carbon resonance envelopes of spectrum A(A'). Spectrum B(B') is obtained from the FID of A(A') after multiplication of the latter by $e^{-\tau/T_A} - Ke^{-\tau/T_B}$. The CD variables T_A, T_B and K are selected empirically to optimize the compromise between enhanced resolution and reduced signal-to-noise (see figure legend). This mathematical operation is equivalent to the subtraction of a broad from a narrow lorentzian line and effectively removes resonances from the spectrum for which $T_2^* << T_B$. Spectrum B(B') retains ca. 25% of the total resonance area measured in A(A'). Variation of the CD parameters to generate spectra which contain 25-100% of the area of A(A') and subsequent integration reveal that f(a) (the fraction of total carbon which is sp^2-hybridized) is nearly independent of the extent of line narrowing (±5%). Structural features emerge from the asymmetric hybridization envelopes by gradual sharpening across the bandwidth, not by preferential signal attenuation of specific spectral regions. This behavior indicates that the relaxation time dispersion is largely independent of the statistical distribution of the carbon types within the sp^2- and sp^3-carbon resonance envelopes. The relaxation time dispersion may result from a random distribution of paramagnetic centers in the sample and/or to regions of the coal matrix which differ in mobility. In either case, the CD spectra appear to be an undistorted representation of the whole coal sample.

Convolution difference spectra do not yield a dramatic increase in resolution within the resonance envelopes but effectively remove

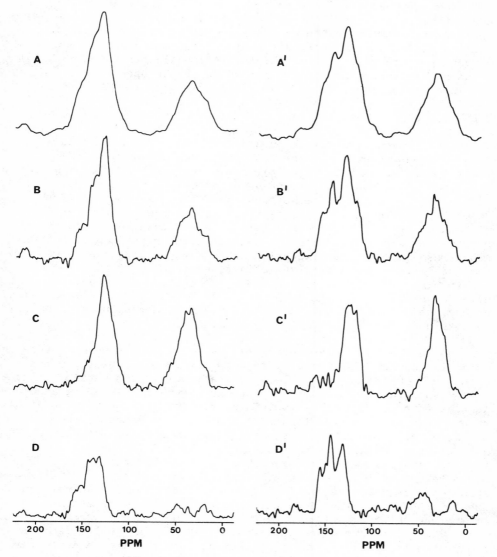

Fig. 1. CP/MAS-^{13}C nmr spectra of bituminous (A-D) and subbituminous coal (A'-D'). Spectrum A(A') was generated with T_{CP} = 1.0 ms and 2.0 s recycle delay; ca. 15,000 transients. Spectrum B(B') was obtained by multiplication of the FID of A(A') by $(e^{-t/.0064} - 0.80\ e^{-t/.0032})$. Spectrum C(C') was generated with T_{CP} = 20 μs and 2.0 s recycle delay; ca. 90,000 transients. The FID was processed as for B(B'). C(C') is shown with an arbitrary vertical scale expansion. Spectrum D(D') is the difference spectrum B - C (B' - C').

the broad resonance tails from the CP/MAS-^{13}C-spectra and allow evaluation of the carbon resonances at the band extrema. The 60-90 ppm spectral region in the coal spectra of Figure 1 is baseline, indicating that the concentration of sp^3-carbons directly bound to oxygen in these coals is negligible. Aliphatic ether and carbonyl bands are prominent and well resolved in the CD spectra of lignite and lower rank coals.

PARTIAL POLARIZATION SPECTRA

Carbon magnetization accrues in the CP experiment by spin polarization transfer from protons to carbon atoms. The ^{1}H-^{13}C cross polarization rate, (T_{CH}^{-1}), maximized at the Hartmann-Hann condition,[21] $\gamma_C H_{1C} = \gamma_H H_{1H}$, is governed by the strength of local static ^{1}H-^{13}C dipolar interactions and follows an inverse sixth power dependence on ^{1}H-^{13}C internuclear distance. Carbon atoms with directly bonded protons polarize faster than non-protonated carbons which must polarize via nearest neighbor protons. Methyl group resonances are an exception to this generalization. Owing to fast internal rotation, the effective strength of the ^{1}H-^{13}C dipolar interaction is reduced and T_{CH} lengthened. Schaefer, et al. have demonstrated that the dependence of T_{CH} on internuclear distance can be used as an assignment criterion and as a resolution enhancement technique.[5]

We have not performed quantitative relaxation time measurements on coals but have estimated average T_{CH} of 50 and 300 µs for protonated and non-protonated carbons, respectively, from partial polarization spectra of the coals recorded in Figure 1. Assuming an exponential form for T_{CH}, these values predict a >80% suppression of resonance area of non-protonated carbons with respect to the normalized protonated carbon resonance area, in spectra recorded with a 20 µs cross polarization contact time. Partial polarization spectra obtained with this CP contact time and processed with CD parameters used for spectrum B(B') are shown in Figure 1 C(C'). These spectra reveal a nearly complete elimination of the resonance area in the spectral region below 148 ppm and a strong attenuation of the resonance area in the central portion of the aromatic envelope. Hence, the resonance area below 148 ppm in B(B') arises from non-protonated carbons, which on chemical shift criteria must bear (predominantly) oxygen substituents. The absence of aliphatic oxygenated carbon moieties indicates that in these coals, the organic oxygen is contained predominantly in phenolic and diaryl ether moieties.

The partial polarization spectra reveal a narrowing of the sp^3-carbon manifold resulting from preferential signal loss from the low and high field extrema of this band, particularly evident in the subbituminous coal (spectrum C').

STRUCTURE ANALYSIS OF COALS

The carbon resonances with long T_{CH} eliminated in C(C') are displayed in D(D'), the difference spectrum obtained by subtraction of C(C') from B(B'). The criterion used in this subtraction is baseline nulling of the area in the aromatic envelope in the spectral region above 115 ppm. Chemical shift considerations of a broad spectrum of aromatic model compounds indicate that this spectral region will contain only aromatic methine carbons. The low intensity bands centered at 47 and 20 ppm in D(D') are due to quaternary (and conceivably R_3C-X, X=N, S) and methyl carbon resonances, respectively. The prominent aromatic band centered at 127-129 ppm in B(B') is a composite of overlapping non-protonated and methine carbon resonances.

Reconstruction of spectrum B by summation of C and D indicates that 70% of the sp^2-carbon resonance area is due to non-protonated carbon resonances. Using chemical shift criteria [phenols, aromatic ethers; 160-145 ppm; alkyl/aryl substituted aromatic carbons (ring junction and sidechain attachment sites); 148-125 ppm], the resonance area in D may be roughly partitioned into components arising from oxygenated and non-oxygenated species in a 1:6 ratio. Hence, the total sp^2-carbon resonance area may be partitioned into components due to carbon substituted by oxygen, carbon and hydrogen in a 1:6:3 ratio, respectively. This ratio statistically corresponds to the distribution of these three aromatic carbon types in a tetra-alkyl substituted naphthol or naphthyl ether.

In the sp^3-carbon region of the spectrum, partial polarization spectra identify non-protonated and methyl carbon resonance contributions to the total resonance area. Utilizing the gross carbon chemical shift dependence on electronegative substituents, each of these carbon types may be further partitioned into subgroups, depending upon the presence or absence of direct oxygen substitution. The remaining and dominant fraction of the sp^3-carbon resonance area in these coals arises from methine and methylene carbons whose resonances extend throughout the 20-50 ppm interval.

The decomposition of the sp^2- and sp^3-carbon resonance envelopes of solid state coal spectra into fractions reflecting immediate carbon bonding environment is summarized schematically in Figure 2. The neglect of organic nitrogen and sulfur moieties and the partially overlapping resonance zones of non-protonated C_{sp^2}-C and C_{sp^2}-O species are current limitations restricting quantitative characterization of coals according to this scheme. The methods nonetheless provide a more detailed description of major carbon types within the hybridization envelopes which can be monitored to map out the principal reaction courses of chemically modified coals.

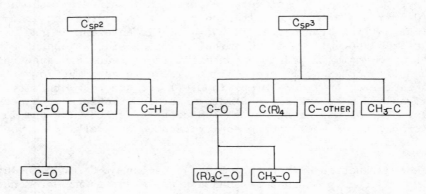

Fig. 2. Decomposition of the C_{sp^2}- and C_{sp^3}-resonance envelopes of CP/MAS-^{13}C nmr spectra of coals.

SPECTRAL ANALYSIS OF THE COAL PRODUCTS FROM THE Na-K REDUCTION OF ILLINOIS NO. 6 VITRINITE IN MIXED GLYMES

The repeated reduction of coal by the action of sodium-potassium alloy in glyme-tryglyme followed quenching with a proton donor or with methyl iodide is an efficient low temperature method for degrading coal by cleavage of aliphatic and aromatic-aliphatic carbon-carbon bonds.[16,17] We have examined the chemically altered products of Illinois No. 6 vitrinite obtained by this reduction scheme utilizing both natural abundance and 20%-^{13}C-enriched methyl iodide as quenching reagent. Multiple reductions were performed by subjecting the total isolated solid product (>95% of the starting material) from one reduction pass to identical reaction conditions following spectral examination.

The CP/MAS-^{13}C-nmr spectrum depicted in Figure 3(A) is of the initial reduction product obtained with a natural abundance methyl iodide quench. The spectrum was acquired using a long cross polarization time (1.0 ms) for which all carbon resonances show maximum intensity. The general character of the spectrum is similar to that of the chemically unaltered coal but with a broadened sp^3-carbon resonance envelope. The changes in this region of the spectrum are apparent in Figure 3(B), the difference spectrum obtained by subtracting the spectrum of the starting material from spectrum 3(A) (arbitrary scale expansion). The criterion used for the subtraction is nulling the aromatic resonance envelope. Birch-Hückel reduction products (di- and tetrahydro aromatic moieties), not expected to result in a linear attenuation in the area across the full width of

STRUCTURE ANALYSIS OF COALS

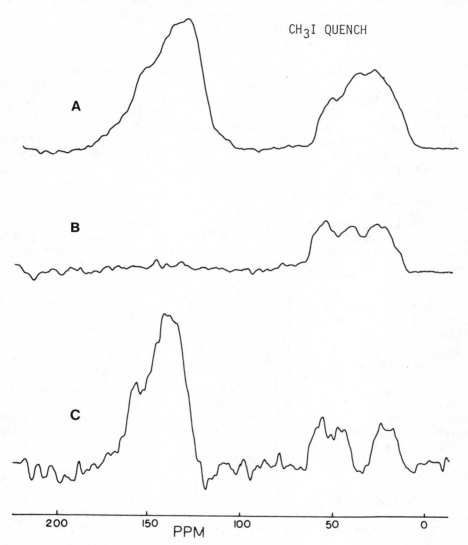

Fig. 3. CP/MAS-^{13}C-nmr spectra of the Na-K reduction product of Illinois No. 6 vitrinite obtained by quenching the reaction with 1.1% ^{13}CH$_3$I. (A) Full spectrum, 1.0 ms CP contact time; (B) Difference spectrum obtained by subtraction of A from the spectrum of the unreacted coal; (C) Partial polarization difference spectrum.

the sp^2-carbon resonance envelope, are not formed in this reaction
in detectable concentrations. Since the generation of 30 dihydro-
aromatic species per hundred individual aromatic rings will cause
a 10% reduction in the total sp^2-carbon resonance area, i.e., a
change estimated to be the minimum significant detectable area difference, cancellation of the aromatic area in these spectra provides an upper limit estimate of the concentration of Birch-Hückel
reduction products.

Two of the three "bands" in the difference spectrum can be
assigned to O-CH$_3$ and C-CH$_3$ resonances on chemical shift basis
alone. Details of the composition of the central band are revealed
in the comparison of spectrum 3B and 3C. The latter is the partial
polarization difference spectrum (equivalent to spectrum D in
Figure 1), which shows that the low field portion of the central
band in 3C is due to quaternary carbon resonances and the balance
to arise from protonated carbon resonances.

Spectra A and B in Figure 4, analogous to spectra A and B in
Figure 3, are of the vitrinite derivative following the second
reduction-quench sequence with 20% ^{13}CH$_3$I. The central aliphatic
resonance "band" in 3B is absent in 4B, indicating that this band
does not contain carbon resonance contributions arising from the
quenching agent. Thus, the methyl iodide consumed in the reaction
formed only O- and C-methyl groups.

Figure 5 is a plot of the O-CH$_3$/C$_{sp^2}$ area ratio vs. reduction
step number. The areas are taken from the spectra of the coal derivatives treated with 20% methyl iodide and acquired with 1.0 ms
CP contact time. The ratios are calculated under the assumption
that the area of the sp^2-carbon resonance envelope remains constant
throughout the six sequential reductions (vide supra). The plot
reveals that by the third reduction, the OCH$_3$/C$_{sp^2}$ ratio has attained its maximum value and remains constant through the remaining
reduction-quench steps. The concentration of aromatic methyl ether
moieties indicated is 9 per 100 aromatic carbon atoms. This value
is in good agreement with the number of oxygen substituted aromatic
carbon centers determined by integration of partial polarization
difference spectra of the chemically untreated coal given above.
The value likely reflects total aromatic oxygen in the coal obtained
by alkylation of the phenolic sites present in the coal and those
generated in the reduction milieu by opening of aromatic ethers.
This point is under further investigation.

The O-methyl resonance band in spectra of the coal derivatives
quenched with 20% ^{13}C-methyl iodide is partially resolved into two
components centered at 55 and 60 ppm. These shift values are characteristic resonances of aromatic methyl ethers in unhindered (one
or both ortho-positions unsubstituted) and hindered (ortho-ortho'-
disubstitution) environments, respectively. The approximate 1:1

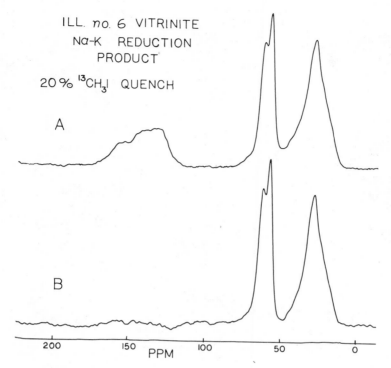

Fig. 4. CP/MAS-^{13}C-nmr spectra of the Na-K reduction product of Illinois no. 6 vitrinite obtained by quenching the reaction with 20% ^{13}CH$_3$I. (A) Full spectrum, 1.0 ms CP contact time; (B) Difference spectrum obtained by subtraction of A from the spectrum of the chemically unaltered coal.

ratio of these signals shown in Figure 4 remains constant throughout the reduction sequence and may reflect the distribution of these species in the coal.

In contrast to the static concentration of O-methyl groups introduced into the coal structure, the number of C-methyl groups — comparable to the O-methyl concentration in the first reduction — continues to increase in each of the six reduction steps. The number of highly substituted sp^3-carbon centers generated in the reaction, i.e., the central "band" in spectrum B, Figure 3, increases with the number of introduced C-methyl groups.

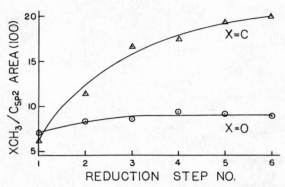

Fig. 5. Plot of the OCH$_3$/aromatic carbon and CCH$_3$/aromatic carbon area ratios vs. reduction step number of Illinois no. 6 vitrinite.

SUMMARY

The application of experimental nmr resolution-enhancement methods to CP/MAS-^{13}C-nmr spectroscopy yields spectra of coals and coal-derived solids which contain structural information reflecting the immediate bonding environment of carbon centers within the hybridization envelopes. These techniques in conjunction with the chemical alteration of coals using natural abundance and ^{13}C-enriched reagents permit the chemical modification of coals to be mapped in the solid state.

REFERENCES

1. Research sponsored by the Division of Chemical Sciences, Office of Basic Energy Sciences, U. S. Department of Energy, under contract W-7405-eng-26 with the Union Carbide Corporation.
2. L. R. Sarles and R. M. Cotts, Double Nuclear Magnetic Resonance and the Dipole Interaction in Solids, Phys. Rev. 111:853 (1958).
3. A. Pines, M. G. Gibby, and J. S. Waugh, Proton-enhanced NMR of Dilute Spins in Solids, J. Chem. Phys. 59:569 (1973).

4. E. R. Andrew, The Narrowing of NMR Spectra of Solids by High-Speed Specimen Rotation and the Resolution of Chemical Shifts and Spin Multiplet Structures for Solids, in "Progress in Nuclear Magnetic Resonance Spectroscopy," J. W. Ensley, J. Feeney, and L. H. Sutcliffe, eds., Pergamon Press, Oxford (1972).
5. J. Schaefer, E. O. Stejskal and R. Buchdahl, High-Resolution Carbon-13 Nuclear Magnetic Resonance Study of Some Solid Glassy Polymers, Macromolecules 8:291 (1975).
6. J. Schaefer and E. O. Stejskal, Carbon-13 Nuclear Magnetic Resonance of Polymers Spinning at the Magic Angle, J. Am. Chem. Soc. 98:1031 (1976).
7. J. Schaefer, E. O. Stejskal, and R. Buchdahl, Magic-Angle ^{13}C NMR Analysis of Motion in Solid Glassy Polymers, Macromolecules 10:384 (1977).
8. C. A. Fyfe, J. R. Lyerla, W. Volksen, and C. S. Yannoni, High-Resolution Carbon-13 Nuclear Magnetic Resonance Studies of Polymers in the Solid State. Aromatic Polyesters, Macromolecules 12:757 (1979).
9. W. L. Earl and D. L. Vanderhart, Observations in Solid Polyethylenes by Carbon-13 Nuclear Magnetic Resonance with Magic Angle Sample Spinning, Macromolecules 12:762 (1979).
10. W. S. Veeman, E. M. Menger, W. Ritchey, and E. deBoer, High-Resolution Carbon-13 Nuclear Magnetic Resonance of Solid Poly(oxymethylene), Macromolecules 12:924 (1979).
11. D. L. Vanderhart and H. L. Retcofsky, Estimation of Coal Aromaticities by Proton-decoupled Carbon-13 Magnetic Resonance Spectra of Whole Coals, Fuel 55:202 (1976).
12. V. J. Bartuska, G. E. Maciel, J. Schaefer, and E. O. Stejskal, Prospects for Carbon-13 Nuclear Magnetic Resonance Analysis of Solid Fossil Fuel Materials, Fuel 56:354 (1977).
13. H. L. Retcofsky and D. L. Vanderhart, ^{13}C-^{1}H Cross-polarization Nuclear Magnetic Resonance Spectra of Macerals from Coal, Fuel 57:421 (1978).
14. K. W. Zilm, R. J. Pugmire, D. M. Grant, R. E. Wood, and W. H. Wiser, A Comparison of the Carbon-13 NMR Spectra of Solid Coals and their Liquids Obtained by Catalytic Hydrogenation, Fuel 58:11 (1979).
15. G. E. Maciel, V. J. Bartuska, and F. P. Miknis, Characterization of Organic Material in Coal by Proton-decoupled ^{13}C Nuclear Magnetic Resonance with Magic Angle Spinning, Fuel 58:391 (1979).
16. H. A. Resing, A. N. Garroway, and R. N. Hazlett, Determination of Aromatic Hydrocarbon Fraction in Oil Shale by ^{13}C N.M.R. with Magic-Angle Spinning, Fuel 57:450 (1978).
17. F. P. Miknis, G. E. Maciel, and V. J. Bartuska, Cross Polarization Magic-Angle Spinning ^{13}C N.M.R. Spectra of Oil Shales, Org. Geochem. 1:169 (1979).
18. I. D. Campbell, C. M. Dobson, R. J. P. Williams, and A. V. Xavier, Resolution Enhancement of Protein PMR Spectra Using

the Difference Between a Broadened and a Normal Spectrum, J. Magn. Res. 11:172 (1973).
19. R. R. Ernst, Sensitivity Enhancement in Magnetic Resonance, in "Advances in Magnetic Resonance," J. S. Waugh, ed., Academic Press, New York (1966).
20. Cf. C. H. A. Seiter, G. W. Feigenson, S. I. Chan, and M.-C. Hsu, Delayed Fourier Transform Proton Magnetic Resonance Spectroscopy, J. Am. Chem. Soc. 94:2535 (1972).
21. S. R. Hartmann and E. L. Hahn, Nuclear Double Resonance in the Rotating Frame, Phys. Rev. 128:2042 (1962).
22. K. Niemann and H.-P. Hombach, Studies in the Chemical Characterization of Coal: Reduction Via Solvated Electrons, Fuel 58:853 (1979).
23. C. J. Collins, H.-P. Hombach, B. Maxwell, M. C. Woody, and B. M. Benjamin, Carbon-Carbon Cleavage during Birch-Hückel-Type Reductions, J. Am. Chem. Soc. 102:851 (1980).

Applications of Nuclear Magnetic Resonance to Oil Shale

Evaluation and Processing

>Francis P. Miknis
>U.S. Department of Energy
>Laramie Energy Technology Center
>Laramie, WY 82071
>
>and
>
>Gary E. Maciel
>Department of Chemistry
>Colorado State University
>Fort Collins, CO 80523

INTRODUCTION

Nuclear Magnetic Resonance (NMR) is playing an increasing role in the characterization of the organic constituents of fossil fuels. The NMR techniques that currently are being applied to fossil fuel characterization utilize both the ^1H and ^{13}C nuclides and can be conveniently divided into those that apply to liquid and those that apply to solid samples. For coals and oil shales, only a small fraction of the organic material can be extracted under the mild conditions that would be expected to retain the primary structural integrity of the organic material. Therefore NMR techniques usually applied to liquids are not applicable to these solid materials; instead NMR techniques developed for solid materials must be applied.

This paper is concerned with the application and refinement of recently introduced ^{13}C NMR techniques for obtaining high resolution spectra of solid materials. These techniques, known as cross polarization (CP)[1] and magic angle spinning (MAS)[2] show promise for the characterization of fossil fuels. One reason is because the CP/MAS techniques provide important information about the organic carbon distribution in intractable solids. The CP/MAS ^{13}C NMR techniques have been applied with great success to a variety of fossil fuel substances, including coals[3-10], oil shales[11-18] and humic materials[19-21]. In this paper, we will confine our discussion to oil shales and the applications of CP/MAS techniques to oil shale evaluation and processing.

THEORETICAL ASPECTS

^{13}C NMR in Solids.

Observation of ^{13}C NMR signals from solid materials is not without its attendant problems, particularly if one is interested in structural information of the type obtained for liquids. The more prominent of these problems are: (a) ^{1}H - ^{13}C dipole-dipole interactions, which cause excessive line broadening in solids, (b) chemical shift anisotropy, which also causes line broadening and unsymmetrical line shapes because of the many different orientations the molecules in an amorphous state can assume in a magnetic field, and (c) long spin-lattice relaxation times, T_1, for ^{13}C in solids ($\cong$ minutes). The T_1 value determines how rapidly a pulsed NMR measurement can be repeated. Factors (a) and (b) are associated with the resolution problem in solids, while factor (c) is important to the sensitivity problem of ^{13}C NMR in solids. In liquids, the normal tumbling motions of the molecules tend to average the dipole-dipole interactions to zero, reduce the chemical shift anisotropy to its isotropic value, and provide a relaxation mechanism to shorten the T_1 from that of the solid. The whole idea of CP/MAS is to overcome each of these problems so that ^{13}C spectra can be obtained for solids, in a reasonable time, and which possess some structural resolution.

In solids line broadening that is caused by ^{1}H-^{13}C dipolar interactions is largely eliminated by high power decoupling techniques. By irradiating the ^{1}H nuclei with a strong radio frequency field at their Larmor resonance frequency, rapid transitions are induced between the ^{1}H energy levels which effectively makes them transparent to the ^{13}C nuclei. Line broadening caused by the anisotropy is not eliminated by decoupling, but is dealt with by another technique. This technique, known as magic-angle spinning, involves rapid rotation of the sample at an angle of 54.7°, relative to the external magnetic field. The chemical shift anisotropy is dominated by a factor $(1-3\cos^2\theta)$. The angle, 54.7°, is the one which makes $(1-3\cos^2\theta)$ vanish. By rapid rotation, we mean rotational rates of the same order of magnitude as the anisotropy. Thus, for aromatic carbons which can have a chemical shift anisotropy of 100-200 ppm, this means the sample must be spun at rates of 1500-3000 revolutions per sec for a 15 MHz ^{13}C frequency.

Because ^{13}C has a low natural abundance, (1.1%), NMR measurements on this nucleus generally require signal averaging. The T_1

determines how rapidly a normal pulse experiment can be repeated. The rule of thumb is to repeat the experiment every $3T_1$ to $5T_1$. Consider the situation in which the T_1 of a ^{13}C nucleus is 10 min. The experiment would then be repeated every 30 to 50 minutes. Suppose 1000 signals must be accumulated to obtain a reasonable signal-to-noise ratio, S/N. For this case (which is not unrealistic for ^{13}C in a solid) the total time for the experiment would be between 21 to 35 days - hardly an attractive feature for a routine method. Fortunately, cross polarization techniques overcome the problems of long T_1's and consequent limited sensitivity.

Cross Polarization.

The key to observing ^{13}C signals from solids is the technique of cross polarization. In this section we describe this process in some detail. More rigorous and thorough discussions about the dynamics of cross polarization can be found in the articles by Pines et al[1], Schaefer at al[2], and Bartuska et al[4].

The cross polarization NMR experiment relies on the presence of a system of abundant nuclear spins (^1H), in order to observe the NMR signal from a dilute spin (^{13}C). The main idea is to provide a mechanism for efficient transfer of polarization (hence signal intensity) from the abundant ^1H spin system to the dilute ^{13}C spin system.

When a sample containing ^1H and ^{13}C nuclei is placed in a magnetic field, H_0, the energy levels of these nuclei are split by an amount which depends upon the field, H_0, and a fundamental constant of the nucleus, the magnetogyric ratio γ (Fig. 1, laboratory frame). The ratio, γ_H/γ_C, is nearly 4 so that for a given field, the ^1H energy levels are split four times greater than the ^{13}C energy levels. For a field of 1.4 T the splittings correspond to frequencies of 60.00 MHz and 15.08 MHz for the ^1H and ^{13}C nuclei, respectively. Because of this "mismatch" in energy levels, any energy given up by the ^1H spin system cannot be effectively utilized by the ^{13}C spin system. This mismatch is indicated schematically in the upper portion of Figure 1. However, it is possible to make a transformation to a rotating reference frame and "match" the ^{13}C and ^1H energy levels by satisfying the Hartmann-Hahn conditions, $\omega_H = \omega_C$. Under these conditions the energy levels of the two spin systems are equalized, and a ^1H spin flip can cause a ^{13}C spin to become aligned with the field, H_0. As a result energy is transferred efficiently between the dilute ^{13}C spin system, and the abundant ^1H spin system. This process is called cross polarization and takes place much more rapidly than spin lattice relaxation processes. Thus, one doesn't have to wait the usual $3-5T_1$ of ^{13}C in order to

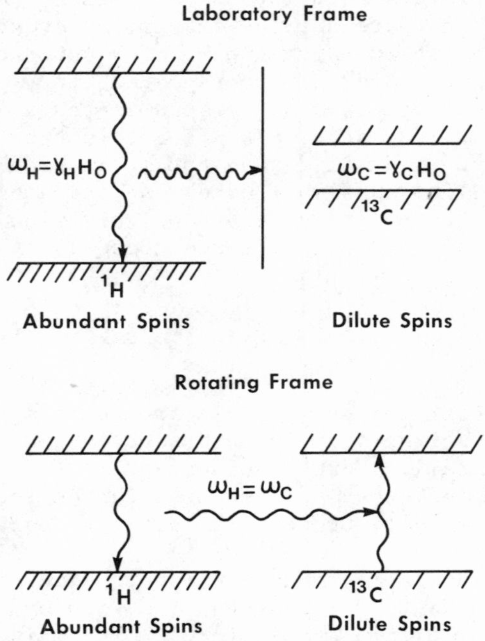

Fig. 1. Schematic diagram of energy level matching for cross polarization under Hartmann-Hahn conditions.

repeat the experiment for purposes of signal averaging. Also, because $\gamma_H/\gamma_C \cong 4$, a fourfold enhancement in ^{13}C signal intensity can be obtained from the cross polarization process.

EXPERIMENTAL

Oil Shales Studied

CP/MAS ^{13}C NMR measurements were made on oil shales that represent a variety of geologic periods (Tertiary to Devonian), depositional environments (lacustrine, marine, and paludal), source locations and elemental compositions. Hence Type I, Type II, and Type III kerogens as described by Tissot and Welte[22] are contained in the sample set. Appropriate information about the samples is given in Table 1. For some of the samples, chemical and petrographic information has been described[23]. These are appropriately noted in the Table. Other samples were obtained from the Laramie Energy Technology Center sample library. All samples were powdered to pass 100 mesh screen size. Most NMR measurements were made on raw oil shales. Kerogen concentrates were available for the French, Scottish and Argentine oil shales and

TABLE 1 SUMMARY OF OIL SHALE ANALYTICAL DATA

Sample	Oil, wt%	Gas, wt%	C(Shale) wt%	C(Oil) wt%	C(Gas) wt%	Fraction of Carbon Converted	Fraction of Aliphatic Carbon, f_{al}	Remarks
1 Alaska	50.5	8.7	57.5	84.75*	55.0†	0.82	0.79	Chandelar Deposit
2 Alaska	32.4	5.7	42.8	84.75*	55.0†	0.71	0.74	Chandelar Deposit
3 Alaska	48.7	6.7	58.5	84.75*	55.0†	0.77	0.79	Chandelar Deposit
4 Argentina	5.2	1.4	7.7	84.75*	55.0†	0.67	0.75	San Juan
5 Australia	8.4	2.5	27.6	85.56	60.0	0.31	0.37	Glen Davis Mine Top Seam
6 Australia	74.9	9.4	81.4	84.75*	70.0	0.84	0.83	Coolaway Mtn
7 Brazil	11.5	3.9	16.5	84.29	55.0†	0.71	0.74	Panal Mine, Sao Paulo, a
8 Colorado	9.5	2.1	11.6	84.61	50.0	0.78	0.80	Green River Formation
9 Colorado	11.9	3.7	15.9	84.69	50.0	0.76	0.78	Green River Formation
10 France	9.5	3.2	22.3	85.16	62.0	0.45	0.49	St. Hilaire a,
11 Kentucky	6.1	2.9	13.6	84.94	55.0†	0.50	0.46	New Albany Formation
12 Kentucky	5.2	2.9	14.8	84.48	55.0†	0.41	0.55	Sunbury Deposit
13 Michigan	3.7	2.6	9.2	86.74	55.0†	0.51	0.50	Antrim Shale
14 Morocco	7.2	3.2	14.8	79.58	55.0†	0.53	0.61	Timahdit Deposit
15 Morocco	8.8	3.6	17.3	84.75*	55.0†	0.51	0.59	Timahdit Deposit
16 Nova Scotia	5.4	2.1	34.3	84.75*	55.0†	0.16	0.24	Stellarton Deposit,a
17 Scotland	8.2	3.0	12.3	84.75*	40.0	0.66	0.69	Westwood Mine a,
18 So.Africa	26.2	4.6	42.2	84.99	61.0	0.60	0.66	Ermelo Deposit a
19 Wyoming	13.7	3.8	18.1	84.75*	50.0	0.75	0.79	Green River Formation

* Average wt% Carbon, based on 29 samples Ref 24   a   Raw oil shale described in Ref 24
† Average wt% Carbon, based on 9 samples Ref 24

NMR measurements were made on these materials. Resulting data, however, were calculated to a raw shale basis. For cases in which compositional data were not available for the shale oils, a value of 84.75 was used for the weight percent carbon in the oil. This represents the average percent carbon of 29 shale oils[23] derived from oil shales from various worldwide deposits. For oil shales in which mass spectrometric analyses of the Fischer assay gases were available, the amount of carbon in the gas was calculated directly. Otherwise a value of 55 was used to represent the percent carbon in the gas. This value was determined by averaging over the available gas compositional data[23].

NMR Procedures

NMR spectra were recorded on a home-built spectrometer based on a 1.4T Varian 30 cm magnet and a JEOL FX60Q* spectrometer, modified for cross polarization and magic angle spinning (Bartuska and Maciel)[24]. Matched spin-lock, single-contact cross polarization experiments were performed using the four-part procedure described by Bartuska et al[4]. Instrumental conditions were: 15.1 MHz ^{13}C frequency, 2 sec. pulse repetition rate, 1 msec cross polarization contact time, 64 msec data acquisition time, 50-60KHz ^1H decoupling field and a spinning rate of 2.5 KHz. Typically, 8,000 to 12,000 transients were recorded for the raw shales. The fraction of aromatic and aliphatic carbons was determined by electronic integration of the signals in the corresponding spectral regions.

For most samples studied, signal strength is a more critical problem than sample availability, so we have chosen to spin as large a sample as possible in experiments. For example, in a raw oil shale the amount of organic carbon is quite small and the distribution of carbon types is wide; therefore, NMR S/N problems are considerable and a large sample is desirable. A Colorado oil shale that assays at 125-146 L/kg (30-35 gal/ton) contains about 12-13 wt% organic carbon. In our experiments on solid fossil fuels, samples of about 1.0 to 1.5 cm^3 volume typically have been employed in spinners machined from Kel-F. Spinner diameters for the two spectrometers used typically are 10 mm and 13 mm. With these large samples, spinning rates of 2.0 to 3.0 KHz have been achieved using a turbine air pressure of 137-207 kPa (20 to 30 psi). This range of spinning rates proves adequate for a 1.4 T magnet system, but much higher speeds are needed to remove serious sideband problems if high-field magnets are employed[17].

---

* Reference to a specific brand name does not imply endorsement by the U.S. Department of Energy.

APPLICATIONS

General Features of CP and CP/MAS ^{13}C Spectra of Oil Shales.

Examples of the type of structural information that is obtainable using CP and CP/MAS techniques are shown in Fig. 2 for a Colorado and Kentucky oil shale. The significance of the Fischer assay and elemental analyses data is discussed later. Although there is a partial resolution of aromatic and aliphatic carbons for the case of non-spinning, the broadening due to chemical shift anisotospy is evident, particularly for the aromatic carbons in the Kentucky oil shale. Magic-angle spinning removes this broadening and provides much better resolution of the aliphatic and aromatic carbons as seen from the example in Fig. 2.

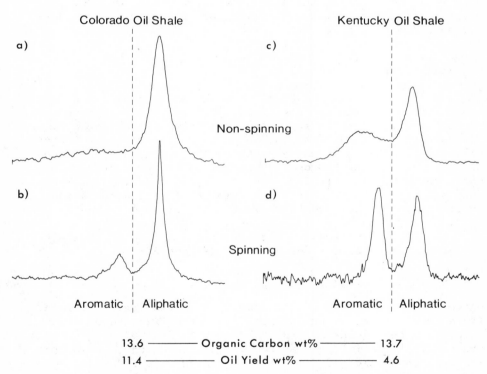

Fig. 2. ^{13}C NMR spectra of a Colorado and Kentucky oil shale obtained by high power ^{1}H decoupling and cross polarization without magic angle spinning (a,c) and with magic angle spinning (b,d). Organic carbon and oil yields of the two oil shales are shown also.

For most oil shales CP/MAS provides only a resolution of aliphatic and aromatic carbons in the ^{13}C NMR spectrum. This common lack of additional spectral detail is due primarily not to limitations of the NMR techniques, but rather to the large number of slightly different aliphatic and aromatic carbons in the kerogen. This gives rise to a multitude of closely spaced resonances that result in broad bands. Regardless of the lack of spectral detail, the major carbon types--aliphatic and aromatic-- are resolved in oil shales.

Even at this level of detail, useful information about the organic carbon distribution in an oil shale can be obtained. If the area to the left of the dashed line in Fig. 2 is denoted as A, the area to the right as B, the carbon aromaticity, f_a, is A/(A+B). Similarly the fraction of aliphatic carbons, f_{al}, is B/(A+B). Actually, carbon resonances from olefins, esters, carbonyl and carboxyl carbons, and possibly inorganic carbonate carbons can contribute to the signal in the aromatic region. Possible interferences from these types of carbons have been described[11,14].

Oil Shale Resource Evaluation

By comparison of the CP/MAS ^{13}C NMR spectra of the Colorado and Kentucky oil shales (Fig. 2b,d) with the corresponding organic carbon and Fischer assay data, it is obvious why the oil yields of Colorado oil shales are greater than those of Kentucky oil shales. This is because of the much more favorable carbon distribution (hence hydrogen availability) of the Colorado oil shale for producing liquids under pyrolysis conditions. These data and spectra suggest that the amount of aliphatic carbon in an oil shale can be used to estimate its oil yield potential.

We recently correlated the oil yield of an oil shale with the amount of aliphatic carbon[12,13], defined as:

$$C_{al} = f_{al} (C_{org}) \qquad (1)$$

where C_{al} is the aliphatic carbon, C_{org} is the organic carbon (total minus mineral carbon) and f_{al} is the fraction of aliphatic carbon in the oil shale. The term, f_{al} is determined by taking the ratio of the integral of the aliphatic region to the total integral (aromatic plus aliphatic). Prior to solid state NMR techniques it was difficult, if not impossible, to obtain a direct measurement of f_{al} for a raw shale. Now such measurements are routine.

The correlation we obtained is shown in Fig. 3, along with the parameters obtained from linear regression analysis of the

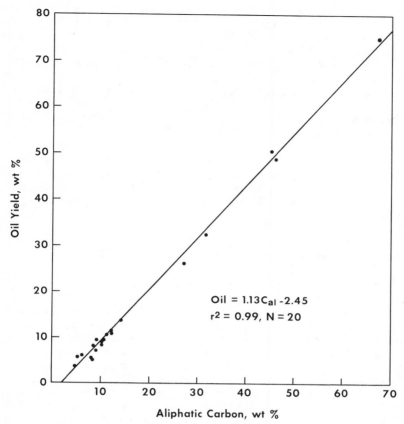

Fig. 3. Correlation between aliphatic carbon and oil yields. The linear regression analysis parameters are given also, r^2 = index of determination, N = number of oil shales.

data. Important aspects of this correlation are that it is linear over a wide range of oil yields, and is independent of the geologic ages, origins, depositional environments, and source locations of the oil shales. Appropriate analytical data about the oil shales are given in Table 1.

The Genetic Potential and Transformation Ratio

Note that C_{al} in eqn. 1 is the product of two terms: C_{org}, which represents the quantity of kerogen; and f_{al}, which is a measure of the quality of the kerogen. An assessment of the quality of an oil shale kerogen for producing liquids and gases, independent of the quantity of kerogen in the oil shale, can be obtained from the transformation ratio of the kerogen. The transformation ratio[22] is the ratio of oil and gas produced with time during heating, to the amount the kerogen is capable of producing. The amount of oil and gas a kergoen will produce given sufficient time and adequate temperature is called the genetic potential[22] of the kerogen. During a Fischer assay, when all the labile organic material is converted to oil and gas, the transformation ratio equals genetic potential.

The genetic potential of an oil shale can be calculated using the Fischer assay and elemental compositional data of Table 1, based on the amount of carbon in the raw shale, converted to carbon in the gaseous and liquid products during the Fischer assay, ie.:

$$\text{genetic potential} = \frac{C(\text{oil}) + C(\text{gas})}{C(\text{raw shale})} \quad (2)$$

By plotting the genetic potential, calculated from eq. 2, against the fraction of aliphatic carbon, f_{al}, in an oil shale, determined solely by NMR, the correlation shown in Fig. 4 has been obtained[18]. Appropriate linear regression parameters are included in Fig. 4. Important aspects about this correlation are that it, too, is independent of the geologic age, origin, depositional environment and source location, and that the dependence upon quantity of kerogen has been removed. For example, one notes that the genetic potential of the Argentine and Scottish oil shales are high, whereas their position on Fig. 3 places them at the lower end of the plot, because of the quantity of organic material present in the raw shale.

Oil Shale Processing

Because the kerogen in oil shale is insoluble in common organic solvents, processes for extracting liquids from such materials necessarily involve the application of heat to thermally decompose the kerogenic material. The fact that CP/MAS techniques are applicable to all types of solid materials suggests that CP/MAS techniques can provide useful information about changes that occur in the carbon distribution of an oil shale upon heating. We are investigating these types of NMR applications with an aim toward a better understanding of the molecular processes involved in the thermal decomposition of oil shales during retorting.

The thermal decomposition of the kerogen in oil shale can be expressed as a general process of the following kind:

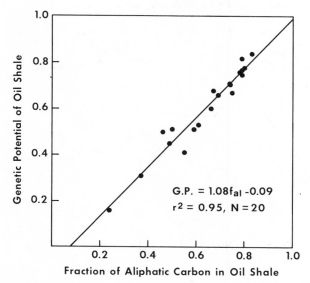

Fig. 4. Correlation between the fraction of aliphatic carbon in oil shale and the genetic potential. The linear regression analysis parameters and giving also, r^2 = index of determination, N = number of oil shales.

$$\text{kerogen} \xrightarrow{\Delta} x \text{ oil} + y \text{ gas} + z \text{ water} + w \text{ carbonaceous residue}$$

where x, y, z and w represent the amounts of indicated constituents resulting from decomposition of one gram of kerogen.

For such a process, not all of the carbon originally present in the raw oil shale is converted to carbon in the gaseous and liquid products. Instead, a certain fraction remains in the carbonaceous residue[25], and can be burned to supply additional energy input to drive the retorting process.

CP/MAS ^{13}C NMR spectra of a 230 L/kg (55 gal/ton) Colorado oil shale heated to various temperatures for 24 hours are shown in Fig. 5. The increase in carbon aromaticity, f_a, with increasing temperature is easily noted. These results further show that the carbon in the residue after retorting an oil shale is almost totally aromatic. The spectra in Fig. 5 are unnormalized so that comparison of the intensities of the bands at different temperatures, 380°C and 250°C for example, are not quantitative. However, at any given temperature, the CP/MAS spectra provide a faithful representation of the organic carbon distribution of the

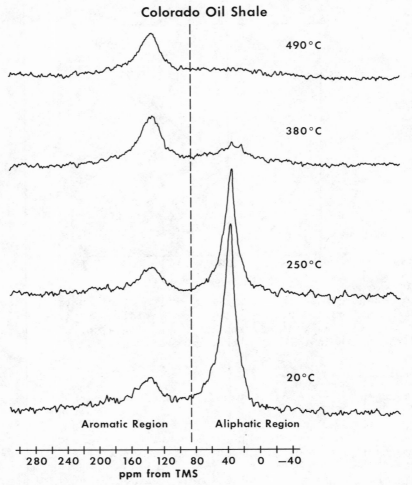

Fig. 5. Cross polarization, magic angle spinning ^{13}C NMR spectra of a Colorado oil shale, heated to various temperatures for 24 hours.

unconverted kerogn. We have recorded CP/MAS ^{13}C NMR spectra of the carbon in spent shales after Fischer assay[26], and in all cases have found the carbon distribution to be aromatic.

In conjunction with the earlier definition of the transformation ratio of an oil shale, and our observation that the fraction of aliphatic carbons correlate with genetic potentials, the results of Fig. 5 suggest that a transformation ratio based on CP/MAS NMR techniques can be defined as:

$$\text{transformation ratio} = \frac{\Delta f_{al}}{f_{al}} \qquad (3)$$

where Δf_{al} is the change in the fraction of aliphatic carbons as a function of temperature.

Work is in progress at the Laramie Energy Technology Center in the development of models for the kinetics of oil shale retorting, based on these NMR data.

SUMMARY

CP/MAS ^{13}C NMR techniques provide a direct measurement of the amounts of aliphatic and aromatic carbon types in oil shales. These measurements are non-destructive and can be made on raw oil shales. From the data derived from NMR measurements, and the organic carbon determined by elemental analysis, it appears feasible to predict the potential oil yield of an oil shale, and the genetic potential of a formation. The transformation ratio, occurring during heating of an oil shale also can be determined. Measurements of the organic carbon distribution of an oil shale by CP/MAS ^{13}C NMR can provide insight into organic structural types of carbons that are important for oil shale processing.

ACKNOWLEDGEMENT

This work was performed under the auspices of the Laramie Energy Technology Center (LETC) as part of its designated role as the U.S. Department of Energy's lead laboratory for research and development on oil shale and non-oil shale in situ processes. The authors gratefully acknowledge support of this work by LETC (Contract number DE-AC20-LC10006). The authors also gratefully acknowledge the technical assistance of V. J. Bartuska.

REFERENCES

1. A. Pines, M. G. Gibby, and J. S. Waugh, Proton-enhanced NMR of dilute spins in solids, J. Chem. Phys. 59:569 (1973).

2. J. Schaefer, E. O. Stejskal, and R. Buchdahl, Magic-angle ^{13}C NMR analysis of motion in solid glassy polymers, Macromolecules 10:384 (1977).

3. D. L. VanderHart, and H. L. Retcofsky, Estimation of coal aromaticities by proton-decoupled carbon-13 magnetic resonance spectra of whole coals, Fuel 55: 202 (1976).

4. V. J. Bartuska, G. E. Maciel, J. Schaefer, and E. O. Stejskal, Prospects for carbon-13 nuclear magnetic resonance analysis of solid fossil fuel materials, Fuel 56:354 (1977).

5. K. W. Zilm, R. J. Pugmire, D. M. Grant, R. E. Wood, and W. H. Wiser, A comparison of the carbon-13 n.m.r. spectra of solid coals and their liquids obtained by catalytic hydrogenation, Fuel 58:11 (1979).

6. H. L. Retcofsky, and D. L. VanderHart, ^{13}C-^{1}H cross polarization nuclear magnetic resonance spectra of macerals from coal, Fuel 57:421 (1978).

7. G. E. Maciel, V. J. Bartuska, and F. P. Miknis, Characterization of organic material in coal by proton-decoupled ^{13}C nuclear magnetic resonance with magic-angle spinning, Fuel 58:391 (1979).

8. B. C. Gerstein, L. M. Tyan, and P. D. Murphy, A tentative identification of average aromatic ring size in an Iowa vitrain and a Virginia vitrain, American Chemical Society, Division Fuel Preprints 24:No. 1, 90 (1979).

9. F. P. Miknis, M. Sullivan, V. J. Bartuska, and G. E. Maciel, Cross polarization magic angle spinning ^{13}C NMR spectra of coals of varying rank, Org. Geochem (in press).

10. P. G. Hatcher, I. A. Breger, N. M. Szeverenyi, and G. E. Maciel, Nuclear magnetic resonance studies of ancient buried wood: II Observations on the origin of coal from brown coal to bituminous coal, Org. Geochem (in press).

11. H. A. Resing, A. N. Garroway, and R. N. Hazlett, Determination of aromatic hydrocarbon fraction in oil shale by ^{13}C NMR with magic-angle spinning, Fuel 57:450 (1978).

12. G. E. Maciel, V. J. Bartuska, and F. P. Miknis, A Correlation between oil yields of oil shales and ^{13}C nuclear magnetic resonance spectra, Fuel 57:505 (1978).

13. G. E. Maciel, V. J. Bartuska, and F. P. Miknis, Improvement in correlation between oil yields of oil shales and ^{13}C NMR spectra, Fuel 58:155 (1979).

14. F. P. Miknis, G. E. Maciel, and V. J. Bartuska, Cross polarization magic-angle spinning ^{13}C NMR spectra of oil shales, Org. Geochem. 1:169 (1979).

15. D. Vitorovic, D. Vucelic, M. J. Gasic, N. Juronic, and S. Macura, Analysis of the organic matter of oil shales by nuclear magnetic resonance, Org. Geochem. 1:89 (1978).

16. D. Vucelic, N. Juranic, and D. Vitorovic, Potential of proton-enhanced ^{13}C NMR for the classification of kerogens, Fuel 58:759 (1979).

17. M. T. Melchior, K. D. Rose, and F. P. Miknis, Artefacts in magic-angle spinning NMR: comments on 'potential of proton enhanced ^{13}C NMR for the classification of kerogens' Fuel 59:594 (1980).

18. F. P. Miknis, J. W. Smith, M. A. Mast, and G. E. Maciel, ^{13}C NMR measurements of the genetic potentials of oil shales, Geochim. Cosmochim. Acta (to be submitted).

19. F. P. Miknis, V. J. Bartuska, and G. E. Maciel, Cross polarization ^{13}C NMR with magic-angle spinning: Some applications to fossil fuels and polymers, Amer. Lab. 11:19 (1979).

20. P. G. Hatcher, D. L. VanderHart, and W. L. Earl, Use of solid state ^{13}C NMR in structural studies of humic acids and humin from Holocene sediments, Org. Geochem 2:87 (1980).

21. P. G. Hatcher, L. W. Dennis, and G. E. Maciel, Prospects of NMR for the study of humic acids, Geochim. Cosmochim. Acta (in press).

22. B. P. Tissot, and D. Welte "Petroleum Formation and Occurrence," Springer-Verlag, New York (1978).

23. H. N. Smith, J. W. Smith, and W. C. Kommes, Petrographic examination and chemical analyses for several foreign oil shales, U. S. Bureau of Mines Rept. Invest. 5504, 34 pp. (1959).

24. V. J. Bartuska, and G. E. Maciel, A magic angle spinning system for bullet type rotors in electromagnets, J. Magn. Reson. (in press).

25. A. B. Hubbard, and W. E. Robinson, A thermal decomposition study of Colorado oil shale, U.S. Bureau of Mines Rept. of Investig. 4744, 24 pp. (1950).

26. F. P. Miknis, N. M. Szeverenyi, and G. E. Maciel, Characterization of the residual carbon in retorted oil shale by solid state ^{13}C NMR, Fuel (to be submitted).

NMR RESULTS RAISE QUESTIONS ON COAL LIQUEFACTION MODEL COMPOUND STUDIES

D. C. Cronauer, R. I. McNeil, D. C. Young, and R. C. Ruberto

Gulf Research & Development Company
P.O. Drawer 2038
Pittsburgh, Pennsylvania 15230

ABSTRACT

To provide a further understanding of the mechanism of coal liquefaction, a wide range of reactions has been studied with model compounds having chemical bonds similar to those believed present in coal. These results have been compared with those obtained by NMR and MS analysis of products from coal liquefaction runs with deuterium labeled solvents. As indicated from the model compound reactions, coal liquefaction products should have much of the deuterium in sites α to aromatic rings and in short chain cracking products. However, the distribution of deuterium in the coal-derived oil and asphaltene fractions is only 40 to 50% in α positions, while 5-15% are at combined β and γ positions, 30-45% in aromatic sites, and 5-10% in C-H sites adjacent to heteroatoms. The distribution is moderately affected by liquefaction temperature but less so reaction time.

INTRODUCTION

Coal has been described by Francis (1954) as a compact stratified mass of mummified plants which have been modified chemically in varying degrees, interspersed with smaller amounts of inorganic matter. While coal is acknowledged to be a heterogeneous mass, several researchers have attempted to present an average molecular picture of the various bonds present. Particular reference is made to the work of Hill and Lyon (1962) and Given (1960). In summary, coal is considered to consist of clusters of condensed aromatic and hydroaromatic groups bonded through alkyl, ether and thioether linkages.

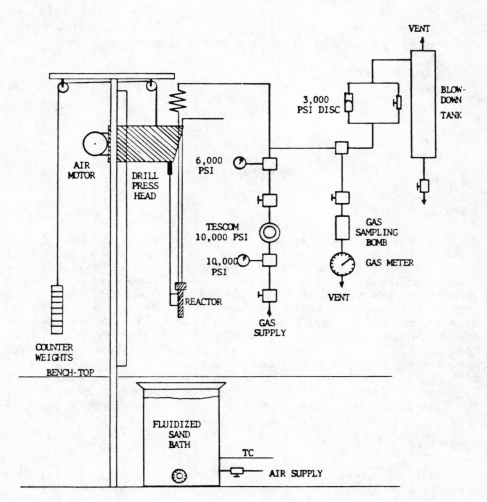

Figure 1. Schematic of the Micro Scale Liquefaction Unit.

Considering the above structure, a series of hydrogen transfer experiments was undertaken at this laboratory using model compounds (acceptors) having C-C linkages and oxygen compounds with functionality likely to be present in coal. The published results (Cronauer, et al., 1979) are summarized in Table 1. It was shown that the cracking of dibenzyl was faster than that of diphenylbutane and 1-phenylhexane at 400-450°C. The primary reaction product was toluene with only low levels of ethylbenzene and benzene. Diphenyl, diphenylmethane and diphenylether were essentially unreactive at 450°C. With oxygen-containing compounds, the relative order of reactivity was: furans < phenols < ketones < aldehydes < chain ethers. It was observed that dibenzylether, benzaldehyde and benzyl alcohol were ultimately converted to toluene at coal liquefaction conditions.

Considering the above experiments, it would be anticipated that coal cracking would strongly occur at carbons α to aromatic rings. To confirm this and to follow the progress of other reactions occurring during liquefaction, a series of runs was made with various coals in deuterated tetralin.

EXPERIMENTAL

Batch hydrogen transfer experiments were made using a small micro-reactor of approximately 10 cc volume. This reactor consisted of a 0.5 inch O.D. by 7-inch tube with Swagelock fittings and a gas sampling line. The reactor was held vertically on a rod which was mounted in a drill press modified to provide up-and-down agitation at about 800 strokes/minute. The reactor was lowered into a pre-heated sandbath to bring its contents quickly to reaction temperature, specifically 2.3 minutes for 450°C. Initial screening experiments indicated that Powhatan coal/solvent mixing was best achieved by placing a small stainless steel ball in the reactor. A schematic of the unit is given in Figure 1.

The liquefaction runs were made at temperatures between 300 and 450°C and reaction times between 0 (allowing 2.3 minutes for heating) and 60 minutes. Runs were made with 1.5 g of one of three bituminous coals characterized in Table 2 and with 3.5 g of either totally- or partially-deuterated tetralin. The average formula of the latter solvent calculated from NMR data was $C_{10}H_{8.2}D_{3.8}$. The level of total hydrogen transfer was calculated from the conversion of tetralin to naphthalene on a basis of equivalent ^1H.

The total reaction product was transferred to a bottle and the tetralin was stripped overnight in a heated oil bath (70°C)

Table 1. Summary of Reaction Kinetics
(Reactions in Excess of Tetralin)

Acceptor	Formula	Half Life $T_{1/2}$ (min) (450°C)	Activation Energy (cal/g mole)
Dibenzyl	$PH-CH_2-CH_2-PH$	55.0	48 100
Diphenyl Methane	$PH-CH_2-PH$	40+ hours	–
Diphenyl Butane	$PH-(CH_2)_4-PH$	64.8	–
Phenyl Hexane	$PH-(CH_2)_5-CH_3$	110	–
Stilbene	$PH-CH=CH-PH$	11.6	28 400
Diphenyl Ether	$PH-O-PH$	40+ hours	–
Dibenzylether	$PH-CH_2-O-CH_2-PH$	2.8	36 000
Benzaldehyde	$PH-\overset{O}{C}-H$	5.8	32 000
Acetophenone	$PH-\overset{O}{C}-CH_3$	161.2	33 500
α-Tetralone	–	101.9	24 750
α-Naphthol	–	301.4	34 000
Benzyl Alcohol	$PH-CH_2-OH$	17.3	26 200
3,5-Xylenol	–	40+ hours	–

Table 2. Analysis of Coal Samples

Coal	Kentucky 9/14	Illinois No. 6	Powhatan No. 5
As-Received, Wt %			
Moisture	5.8	9.3	1.3
Ash	9.0	12.8	9.6
Ultimate, Wt %			
Carbon	70.56	68.99	72.28
Hydrogen	4.74	4.33	5.07
Nitrogen	1.52	1.26	1.47
Oxygen (Difference)	10.13	9.24	7.90
Sulfur	3.49	2.07	3.60
Ash	9.56	14.11	9.68
	100.00	100.00	100.00

with a nitrogen flow. The product was then Soxhlet extracted to give the following fractions (unless specified):

```
oil             = total sample - pentane insolubles
asphaltenes     = pentane insolubles - benzene insolubles
preasphaltenes  = benzene insolubles - tetrahydrofuran
                  insolubles
THF insolubles  = as determined
```

These fractions were examined using a Varian FT-80A NMR (^{2}H-12.21 MHz) equipped with a 10 mm broad band probe. Selected gas samples were analyzed using a CEC-2104 low resolution mass spectrometer.

RESULTS

Coal Conversion

The Kentucky coal liquefaction experiments were done at 450°C in tetralin with and without the addition of various catalysts. Runs were also done with unlabeled octahydrophenanthrene at 450°C. Results from these runs are given in Figure 2.

The addition of either CoMo or NiW catalyst resulted in increased conversion of coal to toluene solubles at reaction times up to 60 minutes. At 60 minutes, the levels were essentially the same. The addition of ash was not as effective as that of commercial catalyst. For reference, the yield of oils followed the same trend. Octahydrophenanthrene was shown to be a very effective solvent.

The levels of hydrogen transfer were also shown in Figure 2. The results of the catalytic and non-catalytic runs were similar with only a minor increase in hydrogen transfer of catalytic runs at intermediate times.

Results of the Illinois coal liquefaction experiments are shown in Figure 3. The primary differences between the results of the two coals are the reduced level of conversion of Illinois coal to toluene solubles and the strong influence of catalyst upon conversion. Suprisingly, the level of hydrogen transfer was not influenced by the catalyst level.

A different approach was taken with Powhatan coal in that temperatures of 300 through 450°C and run times of 0, 10 and

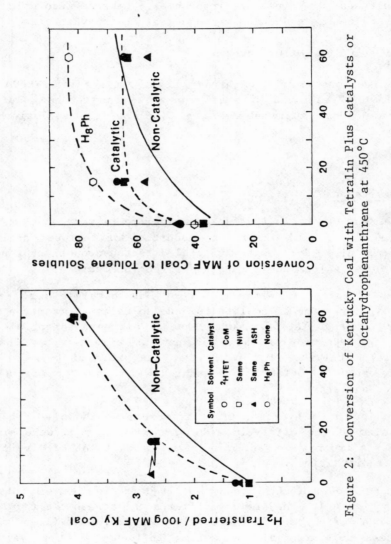

Figure 2: Conversion of Kentucky Coal with Tetralin Plus Catalysts or Octahydrophenanthrene at 450°C

COAL LIQUEFACTION MODEL COMPOUND STUDIES

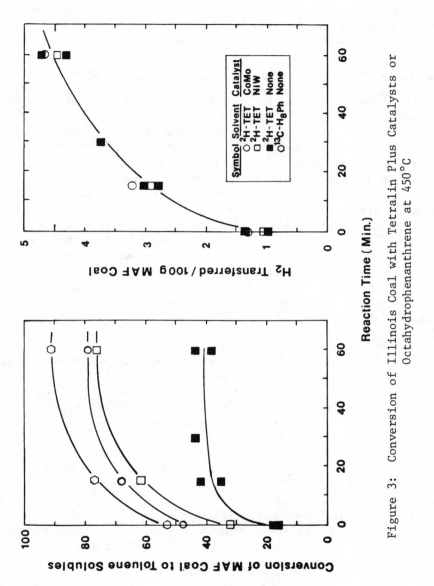

Figure 3: Conversion of Illinois Coal with Tetralin Plus Catalysts or Octahydrophenanthrene at 450°C

30 minutes were used. The results from the runs at 350°C and 450°C are shown in Figure 4. Both low levels of conversion and hydrogen transfer were observed at the lower temperature. The level of hydrogen transfer for the Powhatan runs was lower than that of the other two coals, but the level of conversion to toluene solubles was unexpectedly higher.

Site of Deuterium Transfer (Liquid Products)

From the deuterium spectra of total filtrate product from the Ky liquefaction runs, the deuterated tetralin solvent was the dominant feature. The 400°C, 60-minute run showed a marked decrease in the area of the α carbon deuterium resonance as compared to the starting solvent. The 450°C, 60-minute run product appears to have a distribution more like that of the starting solvent. These results indicate that hydrogen is preferentially withdrawn from the α position of tetralin. However, at higher temperatures, the rate of hydrogen transfer is much greater (namely by a factor of about 8); therefore, there is less selectivity of reaction sites. Scrambling is also more likely to occur at the higher temperature. In addition, any dihydronaphthalene remaining in the reaction solvent is very reactive and the remaining hydrogens would readily be abstracted by coal free radicals.

As previously discussed, samples of the various coals were liquefied in tetralin, stripped of solvent (tetralin and naphthalene) and separated into oils, asphaltenes, preasphaltenes and THF insolubles by Soxhlet extraction. The deuterium spectra of selected fractions were examined for general trends. Figure 5 is an example of a typical deuterium spectrum with the specific zones labeled.

The distribution of deuterium in the toluene solubles derived from the liquefaction of Kentucky and Illinois coals at 450°C with CoMo catalyst is given in Figure 6. The results of the non-catalytic runs were essentially the same. In addition, while the oils contained slightly more hydrogen and deuterium than the total toluene solubles, the above indicated deuterium distributions were generally valid for both.

With respect to the deuterium distributions, it is first pointed out that the profiles are relatively flat in spite of total hydrogen (and deuterium) transfer increasing from about 1 to 4 g/100g MAF coal with increasing reaction time. Positions α to

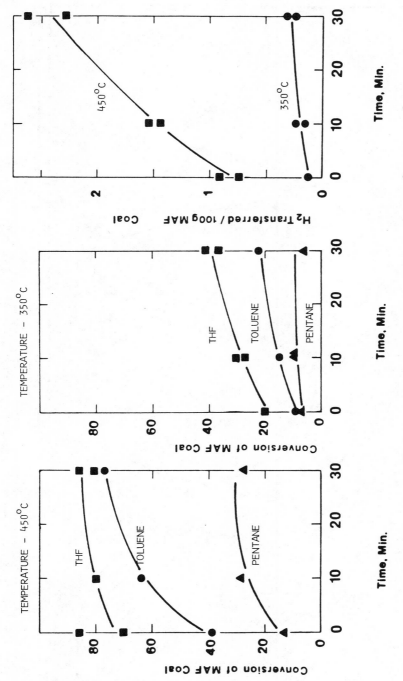

Figure 4: Conversion of Powhatan Coal with Tetralin at 350°C and 450°C

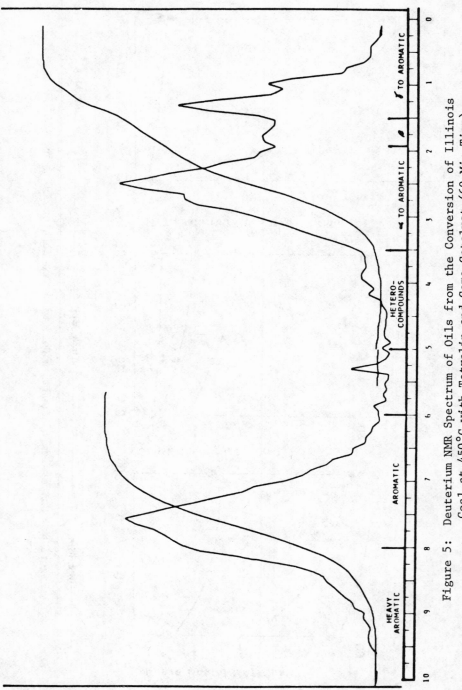

Figure 5: Deuterium NMR Spectrum of Oils from the Conversion of Illinois Coal at 450°C with Tetralin and Como Catalyst (60 Min Time)

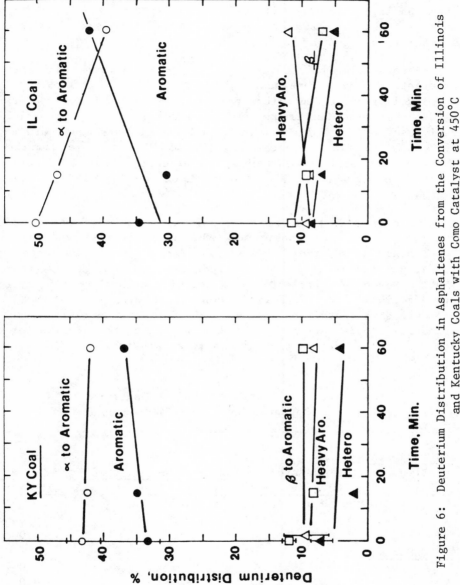

Figure 6: Deuterium Distribution in Asphaltenes from the Conversion of Illinois and Kentucky Coals with Como Catalyst at 450°C

aromatic rings had the greatest fraction of intensity with levels between 40 and 50% tending to decrease with run time. Profiles of β and γ to aromatic position deuteriums were essentially flat with time, each having had levels of about 10%. The level of deuteriums on light aromatics (1-2 rings) was in the range of 30 to 45% with a tendency to increase with time. The level on higher aromatics averaged about 10%.

Resonances indicative of the C-H position adjacent to heteroatoms N or O were assigned to the region of 3.5 to 5.0 ppm. The level of these deuteriums was in the range of 5-10% decreasing with run time. Deuterium distributions of the toluene solubles derived from liquefying Powhatan coal at 350° and 450°C are shown in Figure 7. The results from the 450°C runs were similar to those of the above coals, with the exception that deuteriums in sites α to aromatic fell to the range of 30-40% and β to aromatic increased to 10-20%.

Deuterium distributions in toluene solubles from the 350°C runs are consistent considering the above results and previous experiments with model compounds. First, little hydrogen (and deuterium) was actually transferred, namely one-eighth of that at 450°C. Secondly, little C-C bond breaking probably occurs in coal liquefaction at 350°C; this is consistent with the reduced α to aromatic site distribution, namely 20-30%. The increased level of aromatic site deuteriums may be due in part to the breaking of C-O and C-S bonds, perhaps tied with some scrambling.

Table 3. Levels of Hydrogen and Deuterium in Powhatan Coal Liquefaction Runs

450°C Series: Run Time Min.	^{1}H Transfer g/100 g MAF Coal	^{2}H Transfer, g/100 g MAF Coal	
		Oils + Asph.	Pre-Asph.
0	0.8	0.057	0.0018
10	1.5	-	0.0068
30	2.5	0.103	0.000
350°C Series:			
0	0.13	0.018	0.0016
10	0.24	-	0.0032
30	0.31	0.054	0.0025

Notes: Feed of 1.5 g coal plus 3.5 g D_4 tetralin

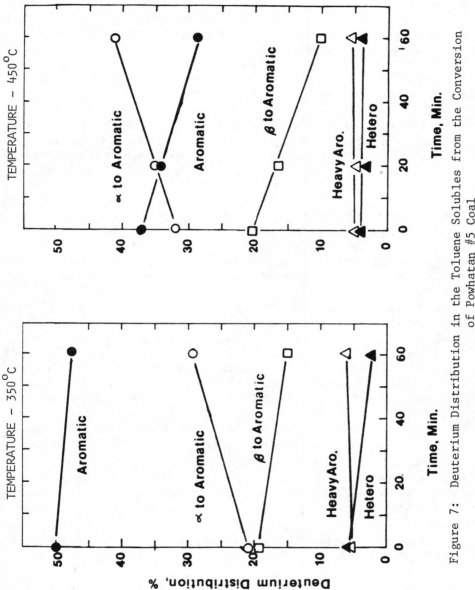

Figure 7: Deuterium Distribution in the Toluene Solubles from the Conversion of Powhatan #5 Coal

Deuterium Balances/Scrambling

In addition to obtaining deuterium distributions, preliminary work has been undertaken to observe absolute amounts of deuterium in the various fractions. The natural abundance of deuterium in the methylene chloride which was used as a solvent for the NMR samples served as reference.

Levels of deuterium in the oil plus asphaltene fractions (toluene solubles) and preasphaltene fractions of the 350° and 450°C runs are given in Table 3. Considering that the tetralin solvent for these runs had about 32% of its "hydrogen" as deuterium, only about one-sixth of the transferred deuterium was contained in the combined oil, asphaltene, and pre-asphaltene fractions. These fractions made up the bulk of the reacted product.

To observe the level of deuterium going into the product gas phase, a liquefaction run was made with totally-deuterated tetralin, Powhatan coal, and an initial hydrogen charge of 500 psig. The reaction temperature and time were 450°C and 30 minutes. About one-half of the "transferred" deuterium ended up in the gas phase with the levels of deuterium of selected components shown in Table 4. Preliminary data from a similar run at 0 time indicated a low level of deuterium incorporation in propane with only about 2% HD and essentially no D_2 present.

It is obvious that HD scrambling is playing a significant role in this process. However, it appears that most of the scrambling occurs on aliphatic carbons. We have observed only 13.3 and 2.3% D_1 and D_2 incorporation in biphenyl and 27.8, 6.6 and 1.0% D_1, D_2 and D_3 incorporation, respectively, in naphthalene at reaction conditions of excess D_{12}-tetralin (1.4/0.4), 450°C and 60 minutes. More deuterium exchange occurs with larger ring-condensed aromatics. These results are consistent with observations of B. M. Benjamin (private communications) in which there was a sizable exchange on the α carbon of butylbenzene, a limited exchange on the remaining aliphatic carbons and essentially none on the aromatic carbons at reaction conditions of 400°C for 8 hours.

Table 4. Deuterium Distribution in "Hydrogen" and Propane from Liquefying Powhatan Coal in D_{12}-Tetralin (450°C, 30 Min., 1.5 g in 3.5 g Tetralin)

Component	$^1H + {}^2H$	Propane
Deuterium Level:		
0	57	32
1	31	18
2	12	17
3		12
4		12
5		9
$\frac{D}{H+D}$ Ratio	.27	.23

DISCUSSION

Considering past experimental results with "representative" model compounds of coal linkages, the results of liquefying coal with deuterium labeled donor solvent only partially conform with anticipated results. Based on model compound experiments and relative bond energies, [see, for example, Benjamin, et al., (1978) and Ross, et al., (1978)], it was expected that little bond breaking would occur between an aromatic ring and an α carbon, while C-C bonds further from the aromatic ring would more readily crack. In addition, HD scrambling was observed to strongly occur on aliphatic carbons and only to a limited extent on aromatic carbons. Therefore, the presence of more deuterium in aromatic sites than α carbons to an aromatic ring is somewhat surprising. Perhaps the high degree of heterofunctionality in coal (specifically phenolics) is strongly affecting the cracking and/or HD scrambling reactions. In addition, the presence of catalytic surfaces such as mineral matter or undissolved coal may influence differently the reactions occurring during liquefaction.

ACKNOWLEDGMENT

The authors are indebted to Drs. A. Bruce King and Sayeed Akhtar for their discussions and to H. K. Little, L. G. Galya, H. T. Best, R. H. Albaugh, and R. G. DeFelice for their participation in the experiments and analyses. This work was supported under Contract No. FE-AC22-80PC30080 of the Department of Energy.

REFERENCES

1. Benjamin, B. M., Raaen, V. F., Maupin, P. H., Brown, L. L., and Collins, C. J., (1978) Fuel 57, 269.

2. Cronauer, D. C., Jewell, D. M., Shah, Y. T., and Modi, R. J., (1979) I&EC Fundamental 18, 153.

3. Francis, W., (1954) "Coal, Its Formation and Composition," Edward Arnold (Publishers) Ltd., 1.

4. Given, P. H., (1960) Fuel 39, 147 (also 1961, Fuel 40, 427).

5. Hill, G. R. and Lyon, L. B., (1962) Ind. Eng. Chem. 54 (6), 36.

6. Ross, D. S., et al., Various Quarterly and Annual Reports reported under DOE Contract No. EF-76-C-01-2202.

CHARACTERIZATION OF SPECIES
IN FOSSIL FUELS

THE IDENTIFICATION OF ORGANIC COMPOUNDS IN OIL SHALE RETORT
WATER BY GC AND GC-MS

D.H. Stuermer, D.J. Ng, C.J. Morris,
R.R. Ireland
Environmental Sciences Division
Lawrence Livermore National Laboratory
Livermore, California 94550

ABSTRACT

A separation scheme is presented for the analysis of oil shale retort water by gas chromatography (GC) and gas chromatography - mass spectrometry (GC-MS). The scheme is based on liquid-liquid extraction, silica chromatography, argentation chromatography and alumina chromatography. Resulting fractions can be resolved by GC and GC-MS and several hundred compounds can be identified. Even so a large fraction of retort water contaminants will require new analytical techniques to handle non-volatile and reactive compounds.

INTRODUCTION

The organic analysis of oil shale retort water is a challenging analytical problem. The composition of these waters is extremely complex and even the impressive separation capability of glass capillary gas chromatography (GC) is inadequate without prior chemical fractionation. Even so, many of the compounds are either not sufficiently volatile or are too reactive to analyze by gas chromatography.

A separation scheme is presented that provides fractions more amenable to GC and GC-mass spectrometry (GC-MS). Several hundred compounds can be identified by a thorough analysis using these techniques. However, new analytical techniques are required to analyze a majority of retort water contaminants.

---

* Work performed under the auspices of the U.S. Department of Energy by the Lawrence Livermore National Laboratory under contract number W-7405-ENG-48.

EXPERIMENTAL

Description of Sample

A representative, time-averaged, 2 l sample of oil shale retort water was collected for analysis from Lawrence Livermore National Laboratory's L-3 retort run. This retort run was designed to simulate a modified in situ oil shale retorting process and used shale from Rio Blanco's Colorado Tract Ca site (108 to 195 l/tonne) and a steam-air mixture for combustion. Approximately 760 l of shale oil and 1700 l of retort water were produced from 6 tonne of oil shale.

All of the retort water from the run was collected, mixed and heated to break oil-water emulsions, and a 2 liter aliquot was taken for analysis. The total organic carbon content was 1813 ppm.

Extraction and Separation

The water was extracted by the liquid-liquid extraction scheme in Fig. 1 to yield acidic, basic and neutral fractions. Briefly, the water was acidified to pH 1.5 with hydrochloric acid and extracted with dichloromethane to yield acidic and neutral components in dichloromethane. The water was then adjusted to pH 12.5 with potassium hydroxide and extracted with dichloromethane to yield the basic components. The acidic and neutral components were separated by back-extracting the acidic components into aqueous base, leaving the neutral components in dichloromethane. This aqueous base was then acidified to pH 1.5 and extracted with dichloromethane to yield the acidic components in dichloromethane.

The overall separation scheme is presented in Fig. 2. The basic components were analyzed directly by GC and GC-MS.

The acidic components required derivitization before acceptable GC performance was achieved. The sample was treated with 14% BF_3 in methanol to form methyl esters of free fatty acids. The esterification step increases the volatility and decreases the polarity of the components, thus making them more amenable to GC and GC-MS analysis.

The neutral fraction was too complex to analyze without further fractionation. The neutral compounds were separated by silica gel liquid-solid column chromatography (Fig. 2). The sample was applied to a 25 cc column (1 cm i.d.) of 1.5% deactivated silica in hexane and fractions were eluted with 25 ml each of the following solvents:

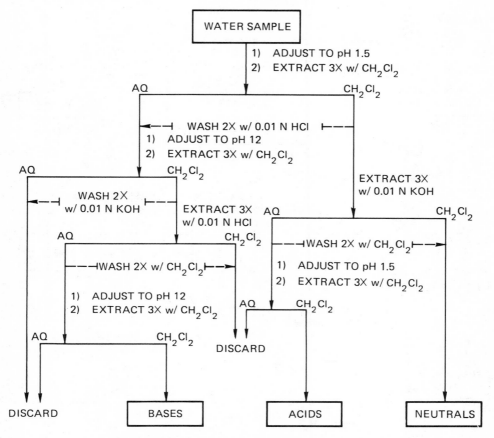

Fig. 1. Liquid-liquid extraction scheme used to extract water samples.

- Fraction 1   100% n-Hexane
- Fraction 2   10% CH_2Cl_2 in n-Hexane
- Fraction 3   20% CH_2Cl_2 in n-Hexane
- Fraction 4   30% CH_2Cl_2 in n-Hexane
- Fraction 5   100% methanol

Fractions 1 and 2 contained mainly saturated hydrocarbons and mainly olefinic hydrocarbons, respectively. To make the separation more complete, these two fractions were combined and then fractionated by argentation chromatography into distinct saturated and olefinic hydrocarbon fractions (Fig. 2). The separation was performed on 25 cc of 1% deactivated 20% $AgNO_3$ on silica gel by elution with a 1:1 mixture of CH_2Cl_2: n-hexane.

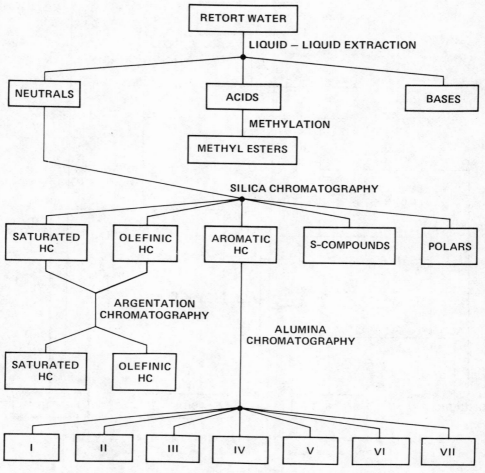

Fig. 2. Overall separation scheme.

The saturated hydrocarbons were eluted in the 0-25 ml fraction and the olefinic hydrocarbons were eluted in the 25-80 ml fraction.

Fraction 3 contained the aromatic hydrocarbons but the gas chromatogram of this fraction (Fig. 3) was too complex to allow identification of individual compounds. Therefore, this fraction was further separated by alumina chromatography. The method was a slight modification of that reported by Giger and Blumer (1974). The sample was applied in hexane to a 14 cc column (1 cm i.d.) of 1% deactivated alumina and then eluted with hexane containing

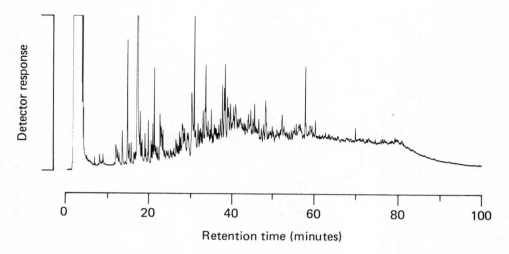

Fig. 3. Gas chromatogram of total aromatic hydrocarbons from silica gel chromatography.

increasing amounts of CH_2Cl_2. Seven fractions were collected according to the scheme in Fig. 4. Each of these fractions was then analyzed by GC and GC-MS.

Fraction 4 from the silica chromatography of the neutral extract was analyzed directly by GC and GC-MS.

Fraction 5 was analyzed by GC but, because of the nature of the resulting chromatogram, no attempt at GC-MS analysis was made.

Analysis

Gas chromatography (GC) was performed on a Hewlett Packard Model 5880 gas chromatograph equipped with a glass capillary injector system (290°C), a 25m, 0.625 mm i.d., SE-54 fused silica WCOT column, and a flame ionization detector (FID) (300°C). The oven temperature was held at 35°C for 5 min, programmed at 3°C/min to 260°C and held at that temperature for 20 minutes.

Combined gas chromatography-mass spectroscopy (GC-MS) was performed on a Hewlett Packard, Model 5985B GC-MS System. GC column and conditions were identical to those described above. Mass spectra were generated by electron impact at 70 eV. Compounds were identified by comparison of mass spectra and retention times with those of known standards.

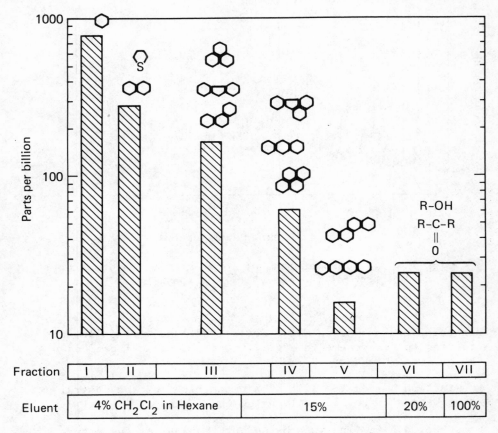

Fig. 4. Alumina chromatography elution scheme for the aromatic fraction from silica chromatography showing compound classes observed in each fraction.

The concentration of compounds amenable to GC analysis in the individual fractions were determined by integration of the FID signal and using detector response factors determined by the analysis of standard mixtures. The total concentration in the individual fractions was determined by weighing the air dried residue of aliquots on a microbalance.

RESULTS

The base fraction of the L-3 oil shale retort water consists mainly of pyridines, quinolines and isoquinolines with minor amounts of anilines and indolines (Fig. 5, Table 1). The total extractable base concentration is 110 ppm.

ORGANIC COMPOUNDS IN OIL SHALE RETORT WATER

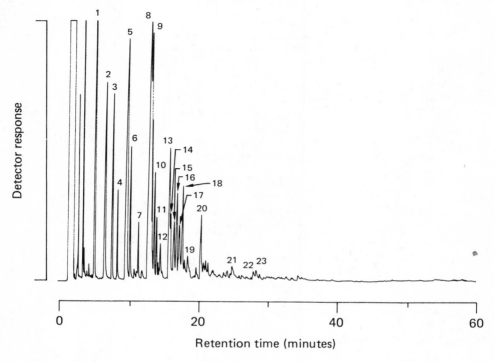

Fig. 5. Gas chromatogram of bases isolated from L-3 retort water. Peak identifications are given in Table 1.

Table 1. Basic compounds Isolated from LLNL L3 Oil Shale Retort Water.

Peak Number	Compound Identification
1	2 - Methylpyridine
2	3- and 4-Methylpyridine
3	Dimethylpyridine (isomer)
4	Ethylpyridine
5	Dimethylpyridine (isomer)
6	Dimethylpyridine (isomer)
7	Ethylmethylpyridine (isomer)
8	Aniline
9	Trimethylpyridine (isomer)
10	Trimethylpyridine (isomer)
11	Ethylmethylpyridine (isomer)
12	Ethylmethylpyridine (isomer)

Table 1. (Continued)

Peak Number	Compound Identification
13	C_9H_9N (isomer)
14	Trimethypyridine (isomer)
15	Ethylmethylpyridine (isomer)
16	Dimethylethylypridine (isomer)
17	Trimethylpyridine (isomer)
18	Dimethylethylpyridine (isomer)
19	Indoline
20	Dimethylethylpyridine (isomer)
21	Quinoline
22	Methylquinoline (isomer)
23	Methylquinoline (isomer)

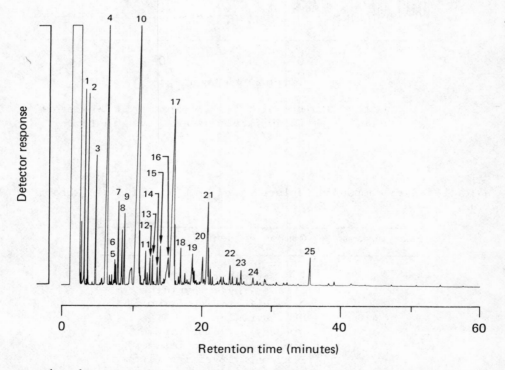

Fig. 6. Gas Chromatogram of methylated acids isolated from L-3 retort water. Peak identifications are given in Table 2.

Table 2. Methyl Esters of acidic compounds Identified in LLNL L-3 Oil Shale Retort Water.

Peak Number	Compound Identification
1	Methyl Isobutyrate
2	Methyl n-Butyrate
3	Methyl Methylbutyrate (isomer)
4	Methyl n-Pentanoate
5	Methyl Ethylbutyrate
6	Methyl Methylpentanoate (isomer)
7	Methyl 2-Methylvalerate
8	Methyl 3-Methylvalerate
9	Methyl Isohexanoate
10	Methyl n-Hexanoate
11	(Methyl Cyclopentane Carboxylate)
12	Methyl Heptanoate (isomer)
13	Methyl Heptanoate (isomer)
14	Methyl Heptanoate (isomer)
15	Methyl Heptanoate (isomer)
16	Methyl Heptanoate (isomer)
17	Methyl n-Heptanoate
18	Methyl Cyclohexane Carboxylate
19	Methyl Benzoate
20	Methyl Octanoate (isomer)
21	Methyl n-Octanoate
22	Methyl Methylbenzoate
23	Methyl n-Nonanoate
24	Methyl Phenylpropionate
25	Methyl Trimethylbenzoate (isomer)

The acid fraction contains mainly normal and branched chain fatty acids with 4- to 9-carbon chain lengths. Minor amounts of cyclic and aromatic acids are also present. These are displayed in Fig. 6 as the methyl ester derivatives and peak identities are presented in Table 2. The total extractable acid concentration is 186 ppm.

The saturated and olefinic hydrocarbons (Fig 7) are the most striking compounds in the gas chromatogram of the L-3 retort water extract even though they constitute only a small fraction of the total extract weight (17 ppm of 1813 ppm total). The major compounds are normal akanes and alkenes, pristane, phytane, and the shorter chained isoprenoid hydrocarbons with minor amounts of branched and cyclic alkanes and alkenes (Table 3).

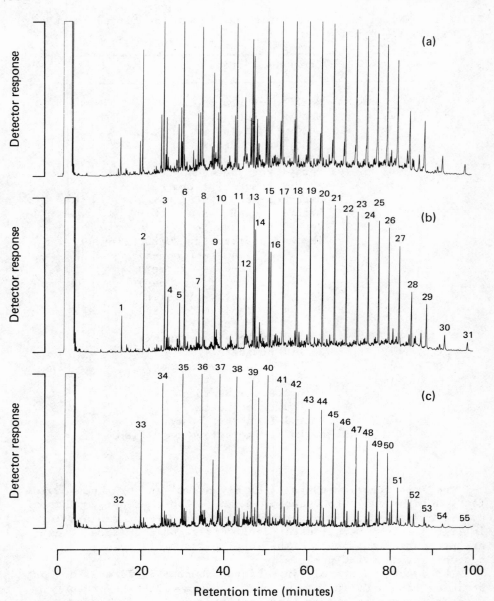

Fig. 7. Gas chromatograms of total aliphatic and olefinic hydrocarbons (a) and the separated saturated (b) and olefinic hydrocarbons (c) obtained by argentation chromatography. Peak identities are given in Table 3.

Table 3. Saturated and Olefinic Hydrocarbons in LLNL L3 Oil Shale Retort Identified water.

Peak Number	Compound Identification
1	Decane
2	Undecane
3	Dodecane
4	2,6-Dimethylundecane
5	2,6,10-Trimethylundecane
6	Tridecane
7	2,6,10-Trimethyldodecane
8	Tetradecane
9	2,6,10-Trimethyltridecane
10	Pentadecane
11	Hexadecane
12	2,6,10-Trimethylpentadecane
13	Heptadecane
14	2,6,10-14-Tetramethylpentadecane (Pristane)
15	Octadecane
16	2,6,10,14-Tetramethylhexadecane (Phytane)
17	Nonadecane
18	Eicosane
19	Heneicosane
20	Docosane
21	Tricosane
22	Tetracosane
23	Pentacosane
24	Hexacosone
25	Heptacosane
26	Octacosane
27	Nonacosane
28	Triacontane
29	Hentriacontane
30	Dotriacontane
31	Tritriacontane
32	1-Decene
33	1-Undecene
34	1-Dodecene
35	1-Tridecene
36	1-Tetradecene
37	1-Pentadecene
38	1-Hexadecene
39	1-Heptadecene
40	1-Octadecene
41	1-Nonadecene
42	1-Eicosene
43	1-Heneicosene

Table 3. (continued)

Peak Number	Compound Identification
44	1-Docosene
45	1-Tricosene
46	1-Tetracosene
47	1-Pentacosene
48	1-Hexacosene
49	1-Heptacosene
50	1-Octacosene
51	1-Nonacosene
52	1-Triacontene
53	1-Hentriacontene
54	1-Dotriacontene
55	1-Tritriacontene

The aromatic hydrocarbons consist of a wide range of structures from one to seven aromatic rings with alkyl and naphthenic substitution. Gas chromatograms of this fraction are presented in Fig. 8.

The sulfur compounds identified in fraction 4 from silica chromatography consist of a series of saturated heterocyclic sulfur compounds with alkyl substituted 5- and 6-membered rings.

The gas chromatogram of the methanol fraction from silica chromatography shows a complex unresolved mixture present in this fraction. Only 883 ppm of the 1340 ppm determined by weighing this fraction was chromatographable indicative of the high boiling and polar nature of the compounds present.

DISCUSSION

The organic compounds identified in the LLNL L3 oil shale retort water are a complex mixture of aliphatic, aromatic and naphthenic hydrocarbons, heterocyclic nitrogen and sulfur compounds, acids, bases and other polar compounds. The relative amounts of these compounds are presented in Table 4 and indicate that the majority of compounds are in the polar fraction eluted from silica gel. The next most abundant classes of compounds are the acids, and bases followed by aliphatic and aromatic hydrocarbons and sulfur compounds.

Chemical separation followed by GC and GC-MS provides a powerful analysis tool for a significant fraction of the compounds in oil shale retort water. Several hundred compounds can be

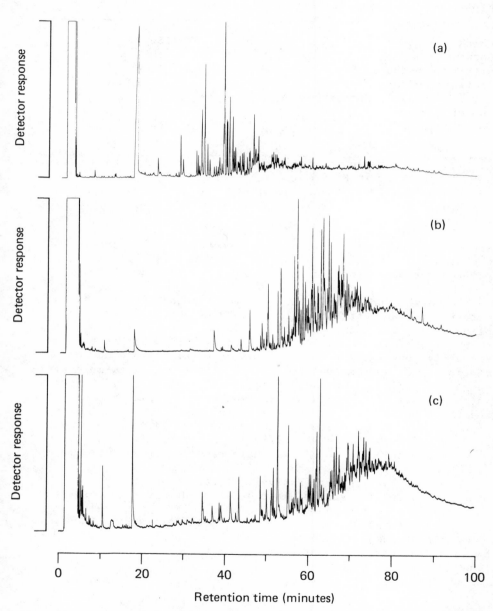

Fig. 8. Gas chromatograms of fractions II (a), IV (b) and V (c) from alumina chromatography in Fig. 2.

Table 4. Concentration of organic compounds identified in L3 oil shale retort water.

Fraction	Concentration (PPM)
Acids	186
Bases	110
Saturated hydrocarbons	12.4
Olefinic hydrocarbons	4.2
Total aromatic hydrocarbons*	1.4
Sulfur compounds**	10.3
Polar compounds***	883

* The concentration of individual aromatic hydrocarbon classes is presented in Figure 4.
** Fraction 4 from silica chromatography.
*** Fraction 5 from silica chromotography.

identified and quantitated by these techniques. However, new techniques are required to analyze the non-volatile and reactive polar fraction of these samples.

The high concentration of organic compounds in oil shale retort water and the composition containing ppm levels of aliphatic hydrocarbons indicates that mechanisms other than true solubility are responsible. Water equilibrated with petroleum generally contains less than 10 ppm of total organic carbon consisting of a relatively simple distribution of low-molecular-weight aromatic hydrocarbons (Stuermer, et al. 1980). Retort water appears to be a mixture or emulsion of shale oil and water and the complexity is representative of the oil itself. It seems likely that methods being developed for the organic analysis of shale oil (Fruchter, et al. 1978; Hertz, et al. 1980) can be applied to the analysis of oil shale retort waters as well.

Even so, the complete analysis of retort waters is not practical when many samples must be studied. In such cases the analysis of specific individual compounds is a much easier and less time consuming task (Hertz, et al. 1980). Particular compounds or compound classes should be identified for analysis based on the particular problems at hand.

REFERENCES

Fruchter, J.S. J.C. Laul, M.R. Petersen, P.W. Ryan and M.E. Turner (1978) "High-Precision Trace Element and Organic Constituent Analysis of Oil Shale and Solvent-Refined Coal Materials" in <u>Anal. Chemistry of Liquid Fuel Sources</u>, P.C. Uden and H.B. Jensen, eds., Adv. Series 170, Amer. Chem. Soc., Washington, D.C. pp. 255-281.

Giger, W., and B. Blumer (1977) Polycyclic Aromatic Hydrocarbons in the Environment: Isolation and Characterization by Chromotography, Visible, Ultravoilet, and Mass Spectrometry. <u>Anal. Chem.</u>, <u>46</u>: 1663-1671.

Hertz, H.S., J.M. Brown, S.N. Chesler, F.R. Guenther, L.R. Hilpert, W.E. May, R.M. Parris, and S.A. Wise (1980) Determination of Individual Organic Compounds in Shale Oil. <u>Anal. Chem.</u>, <u>52</u>: 1650-1657.

McKay, J.F., and D.R. Lathan (1980) High Performance Liquid Chromatographic Separation of Olefin, Saturate and Aromatic Hydrocarbons in High-Boiling Distillates and Residues of Shale Oil. <u>Anal. Chem.</u>, <u>52</u>: 1618-1621.

Stuermer, D.H., R.B. Spies and P.H. Davis (1980) Toxicity of Santa Barbara Seep Oil to Starfish Embryos. <u>Marine Env. Research</u> (in press).

ACKNOWLEDGEMENTS

We gratefully acknowledge the LLNL oil shale group for retort water samples and Ilze Cebers for her labortaory efforts.

PYROLYSIS/HIGH RESOLUTION MASS SPECTROMETRY WITH METASTABLE PEAK MONITORING APPLIED TO THE ANALYSIS OF GREEN RIVER SHALE KEROGEN

Rafael Infante* and Gerhard G. Meisels

Department of Chemistry
University of Nebraska
Lincoln, Nebraska 68588

ABSTRACT

High resolution mass spectra of products resulting from the temperature programmed pyrolysis of Green River shale kerogen in the ion source of a mass spectrometer showed mass peaks to m/z 700. Elemental composition and linked metastable scans, particularly of components containing heteroatoms, were obtained at five second intervals during pyrolysis. This showed that nitrogen and oxygen containing compounds are present in greater abundance than those containing sulfur, and that sulfur always appears in combination with one of the other heteroatoms. Nitrogen is found primarily as a part of stable aromatic ring systems; quinoline and isoquinoline structures appear to be most probable. Oxygen containing species were associated principally with aliphatic hydrocarbon skeletons.

INTRODUCTION

Kerogen from Green River oil shale has been the subject of numerous investigations (Burlingame and Simoneit, 1969; Burlingame and Schones, 1969; Gallegos, 1971 and 1975; Maters et al., 1977; Schmidt-Collerus and Prien, 1974; Yen, 1976); most of these have emphasized the analysis of hydrocarbon compounds. Little information exists on kerogen components containing heteroatoms, however, the role of these species is of primary importance in obtaining a complete picture of kerogen and its binding to the zeolite structure. We have therefore attempted to obtain qualitative and quantitative information about the products resulting from the vacuum pyrolysis

---

*Present address: Department of Chemistry, Catholic University of Puerto Rico, Ponce, Puerto Rico 00731

of kerogen, with particular emphasis on substances containing heteroatoms. Pyrolyses were performed directly in the ion source of a rapid scanning, high resolution mass spectrometer. Carbon number distributions and structural information were derived from elemental maps and from studies of metastable ions using the linked scan method.

EXPERIMENTAL

Green River shale was obtained from the DOE center in Laramie, Wyoming. The elemental composition as well as the Fischer assay analysis for the shale is shown in Table 1.

Table 1. Elemental composition and Fischer assay analysis

Sample	C	H	N	S	Ash	BTU	CO_2
OSCR-76-373	23.07	2.42	0.83	1.06	60.35	3764	18.07

Fischer assay analysis

	Yield of product	
	Weight per cent	Gal per ton
Sample OSCR-76-373		
Oil	13.5	35.2
Water	1.4	3.4

The mass spectrometer was a Kratos MS-5076 coupled to an INCOS data system which acquires and processes the data. The results are then stored on magnetic tape.

The method described by Gallegos (1975) was used for sample preparation. About 30 g of finely crushed Green River shale (80-100 mesh) was extracted for 200 hours using a 50/50 methanol-benzene solution, 10 ml of solution per gram of shale, under reflux conditions. The residue was placed on a capillary tube and brought into the ion source of the mass spectrometer via the direct insertion probe. Data acquisition was initiated and a full scan taken every five seconds. The sample was then heated gradually from 50-400°C by increasing the temperature 25° after every 10th scan.

The high resolution measurements, acquired at a resolving power of 10000, were controlled by the computer and recorded on a magnetic disk followed by subsequent data reduction by the computer. Precise mass measurements were obtained for approximately 700 peaks in each of the spectra studied. Accuracy for the assigned peaks was 1 to 3 millimass units, with a few exceptions. Possible compositions having up to 8 oxygen, 6 nitrogen, and 4 sulfurs were considered in the program used to determine elemental composition.

Linked scan metastable analyses were obtained using the Kratos metastable unit in conjunction with the MS-50 mass spectrometer. The electron energy was set at 70eV and the source temperature was 200°C. A typical scan covering the mass range m to 1/2 m, was obtained in 80 seconds. This procedure identifies metastable and hence fragmentation processes occuring in a single step from an ion of a given mass.

RESULTS AND DISCUSSION

The pyrolysis/mass spectrometer system used in this investigation led to observation of mass peaks higher than m/z 600 from the kerogen samples. The detection of these large pyrolysis products makes this technique suitable for structural studies. However, to reconstruct kerogen from the pyrolysis products, the mechanism by which the pyrolysis products are generated must be known. Ideally one hopes for simple cleavage which allow direct correlation between pyrolysis products and the original sample. Although this did not occur, useful information is obtained and some conclusions can be drawn.

As expected pyrolysis products of higher m/z values are observed as the temperature increases. This can be explained from volatility considerations or from the fact that the higher mass components are held more tightly in the kerogen matrix. In addition, the types of ions observed at different temperatures vary. At lower temperatures, hydrocarbons dominate; as the temperature is increased, oxygen, sulfur and nitrogen containing ions appear. Table 2 summarizes the major products obesved; Table 3 gives their relative abundances.

Hydrocarbons are dominant among the pyrolysis products of Green River shale kerogen. Characteristic ions for hydrocarbons occur over the entire mass range investigated, although they are less abundant in the mass range greater than m/z 200. Saturated hydrocarbons are found in greater abundance. Aromatic hydrocarbons are also observed; these become more prominent as the temperature increases. Naphthenic hydrocarbons start appearing above 200°C.

The hydrocarbon product distribution (Table 4) is very similar to that reported by Schmidt-Collerus and Prien (1974),

Table 2. Major fragmentation products found in the pyrolysis of kerogen

Compound	Formula	Range of n
Alkanes	C_nH_{2n+2}	5-17
Monocycloparaffins	C_nH_{2n}	5-15
Dicycloparaffins	C_nH_{2n-2}	8-14
Tricycloparaffins	C_nH_{2n-4}	11-15, 18
Alkylbenzenes	C_nH_{2n-6}	7-14
Benzocycloparaffins	C_nH_{2n-8}	9-14
Benzodicycloparaffins	C_nH_{2n-10}	10-14, 19
Naphthalenes	C_nH_{2n-12}	11-16
Alkyl Pyridines	$C_nH_{2n-5}N$	7-9
Tetrahydroquinolines or Cycloalkylpyridines	$C_nH_{2n-7}N$	9, 10, 22
Alkyl Indoles	$C_nH_{2n-9}N$	12-14, 23, 24
1-O Compounds	$C_nH_{2n-1}O$	3-6, 14, 16
2-O Compounds	$C_nH_{2n-1}O_2$	4-11

Pyrolysis products containing one to five oxygen atoms as the only heteroatoms are observed. Fragments containing more than one oxygen atom are more abundant at higher temperatures while those containing a single oxygen atom decrease in intensity with increasing temperatures (Table 3). Oxygen is also found in combination with other heteroatoms such as nitrogen and sulfur.

High resolution measurements established that oxygen is contained in two major types of ions: $C_nH_{2n-1}O^+$ and $C_nH_{2n-1}O_2^+$. In both types the hydrocarbon skeleton associated with them is aliphatic.

The low resolution spectra indicated the presence of a number of ions containing two oxygens, such as m/z 60, m/z 74, and m/z 88. High resolution measurements showed that m/z 60 has the composition $C_2H_4O_2^+$; m/z 74, $C_3H_6O_2^+$ and m/z 88, $C_4H_8O_2^+$. These are characteristic rearrangement ions of esters and carboxylic acids. The release of these types of compounds in the pyrolysis of kerogen is supported by the series of odd-mass peaks of general composition $C_nH_{2n-1}O_2$ at m/z 73, 87, 101, etc.

Table 3. Relative intensities for the major pyrolysis products

Compound class	\multicolumn{5}{c}{Temperature}	Total	%Total
	100	175	225	275	300		
Paraffins	150	160	140	110	90	650	22.6
Monocycloparaffins	120	170	130	110	100	630	22.4
Dicycloparaffins	70	40	40	30	20	200	7.0
Tricycloparaffins	30	10	20	20	5	85	3.0
Alkylbenzenes	40	30	60	50	50	230	8.0
Benzocycloparaffins	10	10	20	20	10	70	2.4
Benzodicycloparaffins	1	---	20	20	10	51	1.8
Naphthalenes	---	---	10	10	10	30	1.0
1-O	20	80	70	20	10	200	7.0
2-O	40	50	20	1	10	121	4.2
3-O	---	---	3	4	10	17	0.6
1-N	---	---	10	20	80	110	3.8
N-O	1	20	30	80	70	201	7.0
N-S	---	---	20	60	80	160	5.6
O-S	---	---	---	1	20	21	0.7

Table 4. Hydrocarbon pyrolysis products

Homologous compounds	Empirical formula	Major fragment peaks	Range of n
Paraffins	C_nH_{2n+2}	71,85,99,113,127, 141,155,169,183, 197,211,225	5-17
Monocycloparaffins	C_nH_{2n}	69,83,97,111,125, 139,153,167,181, 195,209	5-15
Dicycloparaffins	C_nH_{2n-2}	109,123,137,151, 165,179,193	8-14
Tricycloparaffins	C_nH_{2n-4}	149,163,177,191, 205,247	11-15,18
Alkylbenzenes	C_nH_{2n-6}	91,105,119,133, 147,161,175,189	7-14
Benzocycloparaffins	C_nH_{2n-8}	117,131,145,159, 173,187	9-14
Benzodicycloparaffins	C_nH_{2n-10}	129,143,157,171, 185,255	10-14,19
Naphthalenes	C_nH_{2n-12}	141,155,169,183, 197,211	11-16

Two oxygen containing ions were studied by the linked scan metastable method. The ion at m/z 264, elemental composition $C_{18}H_{32}O^+$ and the m/z 256 ion, elemental composition $C_{16}H_{32}O_2^+$. On the basis of our observations no final conclusion on the structures of these ions can be reached yet. Comparison with linked scan spectra of reference compounds is currently under way in our laboratory in order to assign the structure.

Nitrogen-containing compounds start appearing once the temperature is in excess of 250°C. They give rise to ions containing one, two, and three nitrogens. Ions containing a single nitrogen are found in greater abundance; their intensity increases with increasing temperature.

A convenient characterization of nitrogen compound types, particularly of nitrogen heterocyclic ring system, is possible by examining the value Z in compositions $C_nH_{2n-Z}N$. Z-number are, of course, not usually sufficient to establish a particular skeleton, since the given value is not necessarily unique for a structural type, and the classification is therefore only suggestive.

Table 5. Nitrogen containing pyrolysis products of Green River shale kerogen

Homologous compounds	Empirical formula	Z	Major fragments peaks	Range of n
Alkyl pyridines	$C_nH_{2n-5}N$	5	107,121,135	7-9
Alkyl indoles	$C_nH_{2n-9}N$	9	172,186,200, 326, 340	12-15,23,24
Tetrahydroquinolines or cycloalkylpyridines	$C_nH_{2n-7}N$	7	133,147,315	9,10,22
Alkyl quinolines or isoquinolines	$C_nH_{2n-11}N$	11	171,185,199	12-14

High resolution mass measurements led to the observations summarized in Table 5. The possible structural types of nitrogen containing ions are the following: alkyl substituted pyridines (Z= 5); tetrahydroquinoline- (Z= 7); alkyl substituted indoles (Z= 9); quinolines or isoquinolines (Z= 11); phenylpyridines (Z= 13); and benzoquinolines (Z= 17).

The metastable scan of the nitrogen containing ion at m/z 225 of composition $C_{16}H_{19}N^+$ is shown in Figure 1. It appears to be a substituted quinoline or isoquinoline. Even mass fragmentation

PYROLYSIS/HIGH RESOLUTION MASS SPECTROMETRY

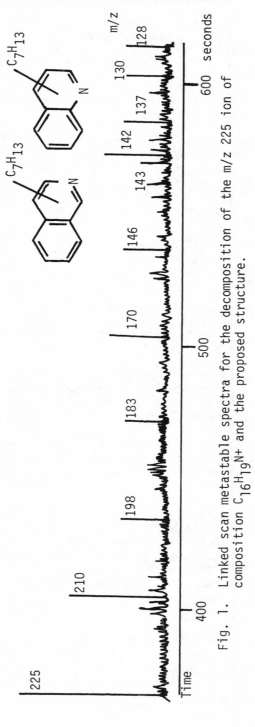

Fig. 1. Linked scan metastable spectra for the decomposition of the m/z 225 ion of composition $C_{16}H_{19}N^+$ and the proposed structure.

ions characteristic of compounds containing an odd number of nitrogen atoms are abundant. The m/z 198 ion may correspond to loss of HCN which is characteristic for this class of compounds. The indication of the side chain length may be given by the m/z 128 ion which corresponds to loss of C_7H_{13}. The m/z 143 ion corresponds to a McLafferty arrangement also characteristic for this class of compounds.

Sulfur is contained in small quantities among pyrolysis products appearing at temperatures over 225ºC. No ions containing sulfur as the only heteroatom were observed. Sulfur was found in combination either with nitrogen or oxygen, with the former combination more intense. No definite conclusions about the way sulfur is bound in these products is possible on the basis of our observation.

CONCLUSIONS

Pyrolysis/high resolution mass spectrometry proved to be succesful in the identification of some of the pyrolysis products of Green River shale kerogen. Pyrolysis products containing oxygen, nitrogen, and sulfur were observed. Nitrogen and oxygen containing components are found in greater abundance than those containing sulfur, which are found in small quantities.

Nitrogen seems to be present mostly in stable ring system. The hydrocarbon skeleton associated with the nitrogen containing ions was aromatic rather than aliphatic. The metastable scan suggested than nitrogen is present in quinoline or isoquinoline structure.

The hydrocarbon skeleton associated with the oxygen containing ions is aliphatic rather than aromatic. Sulfur containing species were difficult to characterize. It is not clear how the sulfur is bonded in these products.

Acknowledgements-- We are greatful for support of this investigation by the Department of Energy under contract number De-AS02-76er02567, and by the University of Nebraska Research Council. The high resolution mass spectrometer was made available as a part of the Regional Facility in Mass Spectrometry at the University of Nebraska supported by the National Science Foundation.

REFERENCES

    Burlingame, A.L., and Simoneit, B.R., 1969, High resolution mass spectrometry of Green River formation kerogen oxidation, Nature, 222:741
    Burlingame, A.L., and Schnoes H.K., 1969, Mass spectrometry in organic geochemistry, in: "Organic Geochemistry,"

G. Eglinton and M.T.J. Murphy, eds., Springer-Verlag, Berlin.

Djuricic, M., Murphy, R.C., Vitorovic, D., Biemann, K., 1971, Organic acids obtained by alkaline permanganate oxidation of kerogen from the Green River (Colorado) shale, Geochim. Cosmochim. Acta, 35:1201.

Gallegos, E.J., 1971, Identification of new steranes, terpanes and branched paraffins in Green River shale by combined capillary gas chromatography and mass spectrometry, Anal. Chem., 43:1151.

Gallegos, E.J., 1975, Terpane-sterane release from kerogen by pyrolysis-gas chromatography-mass spectrometry. Anal. Chem., 47:1524.

Maters, W.L., Meent, D.V.D, Schuyl, P.J.W., deLeew, J.W., Schenk, P.A., Meuzelaar, H.L.C., 1977, "Curie point pyrolysis in org-nic geochemistry", in: "Analytical Pyrolysis," C.E.R. Jones and C.A. Cramers, eds., Elsevier, Amsterdam.

Schmidt-Collerus, J.J., and Prien, C.H., 1974, Investigation of the hydrocarbon structure of kerogen from oil shale of the Green River formation, ACS Div. of Fuel Chem. Preprints, 19:100.

Yen, T.F., 1976, Structural investigation of Green River shale kerogen, in: "Science and Technology of Oil Shale," T.F. Yen, ed., Ann Arbor Science, Michigan.

ADVANCES IN THE CHEMISTRY OF ALBERTA TAR SANDS:

FIELD IONIZATION GAS CHROMATOGRAPHIC/MASS SPECTROMETRIC STUDIES

O.P. Strausz, I. Rubinstein,*, A.M. Hogg,
and J.D. Payzant

Department of Chemistry
University of Alberta
Edmonton, Alberta
Canada T6G 2G2

INTRODUCTION

Tar sands are important future non-conventional sources of liquid fuels because of their total quantity and economy of recovery relative to other non-conventional resources. Tar sand deposits of smaller sizes are of widespread occurrence in the world and can be found on all continents, Figure 1,[1] with the exception of Antarctica.[2] Most of them are without commercial significance but the two largest ones are so huge that they represent by far the largest known oil accumulations in the world, unparalleled by conventional reservoirs. These are located in Venezuela along the Orinoco River and in Alberta, Canada.[1,2]

The in place bitumen content of the Orinoco deposit is 0.7-3 trillion barrels with an estimated recoverable portion of 500 billion barrels.[2] The aggregate amount of bitumen present in the four major tar sand deposits of the Western Canadian sedimentary basin, the Peace River, Wabasca, Athabasca and Cold Lake reservoirs, Figure 2,[3] is in excess of one trillion barrels.[1] Heavy oils are often associated with carbonate facies.[4] In the case of the Alberta tar sand deposits the associated carbonate deposit, in the so-called carbonate trend approximately underlying the Wabasca-Athabasca tar sand accumulations, is estimated to contain 312 billion barrels of bitumen.[1]

---

*Present address: Esso Resources Canada Ltd.,
339 - 50th Avenue S.E., Calgary, Alberta,
Canada T2G 2B3

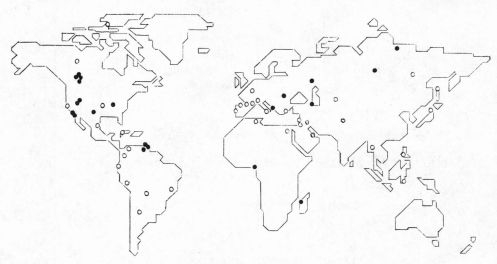

Figure 1. Illustration of major (●) and minor (○) tar sand deposits in the world.[1]

The bitumen that will ultimately be recovered from the Alberta tar sand and carbonate trend reservoirs is estimated to be 213 billion barrels,[1] comprising about 35% of the world's proven conventional oil reserve. Up to 15% of the recoverable oil can be produced by the currently employed surface strip mining and hot water separation technology. Presently there are two tar sand plants in operation, both in the Athabasca area, with a combined production capacity of 180,000 barrels per day of high grade synthetic crude oil. The economy of the process is very inviting and the construction of two additional plants, each with a design capacity of 140,000 barrels synthetic crude per day, is currently being considered. The capital investment required for a tar sand plant of this size is $8-9 billion with a construction time of about four years.

The Alberta oil sands contain about 4-5% water which forms a film surrounding each grain particle of the sand. The bitumen fills the space between the unconsolidated wet sand grains without being in direct contact with the sand itself. This sand is called water-wet, which is an important characteristic of the Alberta tar sands. In contrast, American tar sands are oil-wet. Water-wet oil sands, when placed in water at around 90°C, tend to break up, the oil floating to the surface and the sand sinking to the bottom. This phenomenon forms the basis of the so-called hot water separation technology. The Alberta oil sand bitumen has a high specific gravity, slightly higher than that of water, and flotation is assisted by bubbles of air and gases liberated from the tar sand, which attach themselves to the oil droplets.

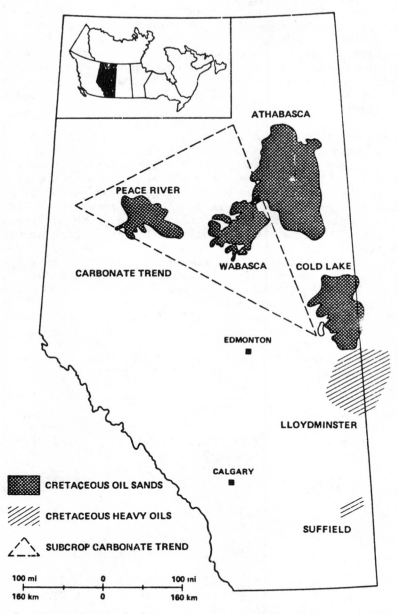

Figure 2. Locations of the major tar sand deposits in Alberta, Canada.[3]

The Alberta tar sands, probably like most tar sands, contain insignificant amounts of kerogen and more than 99% of the organic matter present is soluble in organic solvents. They also contain about 0.5% humic material which is insoluble in organic solvents or water but which can be dissolved with dilute aqueous base solutions. Heteroatoms, N, O, S, are abundant in the bitumen and are concentrated in the asphaltene and resin fractions, Table I. The hydrocarbon content of the bitumen varies between 33% and 50%, the asphaltene content between 14% and 20% and the rest is made up of resins, Table II.

Even though detailed knowledge of the composition of bitumens and crude oils is of considerable interest to the reservoir engineer, refining chemist, petroleum geologist and organic chemist, compositional data on bitumens are scant since analysis of the bitumen is extremely difficult because of the large number of constituents, many of which are poorly characterized or have unknown molecular structures. Even the simplest fraction, comprised of hydrocarbons, presents such a formidable challenge to the analytical chemist that complete analysis of its chromatographically separated subfractions - such as the cycloalkane or aromatic subfractions - could not be satisfactorily achieved with state-of-the-art level analytical methods. These consist largely of some form of mass spectrometry such as combined gas chromatography-electron impact mass spectrometry (GC-EIMS) or chemical ionization mass spectrometry (GC-CIMS) which are powerful tools for

Table I. Elemental Composition of Alberta Tar Sand Bitumens

Fraction	Source*	C	H	N	O	S	H/C	mw
				%				
Whole Bitumen	A	83.98	10.22	0.65	1.97	4.57	1.46	620
	P	81.68	9.98	0.14	2.08	5.60	1.46	520
	C	83.93	10.46	0.23	0.94	4.70	1.49	490
	W	82.44	10.32	0.42	0.82	5.51	1.49	600
Asphaltenes	A	81.31	7.88	1.06	2.79	7.53	1.16	5,920
	P	79.87	8.15	0.78	2.08	8.82	1.21	9,500
	C	80.54	7.39	1.15	1.78	6.51	1.10	8,140
	W	80.46	8.20	0.99	1.16	8.40	1.21	5,760
Deasphaltened Oil	A	84.38	10.63	0.07	0.87	3.91	1.51	435
	P	82.89	10.61	0.08	1.27	5.41	1.53	440
	C	84.19	11.01	0.27	0.61	3.89	1.54	430
	W	82.99	10.74	0.28	1.04	5.25	1.54	530

*A = Athabasca, P = Peace River, C = Cold Lake, W = Wabasca bitumen.

Table II. Average Class Distribution of Alberta Tar Sand Bitumens

Component	Peace R.	Wabasca	Athabasca	Cold L.
		weight %		
Hydrocarbons:	34	33	47	40
Saturated	15	15	18	21
Aromatic	19	18	29	19
Asphaltene:	20	19	17	16
Resins:	44	48	35	44
Acidic	12	10	14	15
Basic	7	6	7	7
Neutral N-Compounds	1	3	1	1
Neutral	24	29	13	21
Total	98	100	99	100

compositional studies of volatile mixtures, but when it comes to mixtures containing many hundreds of hydrocarbons in the $C_{15}-C_{50}$ range, like the cycloalkanes or aromatics of the tar sands, they are inadequate.

In EIMS, hydrocarbon molecules suffer extensive fragmentation giving rise to a large number of fragment ions with more or less intense parent ion peaks. Since crude oil fractions may contain hundreds of hydrocarbons, each yielding many fragment peaks, the resultant mass spectrum is too complex for detailed interpretation. Chemical ionization (CI) and field ionization (FI) are gentler methods for ionizing molecules than the EI mode and with some exceptions they result only in parent ion formation without significant fragmentation. The sensitivity of hydrocarbons with respect to CI shows a marked variation with molecular structure (determined by such molecular parameters as ionization potential, electron affinity, proton affinity, etc.). In CI the reagent gas gives an intense ion current and causes the appearance of the quasi-molecular ions $(M-1)^+$, $(M+1)^+$ and their clusters. FI is largely free of these kinds of complications and the ionization yield is, to a large measure, independent of molecular structure for aromatic hydrocarbons and for alkanes. These characteristics of FI give it a definite advantage over CI in the analysis of petroleum.

We have recently been able to combine gas chromatography with FI mass spectrometry and apply it to the analysis of crude oils and tar sand bitumens. The results of these studies will be reported here.

In FI the sample molecule to be ionized is exposed to the effect of an intense electrical field with field gradients of the same order of magnitude as that experienced by the outermost electrons from the nucleus, which is about 10^8 V cm^{-1}. This results in molecular ion formation *via* a quantum mechanical tunnelling process with little excess vibrational energy imparted to the ion. Therefore the primary molecular ions do not fragment to smaller ions before analysis and direct molecular weight determination can readily be accomplished.

FIMS was pioneered by Beckey[5] and its application for the analysis of hydrocarbons was first reported by Beckey and Wagner[6] who utilized platinum tip emitters to obtain the required electric field. In the present study the emitter source used was the high efficiency carbon needles on tungsten wire type. Beckey and Wagner's experiments showed the potential of FI in the analysis of hydrocarbon mixtures, but at the same time, because of the variation of the relative sensitivity of the various compounds in a mixture with composition, they also pointed to the limitations of FI in quantitative analyses. The problem of sensitivity variation with composition has subsequently been studied by several workers. Thus, Ryska et al.[7,8] found that the relative sensitivities of various hydrocarbons become independent of mixture composition in the presence of ethyl benzene and other aromatics, and Mead[9] reported that the relative sensitivities within major groups (*n*- and *iso*-alkanes, *cyclo*alkanes, alkyl benzenes) are constant in the C_{20}-C_{40} range. Scheppelle et al.,[10,11] on the other hand, reported a dependence of the relative sensitivity on the ion source temperature. They noted, however, that in the temperature range 200-270°C, the relative sensitivities of some 60 aromatic compounds studied were approximately equal. Significantly, they also noted that the relative sensitivities of various hydrocarbons were independent of mixture composition in this temperature range.

In an earlier study from this laboratory, FIMS was applied to the geochemical analysis of oil sand bitumens.[12] The success of this study suggested to us the potential value of combining GC with FIMS.[13] At that time the only reference that existed on this topic in the literature was the work of Damico and Barron[14] who, in 1971, reported the construction of such an apparatus and although it was primitive by today's standards it clearly demonstrated its feasibility and usefulness.

More recently, while our work on the construction of a GC-FIMS was in progress, Milbury and Cook[15] described the GC-FIMS combination on a modern instrument and a considerable improvement in sensitivity was achieved. With their instrument they were able to analyze a number of drugs, pesticides and their metabolites with up to about three dozen components.

The advantages of GC-FIMS in the analysis of crude oils and hydrocarbon mixtures that can be anticipated are the following:

(a) different isomeric molecules with identical molecular weight can be resolved without interference from co-eluting molecules with differing molecular weights,

(b) different homologous series (C_nH_{2n}, C_nH_{2n-14} and C_nH_{2n-28} or C_nH_{2n-6} and $C_nH_{2n-10}S$ etc., can be resolved,

(c) fragmentation can be recognized, corrected for, and used for quantitative analysis and for structural correlations,

(d) since the relative sensitivity varies little with structure within a class of compounds, no calibration is required by individual components for quantitative analysis.

The results presented in our previous article[13] and to be reported here show that these expectations are indeed fulfilled.

EXPERIMENTAL

Materials

Three samples were analyzed, one conventional light oil and two tar sand bitumens. The light oil was from the Leduc oil field in central Alberta (Imperial Oil Ltd.), location 10-50-26-W4, well depth 1600 m (Devonian D3 formation). One of the bitumen samples was a cold bailed material from the Cold Lake formation and the other one was a solvent extracted material from the Athabasca formation.

The crude oil (100 mg) was chromatographed on a 250 g silica gel column and the saturated hydrocarbons were eluted with the first 400 mL aliquot of Skellysolve B. The aromatic fraction was subsequently eluted with 25% dichloromethane in Skellysolve B (400 mL). The polar material was stripped from the column with dichloromethane.[16,17]

The saturated fraction was treated with 5 Å molecular sieves, followed by urea clathration[18,19] to remove the n- and iso-alkanes. The urea non-adduct was thrice clathrated with thiourea to yield the thiourea adduct (TUA) and thiourea non-adduct (TUNA) fractions.[20] The gravimetric results are summarized in Table III.

The procedure for fractionation of the bitumen samples was similar and has been described before.[16,17]

Table III. Separation of Leduc Oil into Various Fractions

Fraction	Weight %	
Saturates:		
Sieve Adduct	63.5	
Urea Adduct	0.5	76
Thiourea Adduct	3.5	
Thiourea Non Adduct	8.5	
Aromatics	23	
Polars	1	
	100 %	

Apparatus

The GC apparatus employed was a Varian 1200 fitted with a 2.5 m x 3 mm stainless steel column packed with 3% OV101 on 100/120 mesh Chromosorb W (AWDMCS). Helium flow was 45 mL min^{-1}, injection temperature 350°C and the oven programmed from 60° to 300° at 2° min^{-1}. Approximately 4 mg of sample were injected for each run. The column exit was coupled to the modified AEI MS-9 mass spectrometer[21,22] by a heated steel capillary transfer line and a Watson-Bieman ceramic separator. The separator outlet was sealed by means of a spring-loaded guide to the source block-counter electrode so that the GC effluent was discharged from a 2 mm orifice in line-of-sight with the emitter wire, 2.3 mm away. All internal surfaces were heated to 250°C or over.

The mass spectrometer was scanned repetitively from m/z 700 to m/z 28 at a scan speed of 5 sec/decade at a repetition rate of 11 sec including reset time. Data acquisition and processing were performed by an on-line AEI DS-50S system.

RESULTS AND DISCUSSION

The compositions of the hydrocarbons from the conventional oil and from the tar sand bitumen are quite different, as seen from the data in Tables II and III. The main component of the conventional oil, the molecular sieve adduct, is the normal alkane, of which the bitumens are almost completely devoid. Since the analysis of the normal alkanes by GC-FIMS is of particular interest because of their extensive fragmentation in FI we first present the results on the conventional oil sample.

Leduc Conventional Oil

(A) Aromatic Fraction. Analysis of the aromatic fractions from crude oil by GC-EIMS is fraught with difficulties, and is especially hampered by the presence of thiophenic components. Direct inlet FIMS analysis, with the spectra containing solely parent ion data, gives information on the distribution of the various classes of compounds within the mixture, and is useful in comparative studies on fossil fuel products.[12] GC-FIMS gives molecular weight data on each individual compound as it elutes from the GC column and thus provides additional, helpful information on the composition of the mixture.[13]

The total ion current (TIC) (in this case, the sum of the ions from m/z = 28 to 600) for the aromatic fraction is shown in Figure 3. A number of well resolved peaks appear at the beginning of the chromatogram; however, as the scan progresses, the complexity of the sample is such that the TIC is manifested as a broad hump with a large number of only partially resolved low intensity peaks.

There are several ways of displaying the resulting GC-FIMS data. One method consists of summing together the molecular ions for a given molecular formula. For example, the naphthalenes (C_nH_{2n-12}) present a homologous series of molecular ions at masses 128, 142, 156, 170, 194, 198... etc. for n = 10, 11, 12, 13, 14, 15... etc. Such a sum is presented in Figure 4. Figure 4 then is

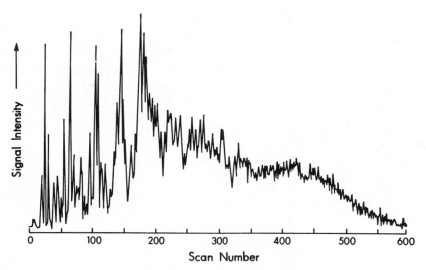

Figure 3. Aromatic fraction. Total ion current (sum from m/z = 28 to 600). Most intense peak is 18 arbitrary units. The scan number is proportional to GC elution time.

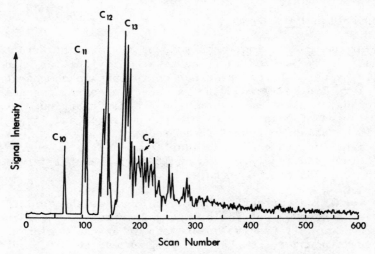

Figure 4. Aromatic fraction. Summation of the molecular ion intensities for the series C_nH_{2n-12} (naphthalenes). Most intense peak is 11 units.

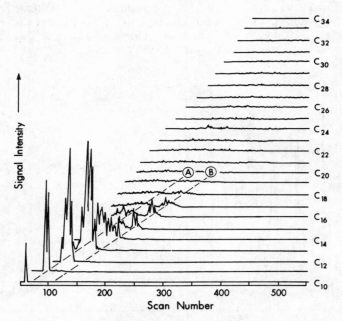

Figure 5. Aromatic fraction. Individual molecular ion intensities for the mass series corresponding to the formula C_nH_{2n-12}. The numbers next to the individual curves are the appropriate values of n in the formula C_nH_{2n-12} (naphthalenes). Most intense peak is 11 units.

a simulated representation of a conventional GC trace which would be observed if all material of the mass series C_nH_{2n-12} had been separated from the remaining aromatic hydrocarbons prior to introduction to the GC column. It is not possible to reconstruct this simulated GC trace from the corresponding GC-EIMS data. This is because of the complexity of the mass scans with their overlapping patterns of fragment and molecular ions. The simple molecular ion spectra of FIMS makes this interpretation possible.

Figure 5 presents an alternative method of displaying this same data, where the scans for the individual molecular ions are positioned in a vertical sequential series. In this manner the various isomers of a given molecular ion may be seen. The trace corresponding to n = 11 (m/z = 142), for example, corresponds to the methyl naphthalenes. The peaks corresponding to n = 13, m/z = 170, are those corresponding to hydrocarbons of the formula $C_{13}H_{14}$, probably a naphthalene carbon skeleton with three additional carbon atom group(s). These group(s) could be three methyl substituents, one methyl plus one ethyl, or n-propyl or iso-propyl groups. The number of isomers is large even for this simple case and they are only partially resolved. As the number of carbon atoms increase many more isomers appear.

The series of formulae C_nH_{2n+2}, C_nH_{2n-12}, C_nH_{2n-26}, C_nH_{2n-40}... etc. all have molecular weights which satisfy the m/z series 128, 142, 158... etc. The formula C_nH_{2n+2} corresponds to saturated acyclic hydrocarbons which have been removed chromatographically from the aromatic fraction. The series C_nH_{2n-40} and the more highly hydrogen deficient series are applicable only for comparatively large values of n and are not expected to be found in significant amounts in a light oil such as the Leduc.

In addition to the two series C_nH_{2n-12} and C_nH_{2n-26} there are a number of series of formula C_xH_yS which fit this m/z series. These formulae are $C_nH_{2n-2}S$, $C_nH_{2n-16}S$, $C_nH_{2n-30}S$, etc. Material of formula $C_nH_{2n-2}S$ is not a major constituent of this oil and $C_nH_{2n-30}S$ is a reasonable formula only for large values of n. The series $C_nH_{2n-16}S$ is the elemental formula for the dibenzothiophenes. The aromatic fraction mass ions at m/z = 184, 198, 212, 226, 240 were examined at high resolution (m/Δm $\sim$ 10,000) using the reservoir inlet system previously described.[12] It was found that these ions were doublets and that the accurate mass measurements were consistent with the assignment of the low mass peak as a sulphur-containing molecule. Thus, the GC mass ion traces of these ions in the naphthalene series must contain peaks due to material of formula $C_nH_{2n-16}S$. The series of peaks (indicated by line B in Figure 5) which appears at longer retention times than the corresponding naphthalenes (line A) are these dibenzothiophenes. Their relative abundance and distribution are consistent with earlier observations made using the reservoir inlet system. The GC retention times relative to the naphthalenes in the Leduc sample were verified with GC retention time studies on model compounds.

Figure 6 displays the ion intensity traces for the series m/z = 120, 134, 148, 162, 176 etc. These masses correspond to values of n = 9, 10, 11, 12, 13... etc. in the formula C_nH_{2n-6} (line A). A series of peaks indicated by line B in Figure 6 was identified as benzothiophenes ($C_nH_{2n-10}S$) which have the mass series m/z = 134, 148, 162... for n = 8, 9, 10... determined using the same criteria as for the case of the dibenzothiophenes. The series of peaks indicated by line C in Figure 6 is probably a series of compounds of formula C_nH_{2n-20} which belong to the series m/z = 176, 190, 204, 218, 232... for n = 14, 15, 16, 17, 18..., since a more hydrogen deficient series (i.e., a poly-condensed aromatic system) would have a longer retention time.

The ion intensity traces for m/z = 120 and 134 are displayed in Figure 7. These masses correspond to n = 9, 10 in the series C_nH_{2n-6}, i.e., a benzene ring with 3 and 4 carbons, respectively, as substituents on the ring. This trace illustrates the potential

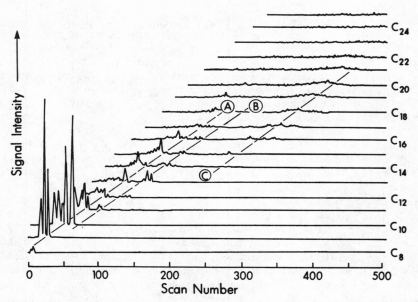

Figure 6. Aromatic fraction. Individual molecular ion intensities for the mass series corresponding to the formula C_nH_{2n-6}. Line A is drawn through the homologous series of benzenes. Line B is drawn through the series of benzothiophenes, $C_nH_{2n-10}S$, and line C through material of formula C_nH_{2n-20}. Most intense peak is 13.5 units.

CHEMISTRY OF ALBERTA TAR SANDS

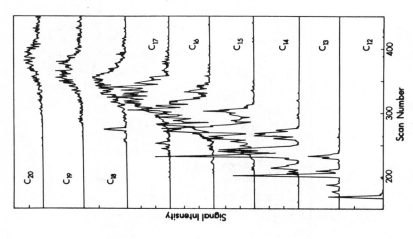

Figure 8. Aromatic fraction. Individual molecular ion intensities for the mass series corresponding to the formula C_nH_{2n-14}. Most intense peak is 3.2 units.

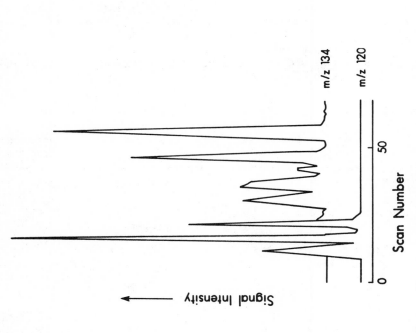

Figure 7. Aromatic fraction. Individual molecular ion intensities for m/z = 120 and 134. These correspond to n = 9 and 10 in the formula C_nH_{2n-6} (alkyl benzenes), respectively. Most intense peak is 13.5 units.

displayed by GC-FIMS for "fingerprint" comparisons between different oils, source-rocks and bitumens. "Fingerprinting" of aromatic fractions is extremely difficult using conventional GC-EIMS techniques.

The ion traces for some members of the C_nH_{2n-14} series are displayed in Figure 8. For each value of n, a number of different isomers are present. In the traces for low values of n, these isomers are partially resolved while for higher values of n the components are not.

As mentioned above, according to Scheppele et al.,[10,11] the relative sensitivities of aromatic compounds to FI are approximately equal. The conditions under which these authors did their measurements are similar to those employed in the present experiments. Assuming then that the relative sensitivities of all aromatic compounds are equal, the relative abundance of the various mass ion series may be obtained by summing the intensities of the appropriate ions for all the scans in the GC run. The results of such a summation are presented in Table IV.

(B) <u>Thiourea Adduct Fraction</u>. The saturated hydrocarbon fractions from fossil fuels have been more extensively investigated by GC-EIMS. The thiourea adduct fraction (TUA) has been of particular interest in the study of the origin and history of fossil fuels. The saturated hydrocarbon components present in this fraction often display very weak parent ions under EIMS conditions, making the determination of molecular weights difficult. GC-FIMS can overcome this problem and present additional information for the identification of the components in the mixture.

Figure 9 shows the TIC of the Leduc TUA fraction, and is equivalent to a GC chromatogram.

Table IV. Mass Series of Aromatic Fraction of Leduc Oil

Simplest formula	Other possible formulae	Masses	Abundance (%)
C_nH_{2n-6}	C_nH_{2n-20}, $C_nH_{2n-10}S$	...190, 204, 218, ...	16.0
C_nH_{2n-8}	C_nH_{2n-22}	...188, 202, 216, ...	16.0
C_nH_{2n-10}		...186, 200, 214, ...	9.0
C_nH_{2n-12}	$C_nH_{2n-16}S$	...184, 198, 212, ...	23.3
C_nH_{2n-14}	C_nH_{2n-28}	...182, 196, 210, ...	14.6
C_nH_{2n-16}		...180, 194, 208, ...	11.2
C_nH_{2n-18}		...178, 192, 206, ...	9.8

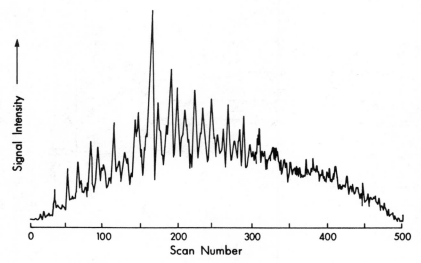

Figure 9. TUA fraction. TIC (sum from m/z = 28-600). Most intense peak is 15.9 units.

Unlike the aromatic fraction where the degree of fragmentation under FI is negligible, the acyclic saturated hydrocarbon molecules (C_nH_{2n+2}) fragment considerably. The amount of fragmentation increases with the number of carbon atoms in a given homologous series, the temperature of the ion source and the temperature of the emitter. Fragmentation also increases with the degree and nature of branching in a molecule of a given elemental formula. The abundance of the fragment ions can range from one tenth to ten times that of the molecular ions, depending on molecular structure and instrument operating conditions.

The literature reports wide variations in the relative sensitivities of saturated hydrocarbons under FI.[5-11] These relative sensitivities were calculated by comparing the ratios of the molecular ions without making appropriate correction for fragmentation. We have examined the fragmentation of a standard mixture of eicosane, pristane and dodecylcyclohexane under a variety of ion source temperatures (140-250°C). Eicosane affords mainly m/z = 29 and a small amount of m/z = 28 as fragment ions and pristane yields mainly m/z = 43 as a fragment ion with a small amount of m/z = 42. The alkyl cyclohexane does not fragment significantly except at high emitter temperatures, >350°C. It was found that the relative ratios of the respective molecular ions in this mixture varied by a factor of 5 with extremes of emitter temperature. However, importantly, the ratios of the sum of the molecular ion and the appropriate fragment ions remained constant within experimental error. The individual molecular ion traces of formula C_nH_{2n+2} for the Leduc TUA are

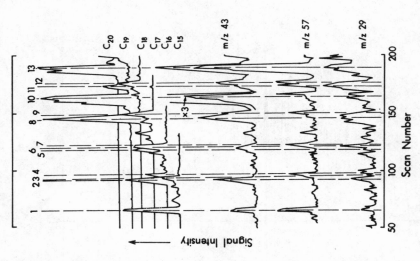

Figure 11. TUA fraction. Expanded display of the molecular ion intensities from n = 15 to 20 in the formula C_nH_{2n+2} with the fragment ion intensities corresponding to m/z = 29, 43, 57. Most intense peak is 3.0 units. See Table V.

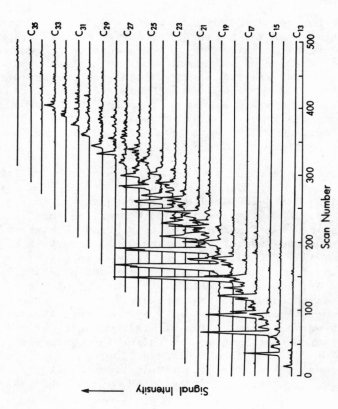

Figure 10. TUA fraction. Individual molecular ion intensities for the mass series C_nH_{2n+2}. Most intense peak is 3.0 units.

CHEMISTRY OF ALBERTA TAR SANDS

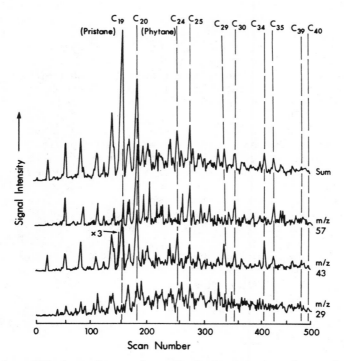

Figure 12. TUA fraction. Sum of the molecular ion intensities for the series C_nH_{2n+2} and the fragment ions m/z = 28, 29, 42, 43, 56, 57 along with the fragment ions m/z = 29, m/z = 43 and m/z = 57. Most intense peak is 13.6 units.

displayed in Figures 10 and 11. Figure 12 displays the sum of the molecular ions for the series C_nH_{2n+2} and the companion fragment ions m/z = 28, 29, 42, 43, 56 and 57. Also displayed are the traces for the fragment ions m/z = 29, 43 and 57.

The fragment ions which were observed under FI in the present experiments are quite different from those which are observed under electron impact. In electron impact the sample molecule is bombarded with 70 eV electrons which have much more kinetic energy than is required to ionize the molecule (typically, 10 eV) or to break a chemical bond (typically, 4 eV). There is considerable excess energy available, depending on the nature of the collision, and usually sufficient energy is imparted to the molecule in the ionization process to induce random fragmentation.

FI is believed to be due to a quantum mechanical tunnelling

process. In the case of an acyclic saturated hydrocarbon, the polarizability of the molecule is considerably greater along the carbon chain than across it. Consequently it will probably approach the surface end on. When FI occurs, the resulting positive charge and any excess energy involved will be localized initially at that end of the molecule nearest the emitter. Experimentally it is observed that the major fragment ion (>90%) from an n-alkane is the m/z = 29 ion. A small amount of m/z = 28 is also observed. Bond rupture therefore occurs between the second and third carbon atoms in the chain. We have observed that the relative abundance of the fragment ion (m/z = 29) increases proportionately with the number of carbon atoms in the molecule. Beckey[5] has explained this phenomenon by arguing that, after ionization, the end of the n-alkane molecule closest to the emitter is repelled from the emitter by the intense electric field. The amount of strain placed on the weakened bond is dependent on the mass and inertia of the remaining part of the molecule. In the case where the mass of the remainder of the molecule is large compared to the fragment, the degree of extension of the weak bond and the inertia of the remainder will be considerably greater than in the case where the fragment ion and the remainder of the molecule are of comparable mass. Thus, the amount of fragmentation within a homologous series increases with increasing number of carbon atoms.

In the case where there is branching at the end of the alkane e.g., pristane

the amount of fragmentation is greater than in the corresponding n-alkane since the resulting fragment ion is thermodynamically more stable.

For pristane the ion m/z = 43 is >90% of the fragment ions and m/z = 42 accounts for most of the remainder; m/z = 29 and 57 are quite small. The molecule phytane,

yields two fragment ions m/z = 43 and 57, with small m/z = 42 and 56 ions corresponding to the fragments from each end of the molecule. Also, m/z = 29 is observed at very low intensity, Figure 12. The m/z = 43 ion is larger than the m/z = 57 ion by a factor of ∼2, suggesting that the greater distance of the methyl branch from end A reduces the amount of fragmentation from this end since the almost symmetrical molecule would not be expected to display a strong preference for orienting one way or the other in the electric field.

Saturated hydrocarbons possessing one or more terminal rings (e.g., dodecylcyclohexane) do not fragment significantly (<2%) under normal FI conditions because as the molecule approaches the emitter its most easily polarizable part will be preferentially attracted to the emitter. In the case of an alkyl cyclohexane molecule, the ring is more readily polarizable than the straight chain attached to the ring. This effect should be more pronounced as the number of rings increases or as the length of the side chain decreases.

If ionization occurs somewhere in the ring and if sufficient excess energy is imparted to rupture a bond then the most likely occurrence is that one of the C-C bonds in the ring will rupture. However, this leaves the resulting ion intact and thus the molecular ion is still detected.

The nature of the fragment ion, because of its unique manner of generation, conveys considerable information about the terminal end of the molecule. For example, the m/z fragments 29, 43 and 57 correlate with molecules having

$\bigwedge\!\bigwedge\!\bigvee^{R}$, $\bigwedge_{R}$ and $\bigvee_{R}$

end groups, respectively.

In the Leduc TUA, structures may be proposed for a number of the alkanes observed. Some of these results are displayed in Figure 11 and Table V. For example, note the incompletely resolved mass ion peaks #2 and 4, Figure 11. Peak #2 correlates with the m/z = 43 fragment while peak #4 correlates with m/z = 57. Many of these proposed structures are based by analogy with the structures of phytane and pristane but they are also in agreement with those proposed on theoretical grounds.[23] Thus, the structure of the terminal moiety of the molecules can be deduced, along with the total number of carbon atoms in the molecules. However, the degree and nature of the branching in the middle of the carbon chain is not determinable.

The sum of the molecular ions for the series C_nH_{2n+2} and the fragment ions m/z = 28, 29, 42, 43, 56 and 57 is displayed in Figure 12. The acyclic hydrocarbons in the TUA display a curious distribution with maxima at C_{19} and C_{20}, C_{24} and C_{25}, C_{29} and C_{30}, C_{34} and C_{35}, and C_{39} and C_{40}, suggesting that these materials were derived from biological polyisoprenoids.

Thus, as can be seen from these data, the disadvantage of extensive cracking in the FIMS analysis of acyclic alkanes becomes a distinct advantage in GC-FIMS, which, in addition to making quantitative analysis of this group of alkanes possible, gives useful structural information on them.

The second largest mass series, Table VI, is that of formula C_nH_{2n}, corresponding to monocyclic alkanes. The profiles of the various mass ions in the series are shown in Figure 13. At least three prominent homologous series of isomeric cyclic molecules are present. The two main series have been identified as alkyl cyclo-

Table V. Structures of Acyclic Alkanes of TUA

Peak #[*]	n[**]	Proposed Structure	Fragment ions found
1	15		43, 57, 29 (?)
2	16		43, 29
3	15		29
4	16		57
5	17		43
6	16		
7	17		57, 29
8	18		43
9	17		
10	19		43
11	19		57, 29
12	18		
13	20		43, 57, 29

[*]See Figure 11. [**]Value of n in formula C_nH_{2n+2}.

Table VI. Thiourea Adduct and Non-Adduct Mass Sums

Mass Series	Abundance (%)	
	Adduct	Non-Adduct
[*]C_nH_{2n+2}	40.1	33.7
C_nH_{2n}	30.6	20.5
C_nH_{2n-2}	15.6	19.0
C_nH_{2n-4}	5.3	12.5
C_nH_{2n-6}	4.0	9.4
C_nH_{2n-8}	2.3	3.8
C_nH_{2n-10}	1.0	1.1

[*]Also includes fragment ions m/z = 28, 29, 42, 43, 56, and 57.

CHEMISTRY OF ALBERTA TAR SANDS

hexanes and alkylmethyl cyclohexanes.

The third largest mass series is that due to molecules of formula C_nH_{2n-2}, i.e. dicyclic alkanes. The molecular ion intensity traces for this series are shown in Figure 14 with odd values of n omitted for clarity. As may be seen, a number of homologous series of compounds is present. This is similar to the C_nH_{2n} series, only more complex.

The other mass series for compounds of formulae C_nH_{2n-4} to C_nH_{2n-10} are similar to those for the C_nH_{2n-2} series, the major differences being the lower quantities present compared to the other series (Table VI) and the poorer resolution of the various isomers. The apparent lack of resolution is due to the much larger number of isomers present in these series of compounds.

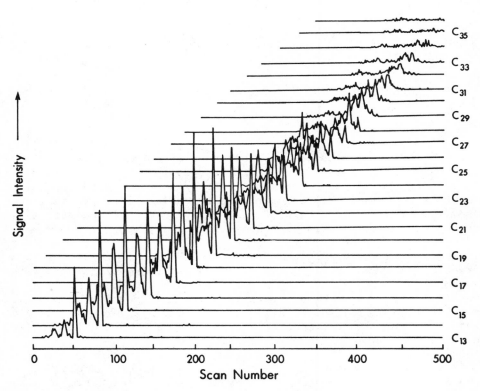

Figure 13. TUA fraction. Molecular ion intensities for the mass series corresponding to the formula C_nH_{2n}. Most intense peak is 4.4 units.

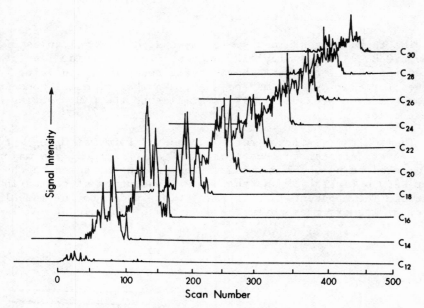

Figure 14. TUA fraction. Molecular ion intensities for the mass series corresponding to the formula C_nH_{2n-2}. Most intense peak is 1.2 units.

The other mass series for compounds of formulae C_nH_{2n-4} to C_nH_{2n-10} are similar to those for the C_nH_{2n-2} series, the major differences being the lower quantities present compared to the other series (Table VI) and the poorer resolution of the various isomers. The apparent lack of resolution is due to the much larger number of isomers present in these series of compounds.

(C) Thiourea Non-Adduct Fraction. The TIC of the Leduc TUNA fraction is shown in Figure 15. It consists of a broad hump with a number of peaks at the beginning (scans 50-250) corresponding to materials of formula C_nH_{2n+2}. They are a mixture of n-alkanes and acyclic isoprenoids which have not been completely removed by clathration. These are the same compounds which are in the molecular sieve and thiourea adduct fractions.

The molecular ion intensities for material of formula C_nH_{2n} are displayed in Figure 16. On comparison with the corresponding traces for the TUA in Figure 13 it can be seen that the TUNA contains more isomers than the TUA and consequently the spectrum is only partially resolved by GC.

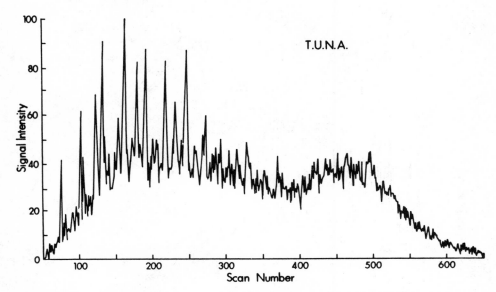

Figure 15. TUNA fraction. TIC (sum from m/z = 28-600). Most intense peak is 10.5 units.

The molecular ion traces for material of formula C_nH_{2n-2} are not shown for the TUNA but they are similar to those of corresponding plots for the TUA, Figure 14.

The molecular ion intensities for material of formula C_nH_{2n-6} in the TUNA are displayed in Figure 17. This class contains the important biological markers, the steranes, and as seen the most abundant of these are the n = 27-31 members.

An alternative approach to displaying the data is to sum together all the ions for all the scans in each GC trace. This then gives a distribution curve for each homologous series according to the carbon number. The representative plots displayed in Figure 18 have been generated from such summations and give the distribution of materials according to carbon numbers for C_nH_{2n-6} and C_nH_{2n-8} in the TUA and TUNA fractions.

The C_nH_{2n-6} series contain the steranes and the C_nH_{2n-8} series, the pentacyclic triterpanes, hopanes and their isomers. Some aromatic

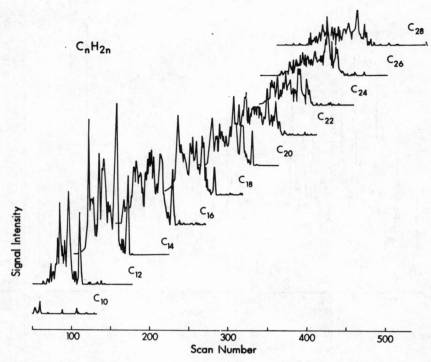

Figure 16. TUNA fraction. Molecular ion intensities for the mass series corresponding to the formula C_nH_{2n}. Most intense peak is 1.4 units. Only the even values of n are plotted for clarity.

material that was not removed chromatographically may also be present. The intensities are normalized, and the peak heights in the figure are proportional to the quantities of material present in each sample. It is then seen that most of the tetra and pentacyclic triterpanes remain in the TUNA fraction.[24]

Athabasca and Cold Lake Bitumens

Similar studies were carried out on the aromatic and saturate fractions of the Athabasca and Cold Lake Bitumens.

The results compiled in Table VII on the aromatic fraction appear to suggest no great differences between the two bitumens with

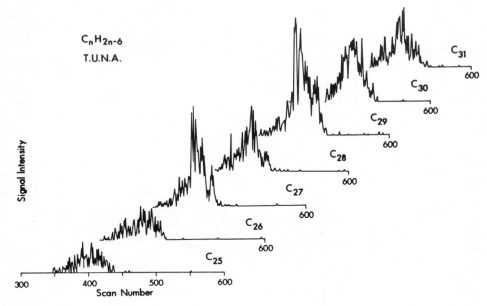

Figure 17. TUNA fraction. Molecular ion intensities for the mass series corresponding to the formula C_nH_{2n-6}. Most intense peak is 1.5 units.

respect to the distribution of their constituent homologous series. The dinuclear series comprise 35-38%, the mononuclear 27-32% and thiophenes, 13-17% of the aromatic fraction. The C_nH_{2n-14} and C_nH_{2n-16} series may contain biphenyl type compounds which would more appropriately belong to the mononuclear aromatic series, therefore the concentration given for the dinuclear series is an upper limit and for the mononuclear series, a lower limit. The average ratios of mono, di and trinuclear aromatics, 30:36:7, is not too different from the values obtained for the Leduc conventional oil, 41:49:10, Table IV, however the thiophenic content of the bitumen is much higher.

The distribution of homologous series of hydrocarbons from acyclic to hexacyclic was determined in the TUA and TUNA fractions of the Cold Lake saturates. The results are tabulated in Table VIII. In both fractions the bulk of the material is made up of the mono to tricyclics, comprising 70-80% of the total.

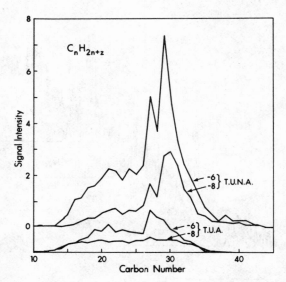

Figure 18. Plots of the distribution of material of formula C_nH_{2n+z} (z = -6,-8) as a function of n for the TUA and TUNA fractions, generated from the summation of all the mass scans of the respective g/c runs. They have been offset for clarity. See text for further information.

Table VII. Mass Series of Aromatic Fraction of Cold Lake (CL) and Athabasca (A) Bitumen

Component		CL	A
		weight %	
C_nH_{2n-6}		4.9	3.7
C_nH_{2n-8}		14.7	10.8
C_nH_{2n-10}		12.0	12.2
C_nH_{2n-12}		14.5	11.5
C_nH_{2n-14}		14.5	13.2
C_nH_{2n-16}		9.0	10.5
C_nH_{2n-18}		5.8	7.6
benzothiophene		≲ 11.0	≲ 8.9
dibenzothiophene		≲ 5.7	≲ 4.5

Table VIII. Mass Series of Saturates in the Cold Lake Bitumen (wt. %).

Formula	Series	TUA	Whole Bitumen	TUNA	Whole Bitumen
C_nH_{2n+2}	acyclic	5	0.12	3	0.54
C_nH_{2n}	monocyclic	27	0.68	23	4.14
C_nH_{2n-2}	dicyclic	36	0.90	28	5.04
C_nH_{2n-4}	tricyclic	17	0.43	19	3.42
C_nH_{2n-6}	tetracyclic	9	0.22	15	2.70
C_nH_{2n-8}	pentacyclic	4	0.10	9	1.62
C_nH_{2n-10}	hexacyclic	2	0.05	3	0.54
Total		100	2.5	100	18.0

It is interesting to compare here the spectra of the tricyclic components of the TUA and TUNA fractions, Figures 19 and 20. The number of isomers at each carbon number is large and their resolution is low in both cases but the TUNA fraction is distinguished by the appearance of a series of prominent, sharp peaks in the C_{20}-C_{30} range which is absent in the TUA fraction. These compounds are tricyclic

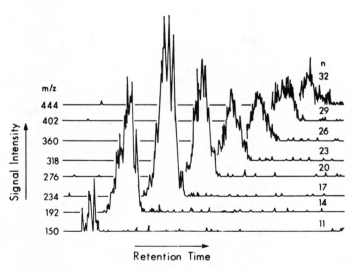

Figure 19. FIMS scan of the C_{11}, C_{14}, C_{17}, C_{20}, C_{23}, C_{26}, C_{29} and C_{32} tricyclic components of the TUA fraction of Cold Lake bitumen.[13]

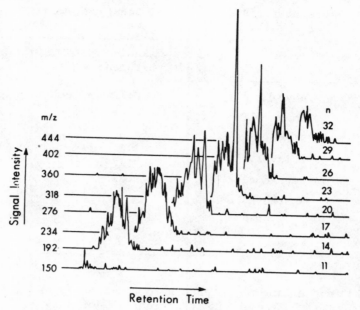

Figure 20. FIMS scans of the C_{11}, C_{14}, C_{17}, C_{20}, C_{23}, C_{26}, C_{29} and C_{32} tricyclic components of the TUNA fraction of Cold Lake bitumen.[13]

terpanes which have been identified in a subsequent organic geochemical study.[25]

Figures 21-23 illustrate the relationship between the GC-EIMS and GC-FIMS data on the geochemically important tetra- and pentacyclic triterpane fractions of the Cold Lake saturates. The m/z 217 characteristic ion fragmentogram for the 5α steranes of the TUA and the FIMS scans for the C_{26} and C_{27} components of the same TUA are shown in Figure 21. Similar data for the 5β steranes and the pentacyclic triterpanes of the TUNA are shown in Figures 22 and 23, respectively. As can be seen from the data, and discussed before,[13] the FIMS ion scans reveal the presence of more isomers than indicated by the EIMS scans and also lend themselves more readily to quantitative analysis.

CONCLUSIONS

It has been shown that the GC-FIMS method of analysis of the hydrocarbon fractions of a conventional crude oil and of two tar sand bitumen samples can yield qualitative and quantitative compositional data not available by other means. The combination of GC FIMS and GC-EIMS provides the most powerful tool for the analysis of the volatile fraction of petroleum.

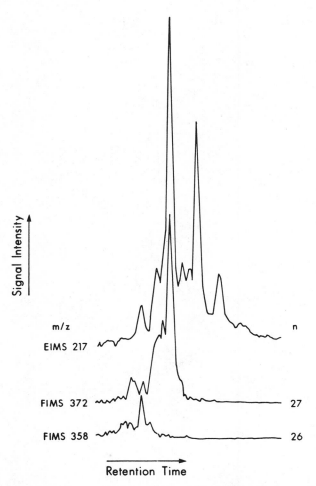

Figure 21. Comparison of the m/z = 217 GC-EIMS fragmentogram of the 5α steranes in the TUA fraction of the Cold Lake bitumen with the $C_{27}H_{48}$ and $C_{26}H_{46}$ GC-FIMS chromatograms.[13]

ACKNOWLEDGMENTS

The authors thank the Alberta Oil Sands Technology and Research Authority and the Natural Sciences and Engineering Research Council of Canada for financial support and Dr. D.S. Montgomery for many helpful discussions.

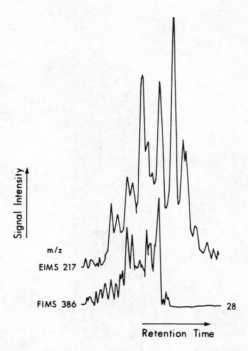

Figure 22. Comparison of GC-EIMS and GC-FIMS scans of the 5β steranes in the TUNA fraction of the Cold Lake bitumen.[13]

REFERENCES

1. Alberta Oil Sands Technology and Research Authority Fifth Annual Report and Five-Year Overview, 1975-1980.
2. R.F. Meyer and W.D. Dietzman, World Geography of Heavy Crude Oils, *in* "The Future of Heavy Crudes and Tar Sands", R.F. Meyer and C.T. Steele, eds., McGraw-Hill, Inc., New York (1981); A. Janisch, Oil Sands and Heavy Oil: Can They Ease the Energy Shortage?, *ibid*.
3. G.D. Mossop, J.W. Kramers, P.D. Flach and B.A. Rottenfusser, Geology of Alberta's Oil Sands and Heavy Oil Deposits, *ibid*.
4. I.A. Breger, Geochemical Considerations Regarding the Origin of Heavy Crude Oils; Suggestions for Exploration, *ibid*.
5. H.D. Beckey, "Principles of Field Ionization and Field Desorption Mass Spectrometry", Pergaman Press, New York (1977).
6. H.D. Beckey and G. Wagner,"Analytische Anwendungsmöglichkeiten des Fieldionen-Massenspektrometers", *Z. Anal. Chem.* 197:58 (1963)

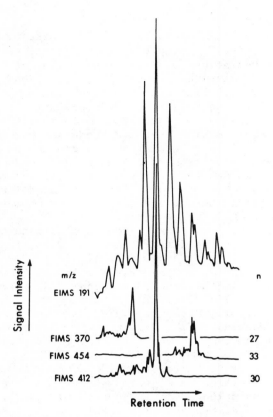

Figure 23. Comparison of GC-EIMS and GC-FIMS scans of the pentacyclic triterpanes in the TUNA fraction of the Cold Lake bitumen.[13]

References cont'd.

7. M. Ryska, M. Kuras and J. Mostecky, "Phenomenology of Adsorption Processes on Emitters in the Field Ionization of Hydrocarbon Mixtures", *Int. J. Mass Spec. and Ion Phys.* 16:257 (1975).
8. M. Kuras, M. Ryska and J. Mostecky, "Analysis of Saturated Hydrocarbons by Field Ionization Mass Spectrometry", *Anal. Chem.* 48:196 (1976).
9. W.L. Mead, "Field Ionization Mass Spectrometry of Heavy Petroleum Fractions-Waxes", *Anal. Chem.* 40:743 (1968).
10. S.E. Scheppele, P.L. Grizzle, G.J. Greenwood, T.D. Marriott and N.B. Perreira, "Determination of Field-Ionization Relative Sensitivities for the Analysis of Coal-Derived Liquids and their Correlation with Ion-Voltage Electron-Impact Relative Sensitivities", *Anal. Chem.* 48:2105 (1976).

References cont'd.

11. S.E. Scheppele, G.J. Greenwood and P.A. Benson, "Field Ionization Relative Sensitivities for Analysis of Coal-Derived Liquids Determined as a function of Ion-Source Temperature and Binary-Mixture Composition", *Anal. Chem.* 49:1847 (1977).
12. J.D. Payzant, I. Rubinstein, A.M. Hogg and O.P. Strausz, "Field Ionization Mass Spectrometry: Application to Geochemical Analysis", *Geochim. Cosmochim. Acta* 43:1187 (1979).
13. J.D. Payzant, I. Rubinstein, A.M. Hogg and O.P. Strausz, "Analysis of Cold Lake Bitumen Hydrocarbons by Combined GLC-Field Ionization Mass Spectrometry and GLC-Electron Impact Mass Spectrometry", *Chem. Geol.* 29:73 (1980).
14. J.N. Damico and R.P. Barron, "Application of Field Ionization to Gas-Liquid Chromatography-Mass Spectrometry (GLC-MS) Studies", *Anal. Chem.* 43:17 (1971).
15. R.M. Milberg and J.C. Cook, "Some Applications of High Sensitivity Combined Field Ionization Gas Chromatography-Mass Spectrometry", *J. Chrom. Sci.* 17:43 (1979).
16. I. Rubinstein and O.P. Strausz, "Ion Subtraction – A Computer-Aided Step in the Geochemical Analysis of Fossil Fuels", *Chem. Geol.* 25:327 (1979).
17. I. Rubinstein, O.P. Strausz, C. Spyckerelle, R.J. Crawford and D.W.S. Westlake, "The Origin of the Oil Sand Bitumens of Alberta: A Chemical and a Microbiological Simulation Study", *Geochim. Cosmochim. Acta* 41:1341 (1977).
18. J.G. O'Connor, F.H. Burrow and M.S. Norris, "Determination of Normal Paraffins in C_{20} to C_{32} Paraffin Waxes by Molecular Sieve Adsorption", *Anal. Chem.* 34:82 (1962).
19. B.J. Kimble, J.R. Maxwell, R.P. Philp, G. Eglinton, P. Albrecht, A. Ensminger, P. Arpino and G. Ourisson, "Tri- and Tetraterpenoid Hydrocarbons in the Messel Oil Shale", *Geochim. Cosmochim. Acta* 38:1165 (1974).
20. I. Rubinstein and O.P. Strausz, "Geochemistry of the Thiourea Adduct Fraction from an Alberta Petroleum", *Geochim. Cosmochim. Acta* 43:1387 (1979).
21. A.M. Hogg and J.D. Payzant, "Design of a Field Ionization/Field Desorption/Electron Impact Ion Source and its Performance on a Modified AEI MS9 Mass Spectrometer", *Int. J. Mass Spec. and Ion Phys.* 27:291 (1978).
22. A.M. Hogg, J.D. Payzant, I. Rubinstein and O.P. Strausz, "Application of GC/FIMS to Hydrocarbon Analysis", *27th Annual Conference on Mass Spectrometry and Allied Topics, Seattle, U.S.A.* 1979.
23. D.W. Waples and L. Tornheim, "Mathematical Models for Petroleum-Forming Processes: n-Paraffins and Isoprenoid Hydrocarbons", *Geochim. Cosmochim. Acta* 42:457 (1978).

References cont'd.

24. T.C. Hoerin, Carnegie Inst. Yearbook, p.303, 1969; E. Gelpi, P.C. Wszolek, E. Yang and A.L. Burlingame, "Evaluation of Chromatographic Techniques for the Preparative Separation of Steranes and Triterpanes from Green River Formation Oil Shale", *J. Chromat. Sci.* 9:147 (1971); I. Rubinstein and O.P. Strausz, "Thermal Treatment of the Athabasca Oil Sand Bitumen and its Component Parts", *Geochim. Cosmochim. Acta* 43:1887 (1979).
25. C.M. Ekweozor and O.P. Strausz, to be published.

SULFUR FORMS DETERMINATION IN COAL USING MICROWAVE

DISCHARGE ACTIVATED OXYGEN AND MÖSSBAUER SPECTROSCOPY

James L. Giulianelli and D.L. Williamson

Department of Chemistry/Geochemistry and
Department of Physics
Colorado School of Mines, Golden, Colorado   80401

ABSTRACT

   The use of a microwave discharge to activate oxygen for the low temperature ashing of coal is being tested. The effect of low temperature ashing (LTA) by the normal radiofrequency procedure and this new microwave procedure upon the Fe-S compounds is monitored by the use of ^{57}Fe Mössbauer spectroscopy. Results show that the microwave technique, like the normal LTA procedure, effectively ashes the coal without the oxidation of pyritic sulfur. Dehydration of iron sulfates and oxidation of Fe^{2+} to Fe^{3+} in sulfates occurs in both ashing procedures under the conditions tested. Separate experiments show that dehydration is caused by sample heating rather than by reaction with the activated oxygen. The Mössbauer recoilless fractions of the Fe-sulfates observed in this study are found to be only about half that of pyrite.

INTRODUCTION

   The low temperature ashing (LTA) of coal in an oxygen plasma has been used for years as a means of isolating the coal minerals with little alteration in their structures or amounts (Gluskoter, 1965; Frazer and Belcher, 1972). In all commercial ashers, the ground coal is placed in a thin layer in a sample boat and bathed in an oxygen plasma created by a radiofrequency (13.56 MHz) discharge. With most equipment, the complete ashing of a gram of coal requires 70-90 hours and intermittent raking to expose fresh coal surfaces to the plasma. Surface temperatures ordinarily do not exceed 150°C.

Recently, several attempts (Paris, 1977; Hamersma and Kraft, 1979) have been made to incorporate into commercial LTA equipment the quantitative capture of the evolved sulfur gases as a means of directly determining the organic sulfur in coal. A number of difficulties have arisen in these attempts to selectively separate sulfur forms by the use of low temperature plasma ashing. One problem has been the inability to quantitatively recover all of the SO_x gases which apparently adhere to the glass chamber and exit tubing. Additional difficulties stem from (a) the inherent slowness of the process under usual conditions, (b) the oxidation of pyrite (FeS_2) at high powers and the apparent tendency for some pyrite particles to spontaneously and unpredictably oxidize in the plasma, and (c) the phenomenon of the fixing of the organic sulfur as sulfate in low-rank coals containing carboxylate salts of calcium and sodium (Miller, Yarzab, and Given, 1979). These problems as they relate to the isolation of unaltered minerals have recently been reviewed (Given and Yarzab, 1978; Jenkins and Walker, 1978).

In order to determine whether or not these difficulties might be overcome by using a specially designed asher for separation of the sulfur forms in coal, work is underway in which the ashing of coal is by atomic oxygen generated in a microwave cavity. The asher will incorporate the features of rapid ashing, compact chamber construction, temperature control, fewer ionic species within the reaction chamber, and more rapid exit of product gases from the chamber. As a first step in evaluating the feasibility of this design, the occurence and selectivity of the ashing process using microwave (MW) LTA is demonstrated in this paper. This MW procedure will be referred to as <u>indirect</u> MW ashing of coal to distinguish it from the more commonly employed use of MW in direct heating (pyrolysis) of materials located within the cavity.

Furthermore, motivated by the need to compare the degree of mineral alteration by the usual (RF) and proposed (MW) ashing procedures, changes in the iron minerals due to ashing are analyzed by the use of Mössbauer spectroscopy. This nuclear gamma-ray resonance technique has been employed in coal and coal product analysis in this (Williamson, Guettinger, Dickerhoof, 1980) and other laboratories (Montano, 1979; Huffman and Huggins, 1978; Saporoschenko, et. al., 1980; Jacobs and Levinson, 1979; Montano, 1980) where it has been shown to be a powerful tool which complements chemical analysis for identifying and monitoring changes in the iron minerology. For example, prior to the application of Mössbauer spectroscopy, the only reported effects of low temperature ashing at low powers (50W) for short times (2-3 days) was the partial dehydration of gypsum ($CaSO_4 \cdot 2H_2O$) to hemihydrate ($CaSO_4 \cdot 1/2H_2O$) by Gluskoter (1975) and Frazer and Belcher (1973). Although there has been no systematic Mössbauer study of the influence of LTA upon the iron compounds, it has since been shown that chemical changes during ashing can occur in Fe-S compounds and in the Fe^{2+} in clays (Montano, 1977; Montano, 1979; Huggins and Huffman, 1979). Because it is important to understand these changes which can occur during LTA if the latter technique is to be applied to quantitative analysis, this paper will emphasize our Mössbauer results on several raw and ashed coals.

EXPERIMENTAL

Coal Samples

Most coals used in this study were obtained from the Penn. State Coal Bank. About 50 grams of sample were split and ground to pass 100% through a -100 mesh (Tyler) brass sieve and splits of this were used without further treatment for each analysis. Forms of sulfur were analyzed by ASTM Procedure D2492, in triplicate, by the "nonsimultaneous" method (Given and Yarzab, 1978) of analyzing the HCl acid leach from the same sample as used for pyritic sulfur determination. This procedure was modified so as to include titration of the HCl-soluble iron fraction (Gladfelter and Dickerhoof, 1976).

LTA in an RF (13.56 MHz) Asher

Coals were ashed in a single chamber Tegal Plasmod RF Low Temperature Asher at 50 watts, employing a flow rate of about 50 sccm and 1-2 torr chamber pressure. Unlike in some commercial ashers, in this model, the plasma is generated directly at the sample location since the RF inductive coil surrounds the entire chamber, not just the inlet. To ash one gram of thinly-spread coal in either a Pyrex petrie dish or aluminum tray required anywhere from 24 to 48 hours, including periodic interruptions for raking to expose new surface to the plasma.

LTA in an MW (2,450 MHz) Asher

Figure 1 depicts the oxygen flow line, microwave cavity, and sample chambers constructed for these experiments. No attempt was made to trap the evolved gases. A Holiday Industries Microwave Power Source with variable power control (0-300W) was coupled via coaxial cable to an air-cooled Opthos Evenson Type N microwave cavity surrounding a one-half inch pyrex inlet tube to the reactor chamber. Forward power levels were maintained in the 90-120W range by occasional retuning of the cavity, keeping the reflected power at a minimum. Discharge was initiated with a spark from a Tesla coil, although this was not always necessary.

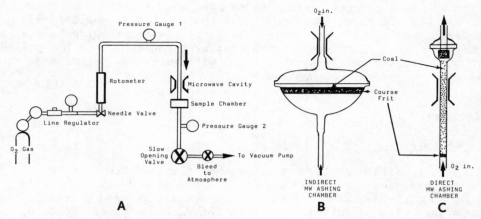

Fig. 1. A. Schematic representation of flow system for low temperature microwave ashing of coals; B. Sample chamber for indirect low temperature microwave ashing of coals; C. Sample chamber for direct microwave heating of fluidized bed of coal.

By adjusting the needle valve at the flow meter and by throttling the flow before the vacuum pump the pressure and flow rates were generally maintained between 1.5-4.0 torr and 180-220 sccm, respectively. Preliminary experiments using a pure graphite rod placed in a horizontal tube one inch from the cavity dramatically illustrated the effects of increased pressure (ca. 30 torr) and high flow rates (ca. 2800 sccm) in causing sparks to fly from the end of the rod. Since the gas is heated to a greater extent at high pressures, this rapid oxidation is most likely thermal, that is, due to molecular oxygen, and would also probably oxidize the pyrite in the coal. Under the actual conditions used for the indirect MW ashing of the PSOC-284 coal discussed below, the temperature of the coal bed probably did not exceed 150°C, as is the case for RF ashing, although this point was not verified directly.

The sample chamber for the indirect ashing (outside of the MW cavity) of coal is shown in Figure 1B. It is constructed from a 150ml Buchner coarse grade funnel. The coal is spread evenly on the porous plate about one inch from the end of the Evenson cavity. To date little effort has been made to optimize the ashing rates for this reactor. Under the conditions employed, the 170 mg of this Illinois coal required about 5 hours for complete ashing, whereas 2.3 grams required 35 hours and frequent raking. At 2 torr and above and 180 sccm ashing is concentrated in the center portion of the bed.

SULFUR FORMS DETERMINATION IN COAL

The reaction chamber shown in Fig. 1C was used in only one experiment intended to demonstrate the result of the direct MW ashing of coal which enters the MW cavity. Oxygen flowing upward through a frit supporting the sample raises the coal in a fluidized bed and the direct combustion is evidenced by the sparking of the coal particles as they transverse the cavity. Figure 2 is the Mössbauer spectra of an Illinois No.6 coal treated in this way. It clearly shows the six line spectrum due to hematite (Fe_2O_3) which is produced from the pyrite in the direct MW treatment in an oxygen atmosphere and cautions against the inadvertent introduction of coal particles into the cavity in an indirect ashing experiment.

Mössbauer Analysis

Mössbauer data were acquired for each sample in the standard transmission mode using a conventional constant acceleration spectrometer employing a $^{57}Co:Rh$ source of about 25 mCi strength. Absorbers of coal were prepared by compressing either 0.500 gm or 1.000 gm of -100 mesh coal into a 2.54 cm diam. pellet. Where possible, coal absorbers were prepared such that the density of ash in the γ-ray beam was about the same as for the original (unashed) coal

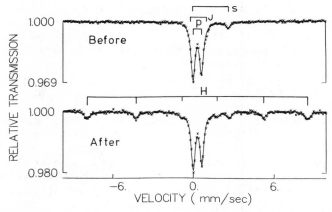

Fig. 2. Mössbauer spectra of Illinois Number 6 coal showing formation of hematite from pyrite due to direct microwave treatment. p = pyrite; j = Fe^{3+} (similar to jarosite); s = szomolnokite; H = hematite.
Note: The zero of velocity in this and all subsequent spectra is the velocity corresponding to the center of the α-Fe calibration spectrum.

absorber. This procedure results in Mössbauer spectra with comparable total resonance intensities for ashed and unashed samples and allows a more accurate analysis of subtle changes.

Each spectrum was computer-fitted with a sum of Lorentzian-shaped resonance lines. For all of the spectra acquired in this study, a fitting constraint of equal intensities on the two lines of each quadrupole pair was imposed since it is theoretically expected from powder absorbers with no grain texture. Such a constraint allows the computer decomposition of rather complicated multicomponent spectra provided only one of the quadrupole lines of each component is partially resolved. For identification of the various iron-containing compounds, the isomer shift and quadrupole splitting of each spectral component were compared to literature values of these parameters for known compounds (Huggins and Huffman, 1979; Montano, 1980).

The integrated intensities (areas) from the computer fits were used to give the concentration of Fe in the various mineral phases. The details of the quantitative analysis method are described elsewhere (Williamson et al, 1980). In this work, however, we account for differences in the Mössbauer recoilless fractions (f-values) of pyrite and Fe-sulfates. Coal absorbers containing pyrite and various Fe-sulfates were run at both room temperature and liquid nitrogen temperature. Increases in the intensity of the sulfate resonances compared to the pyrite resonance were observed upon cooling, thereby indicating substantially lower f-values for the Fe-sulfates. This effect is illustrated in Fig. 3 for a coal containing pyrite (FeS_2), rozenite ($FeSO_4 \cdot 4H_2O$), and a Fe^{3+} compound which has parameters similar to coquimbite ($Fe_2(SO_4)_3 \cdot 9H_2O$). Comparison of the experimental relative resonance areas with Debye model calculations yielded ratios of f-values for the sulfates to that of pyrite. Within our limits of error, all Fe-sulfates studied by this procedure ($FeSO_4 \cdot H_2O$, $FeSO_4 \cdot 4H_2O$, $Fe_2(SO_4)_3 \cdot xH_2O$) were found to have the same f-value: $f(sulfates)/f(pyrite) = 0.55 \pm .15$ at room temperature.

As discussed previously in detail (Williamson et al, 1980), the absolute f-value of pyrite is not well established because of the granular nature of pyrite absorbers prepared from coal or from pure pyrite powders. However, a calibration using coal absorbers containing pyrite yielded f(pyrite)=0.53 and it was observed that this value gave agreement on the average between chemical ASTM analysis and Mössbauer analysis for several coals. Taking f(pyrite) = 0.53 for this work, the above ratio yields f(sulfates)=0.29. Although these may not be the true f-values because of granularity problems, reasonable agreement with ASTM analysis on total Fe and excellent agreement on the ratio of pyritic Fe to total Fe is obtained. Because of this low f-value for the sulfates compared to pyrite, we caution the reader that visual inspection of relative intensities in a Mössbauer spectrum can be misleading; for example,

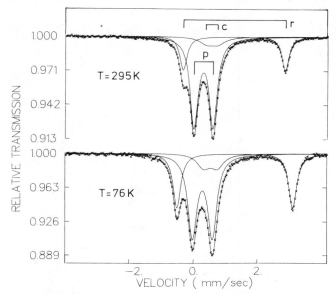

Fig. 3. Mössbauer spectra of Pennsylvania LVB coal (PSOC-319) at room temperature and liquid nitrogen temperature. p = pyrite; r = rozenite; c = Fe^{3+} (similar to Fe^{3+} in coquimbite or kornelite). Note the significant increase in the r and c resonances relative to those of p upon cooling.

the room temperature spectrum of Fig. 3 suggests that the ratio Fe as pyrite/Fe as rozenite ≃3 whereas the correct ratio is less than 2.

RESULTS

Quantitative chemical and Mössbauer analysis results on iron and sulfur forms for the four untreated, as-received coals used in this work are shown in Table 1. These coals represent a wide range of sulfur forms and concentrations, rank, and origin. The high sulfate concentrations in three of these coals reflect the high degree

Table 1. Chemical and Mössbauer Analysis of Iron and Sulfur Forms in Several Coals Prior to Ashing.

COAL SAMPLE (-100 Mesh)	% SULFUR[a]				% IRON[a]		Pyritic Fe / Total Fe	
	Pyritic		Sulfatic	Organic	Non pyritic			
	ASTM[b] D2492	Mössbauer[c]	ASTM D2492	(d)	ASTM (Modified)[e]	Mössbauer[f]	ASTM	Mössbauer
Lower Dekoven Ill., HVAB (PSOC-284)	2.42	2.43	1.06	1.12	1.27	1.44	0.62	0.60
Upper Kittanning Penn., LVB (PSOC-319)	2.96	2.53	1.63	0.078	2.07	1.77	0.55	0.55
Rosebud Mont., SUBB (PSOC-230)	0.28	0.26	0.07	0.038	0.04	0.02	0.9	0.9
Ill. No.6 Ill., HVCB	1.73	2.00	-	-	0.74	0.80	0.67	0.68

a. Weight per cent, on an as-received basis.
b. Both acid extractions performed on the same coal sample; done in triplicate.
c. Pyrite recoilless fraction f = 0.53.
d. Organic sulfur values taken from Penn State Coal Data Base analysis, corrected to wet basis from moisture contents determined in this laboratory.
e. Iron which is soluble in 30 minutes in boiling 5M HCL -- determined by titration.
f. Nonpyrite recoilless fraction f = 0.29 (see text)

of "weathering" incurred prior to their use in these experiments. Values for organic sulfur are approximate and are presented to illustrate the range for the coals used.

Satisfactory agreement was obtained for absolute values of both pyritic and nonpyritic iron as determined by chemical ASTM analysis and by Mössbauer spectroscopy. However, a more important indicator of the quantitative nature of the Mössbauer results is shown by comparison of the numbers in the last two columns of the table. The excellent agreement obtained for the ratio of pyritic to total iron when calculated using chemical and Mössbauer data is a strong indication that all of the iron species are being accounted for in each method and also substantiates the lower f-values for the iron sulfates determined in this work (see EXPERIMENTAL section).

Table 2. Mössbauer Analysis of Effects of Low Temperature Ashing on the Iron-Containing Minerals of Several Coals.

Coal Sample	Treatment	Per Cent of Total Fe as Mineral Indicated[a]					
		Pyrite (FeS_2)	Rozenite ($FeSO_4 \cdot 4H_2O$)	Szomolnokite ($FeSO_4 \cdot H_2O$)	Fe^{3+} (b)	Fe^{3+} (c)	Fe^{2+} (Unknown)
Lower Dekoven Ill., HVAB (PSOC-284)	None	60	26	11	3	0	0
	LTA(RF)	62	0	18	6	7	7
	LTA(MW)	57	0	21	11	6	5
	Heated at 130°C[d]	57	0	15	22	0	6
	Heated at 1650°C[d]	56	0	10	27	0	7
	HCl[e]	88	0	0	12	0	0
	HCl-LTA(RF)	89	0	0	11	0	0
	HCl-LTA(MW)	87	0	0	13	0	0
Upper Kittanning Penn., LVB (PSOC-319)	None	55	33	0	0	12	0
	LTA(RF)	52	0	4	0	35	9
	Heated at 130°C[d]	54	0	17	0	18	11
Rosebud Mont., SUBB	None	92	8	0	0	0	0
	LTA(RF)	91	0	0	9	0	0
Ill. No. 6 Ill., HVCB	None	68	0	16	16	0	0
	LTA(RF)	66	0	10	7	17	0

a. All uncertainties are approx. ±3% except for the Rosebud coal, where, because of the low Fe content, the uncertainties are approx. ±6%. A ratio of recoilless fractions of f(sulfates)/f(pyrite) = 0.55 was used, based on the liquid nitrogen measurements described in the text. Other Fe compounds were assumed to have this same ratio compared to pyrite.
b. This component is similar to a substituted jarosite, e.g. $H_3OFe_3(SO4)_2(OH)_6$, or an Fe^{3+}-containing clay or a mixture of both.
c. This component overlaps the higher velocity pyrite line and is similar to coquimbite, $Fe_2(SO_4)_3 \cdot 9H_2O$, or kornelite, $Fe_2(SO_4)_3 \cdot 7H_2O$, or Fe^{3+} in clays.
d. In air, for 24 hours.
e. A split of the PSOC-284 coal, boiled in 5M HCL for 30 min.

The four coals were low temperature ashed by the normal RF procedure and one of these, PSOC-284, by the indirect MW procedure. The Mössbauer spectra of the Illinois HVAB coal (PSOC-284) and its low temperature ashes produced by RF and indirect MW activated oxygen are presented in Fig. 4. Mössbauer results for the other coals and their LTA(RF) products are presented in Fig. 5. The various iron-containing phases have been indicated by stick diagrams in these figures and their relative fractions, deduced from analysis

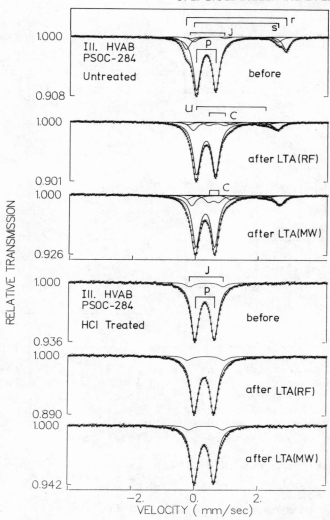

Fig. 4. Comparison of effect of LTA(RF) and LTA(MW) on Lower Dekoven, Illinois coal, as-received and after treatment with boiling 5M HCl. r = rozenite; s = szomolnokite; j = Fe^{3+} (similar to Fe^{3+} in jarosites or clays); p = pyrite; u = unknown Fe^{2+} compound; c = Fe^{3+} (similar to Fe^{3+} in coquimbite or kornelite).

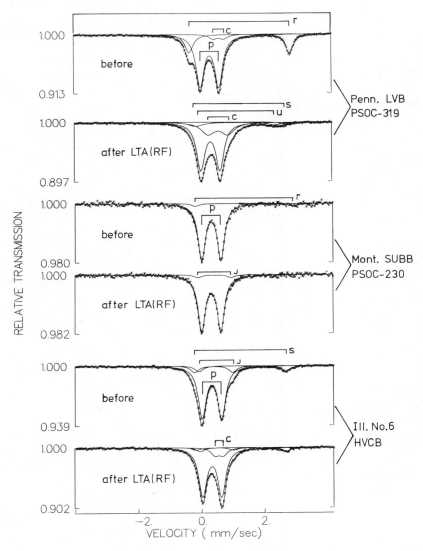

Fig. 5. Effect of LTA(RF) ashing on various coals. See Fig. 4 caption for identification of stick diagram labels.

of the Mössbauer spectra, are summarized in Table 2. Results are also presented on the ashing of an acid-treated coal and on the effects of heat treatment on the iron species.

DISCUSSION

LTA by MW Activated Oxygen

The ashes formed by LTA in the normal RF procedure and in the indirect MW procedure were identical in appearance and represented approximately the same yield (<u>ca</u>. 24 weight percent) of the original Illinois HVAB coal. Within the estimated experimental error of ± 3 per cent defined in Table 2, the MW-and RF-LTA treatments yielded the same iron-sulfur compounds (Figure 4) in the same concentration in the ash. Chemical analysis of the PSOC-284 coal and its ashes produced by each method, presented in Table 3, confirms these Mössbauer results and also shows that, within experimental error, no additional sulfates have been formed.

In order to isolate the effect of the LTA treatments upon the pyritic fraction only, this Lower Dekoven coal was washed in boiling 5M HCl for thirty minutes, leaving only the FeS_2 and a small amount of Fe^{3+} in the coal. The coals were then low temperature ashed by

Table 3. Chemical Analysis of an Illinois Coal and Its Low Temperature Ashes Produced by RF and Indirect MW Treatment (*)

Coal	% Sulfur		% Iron
	Pyritic	Sulfatic	Nonpyritic
Untreated PSOC-284	2.42(1)	1.06(4)	1.27(1)
RF Ash (23.0% LTA)	2.39(1)	1.18(4)	1.33(2)
MW Ash (25.0% LTA)	2.37(5)	1.16(4)	1.5(2)

(*) Uncertainties, given in parentheses as the error in the last figure, are estimated from triplicate and duplicate analyses for the raw coal and its RF ash, respectively. For the MW ash, the larger estimated error results from a greater uncertainty in the weight of ash recovered and the fact that the result is from a single determination.

each procedure. Figure 4 and Table 2 show that no other Fe-S compounds were produced as a result of either LTA treatment on the acid washed coal (within a detection limit of about 3% of the total Fe). This confirms that the pyrite is not affected by either process and suggests that the Fe^{3+} phase sometimes reported (Montano, 1977; Huggins and Huffman, 1979) as formed by LTA from oxidation of pyrite results from a harsher set of LTA conditions or at least is not a general occurrence.

Effect of LTA on Iron-bearing Minerals

A comparison of the iron phases present in the low temperature ashes with those of the original coals leads to the following conclusions about the effect of the LTA upon the iron minerals in these coals (see Figs. 4 and 5 and Tables 2 and 3).

(1) No oxidation of pyrite has occurred (within the approximate three percent experimental uncertainty). This Mössbauer result is confirmed by chemical analysis on the two coals (PSOC-284 and Illinois No. 6) analyzed.

(2) The nonpyritic iron-sulfate minerals are altered differently in different coals upon LTA. This can be seen in Table 2 where the final products in the Lower Dekoven, Illinois coal and the Upper Kittanning, Pennsylvania coal have dramatically different distributions even though rozenite was the precursor in both cases.

(3) Rozenite, $FeSO_4 \cdot 4H_2O$, is completely removed in the LTA process. Dehydration to szomolnokite, $FeSO_4 \cdot H_2O$, occurs and further alterations take place to yield an unidentified Fe^{2+}-bearing compound and at least two Fe^{3+}-bearing compounds.

Although the Fe^{3+} resonances observed from the LTA samples may be due in part to clays (Montano, 1980), the most reasonable oxidation product(s) from a ferrous sulfate consistent with the data, is one or more of the ferric sulfates such as coquimbite, $Fe_2(SO_4)_3 \cdot 9H_2O$, kornelite, $Fe_2(SO_4)_3 \cdot 7H_2O$, or one of the substituted jarosites, e.g. $H_3O\ Fe_3(SO_4)_2(OH)_6$. Other workers have also observed the dehydration of rozenite (Montano, 1979) and the formation of coquimbite (Huggins and Huffman, 1979; Rao and Gluskoter, 1973) as a result of LTA. Oxidation of Fe^{2+} in clays during LTA has been reported (Huggins and Huffman, 1979).

Experiments in which the original coal was simply heated in air at either 130°C or 165°C without ashing (see Table 2), also showed the conversion of rozenite to szomolnokite and is consistent with the dehydration temperature of 90°C for rozenite (Weast, 1977). Oxidation to coquimbite or kornelite in coal sample PSOC-319 and a substituted jarosite in PSOC-284 was also observed upon heating. This indicates that the oxidation of szomolnokite during LTA is also probably thermal in origin and not directly caused by the plasma. Thus, even though the presence of the sulfates in this study complicates the analysis, the results suggest that the temperature dependent conversions of these sulfates in a particular coal offers a sensitive way of monitoring possible differences in ashing conditions in the RF and MW processes.

The unidentified Fe^{2+}-compound is readily detected in the LTA from two different coals by the high velocity resonance line at 2.3 ± 0.1 mm/sec (relative to α-Fe). This velocity does not agree with that reported for the Fe^{2+}-site in roemerite, $FeSO_4 \cdot Fe_2(SO_4)_3 \cdot 14H_2O$ (Huggins and Huffman, 1979) which is a common decomposition product of szomolnokite (Palache, 1951). The possibility that it is a ferrous sulfate yet to be characterized by Mössbauer spectroscopy cannot be excluded.

SUMMARY

Low temperature ashing of coal by oxygen gas dissociated by microwave power has been demonstrated. The degree of conversion of the iron-sulfur compounds, as characterized by Mössbauer spectroscopy, is the same for both RF and MW processes and no oxidation of pyrite due to ashing can be detected in either LTA procedure.

Mössbauer spectroscopy was used to determine the following sequence of events during LTA of the samples studied:

$$FeSO_4 \cdot 4H_2O \rightarrow FeSO_4 \cdot H_2O \rightarrow Fe^{3+}$$

The combination of the Mössbauer method and chemical analysis has allowed the determination of the recoilless fraction ratio $f(sulfate)/f(pyrite) = 0.55 \pm .15$, putting the Mössbauer results on a more quantitative basis than before. To our knowledge this is the first quantitative work on coal and coal products which has accounted for differences in f-values for different Fe compounds.

Future work will attempt to improve on the reaction rate and chamber design in the MW procedure. Experiments will include coals of lower rank and attempts at the complete recovery of the SO_x gases emanating from the chamber.

ACKNOWLEDGMENTS

We are grateful for the financial support provided to one of us (J.L.G.) by the Beckham Fund and are especially thankful to two Colorado School of Mines students, Tim Lee and Larry Peppers, for their assistance.

REFERENCES

Frazer, F.W. and Belcher, C.B., 1973, Quantitative determination of the mineral-matter content of coal by a radiofrequency-oxidation technique, Fuel, 52:41.

Given, P.H. and Yarzab, R.F., 1978, Analysis of the organic substance of coals: problems posed by the presence of mineral matter, in: "Analytical Methods for Coal and Coal Products," Vol.II, C. Karr, Jr., ed., Academic Press, New York.

Gladfelter, W.L. and Dickerhoof, D.W., 1976, Determination of sulfur forms in hydrodesulfurized coal, Fuel, 55:360.

Gluskoter, H., 1965, Electronic low-temperature ashing of bituminous coal, Fuel, 44:285.

Hamersma, J.W. and Kraft, M.L., 1979, Coal Sulfur Measurements, EPA-600/7-79-150.

Huffman, G.P. and Huggins, F.E., 1978, Mössbauer studies of coal and coke: quantitative phase identification and direct determination of pyritic and sulphide sulphur content, Fuel, 57:592.

Huggins, F.E. and Huffman, G.P., 1979, Mössbauer analysis of iron-bearing phases in coal, coke, and ash, in: "Analytical Methods for Coal and Coal Products," Vol. III, C. Karr, Jr., ed., Academic Press, New York.

Jacobs, T.S. and Levinson, L.M., 1979, Enhancement of magnetic separability in coal liquefaction residual solids, J. Appl. Phys., 50:2422.

Jenkins, R.G. and Walker, P.L., Jr., 1978, Analysis of mineral matter in coal, in: "Analytical Methods for Coal and Coal Products," Vol. II, C. Karr, Jr., ed., Academic Press, New York.

Miller, R.N., Yarzab, R.F., and Given, P.H., 1979, Determination of the mineral-matter contents of coals by low-temperature ashing, Fuel, 58:4.

Montano, P., 1979, Characterization of iron bearing minerals in coal, in:"Preprints Amer. Chem. Soc.-Div. Fuel Chem.," 24:218.

Montano, P.A., 1980, Application of Mössbauer spectroscopy to coal characterization and utilization, in: "Recent Chemical Applications of Mössbauer Spectroscopy," A.C.S. Symposium Series, C.K. Shenoy and V.G. Stevens, ed., American Chemical Society, Washington, D.C., in press.

Palache, C., Berman, H., Frondel, C., 1951, "Dana's System of Mineralogy," Vol. II, John Wiley and Sons, New York.

Paris, B., 1977, The direct determination of organic sulfur in raw coals, in: "Coal Desulfurization, Chemical and Physical Methods," A.C.S. Symposium Series 64, T.D. Wheelock, ed., American Chemical Society, Washington, D.C.

Saporoschenko, Mykola, Hinckley, C.C., Smith, G.V., Twardowska, H., Shiley, R.H., Griffin, R.A., and Russell, S.J., 1980, Mössbauer spectroscopic studies of the mineralogical changes in coal as a function of cleaning, pyrolysis, combustion and coal-conversion processes, Fuel, 59:567.

Weast, Robert C., ed., 1977-78, CRC Handbook of Chemistry and Physics," 58th edition, CRC Press, Inc., Florida.

Williamson, D.L., Guettinger, T.W., and Dickerhoof, D., Quantitative investigations of coal, *in*: "Recent Chemical Applications of Mössbauer Spectroscopy," A.C.S. Symposium Series, C.K. Shenoy and V.G. Stevens, ed., American Chemical Society, Washington, D.C., in press.

ANALYSIS OF RADIONUCLIDES IN AIRBORNE EFFLUENTS

FROM COAL-FIRED POWER PLANTS

G. Rosner, B. Chatterjee, H. Hötzl, R. Winkler

Gesellschaft für Strahlen- und Umweltforschung mbH
München, Institut für Strahlenschutz,
D-8042 Neuherberg, Fed. Rep. Germany

ABSTRACT

In order to assess the level of radioactivity emitted by coal-fired power plants in detail, specific activities of several radionuclides have been measured in samples from a coal-fired and a brown coal-fired plant in the Federal Republic of Germany. Samples measured included coal, brown coal, bottom ash, collected fly ash from the various electrostatic precipitator stages and sieve fractions of collected fly ash as well as samples of escaping fly ash taken from the exhaust stream, all taken simultaneously on three operating days. Nuclides measured were U-238, U-234, Th-232, Th-230, Th-228, Ra-226, Pb-210, Po-210 and K-40. Methods applied included (i) direct gamma spectrometry, (ii) radiochemical separation with subsequent alpha spectrometry and (iii) direct alpha spectrometry. Methods are described and discussed. Finally, annual emission rates of airborne radionuclides are calculated for both plants.

INTRODUCTION

A comparison of the risks arising from conventional and nuclear power production is a subject of considerable public interest. One part of this comparison concerns the radiation exposure of the population living in the vicinity of power plants due to airborne effluents. In this context, measurements of the radionuclide content of samples from coal-fired power plants are required. For this reason the Analytics and Ecology of Radionuclides Group of the Institute for Radiation Protection of the GSF, Munich, is engaged in the development of suitable methods and in the subsequent performance of such measurements. These methods and measurements are described in this paper.

In fact, coal contains several naturally occurring radionuclides, all of which can be present in the effluents of power plants. From the radiological point of view, the most important radionuclides are the longer-lived members of the U-238 and Th-232 decay series and Potassium-40. The pathway of radionuclides from the coal to the airborne effluents has been investigated in several studies[8,21] and may be sketched briefly as follows: During combustion, the radionuclides, as well as other constituents of the coal, have a tendency to volatilize, more or less, according to the vapor pressure of their respective compounds. Volatilized compounds, on their way to the stack, are on cooling condensed onto the surface of fly ash particles. The most volatile compounds are enriched in the smaller size particle fractions of the fly ash, since the smaller particles have a larger surface-to-volume ratio than the larger ones. Radioactive gases (Rn-220, Rn-222) leave the plant as such.

A small fraction of the fly ash is not retained by the plant's emission control devices, but is released to the atmosphere via the stack as an airborne effluent. Knowledge of the radioactivity content of this material is required for the assessment of the radiation exposure of the population in the vicinity of coal-fired power stations. Directed towards this aim, the present paper describes techniques used in our analyses of samples related to coal-fired power plants. Though there exist papers dealing with similar measurements [1,7], their conclusions seem to be drawn in most cases from measurements of the imput materials and of fly ash samples collected from the emission control devices rather than from measurements of the released fly ash itself. In the present study, in addition to bulk materials such as coal and collected fly ash, samples of escaping fly ash were included. Also, an analysis was carried out for the pure alpha emitter Po-210, for which only a few experimental data[3,8] were available.

SAMPLES ANALYZED

Samples analyzed comprised coal, brown coal, bottom ash, collected fly ash from the various electrostatic filter stages and four sieve fractions (<4µm, 4-6µm, 6-10µm, and >10µm) of collected fly ash, separated by wind sighting, - which are referred to, collectively, in the following as "bulk materials" -, and samples of escaping fly ash collected from the exhaust stream. The latter were deposited onto quartz wool according to a German guideline[9], with a filter head device for the determination of the dust content of the exhaust stream[10], and with an Andersen Mark III impactor for particle size analysis[11]. All samples were taken simulataneously on three operating days early in summer 1979 from two power stations in the FRG, which are fired by coal and by brown coal. Sampling was done by the Technischer Überwachungsverein Rheinland e.V.

GAMMA SPECTROMETRY

The choice of the analytical method depends strongly on the nuclide to be determined and on the quantity of sample material available. Gamma spectrometry can be applied if the nuclide to be determined is a gamma emitter, or if it has a suitable gamma emitting daughter nuclide and if the sample quantity available is large enough. In our study, gamma spectroscopy was used for the determination of U-238, Th-232, Ra-226, Pb-210 and K-40 in the bulk materials. U-238 was determined via the gamma lines of Th-234, Th-232 via the gamma lines of Ac-228, Pb-212 and Po-212, and Ra-226 via the gamma lines of Pb-214 and Bi-214. Indirect determination of mother nuclides via their respective decay products involves that radioactive equilibrium be established between the members of the decay chains. Direct measurement of single radionuclides is therefore preferable and, using gamma spectrometry, this is practicable for K-40 and Pb-210 in bulk materials. For other radionuclides and samples, the procedures are described below.

Measurements of gamma energies above 100 keV were carried out with two coaxially drifted Ge(Li) detectors with a resolution of 2.45 (2.8) keV, relative full energy peak efficiency of 13.7 (8.0)% and peak-to-compton ratio of 29:1 (19:1), all of which refer to the 1.33 MeV line of Co-60. Detector shielding consisted of lead rings of 5 cm wall thickness, lined with copper and cadmium foil. Detectors were connected via standard nuclear electronics to a multichannel analyzer ND 4420. Spectra were taken simultaneously in 2048 channels each at an energy scale of approximately 1 keV/channel.

Measurements of gamma energies from 10 to 200 keV were carried out with a high purity Germanium detector (Ortec Mod. 1013-25360) of 25 mm dia x 10 mm height, of 310 eV FWHM at 5.9 keV, connected to a ND 600 analyzer with 1024 channels at an energy scale of 187 eV/channel. Lead shielding and counting geometry were identical for all 3 detectors used.

The full energy peak efficiency was determined with a series of standard solutions[15,16], which were diluted and brought into a standard counting geometry (35 ml solution in a 50 ml polyethylene flask), and with spiked fly ash samples. The curves of photopeak efficiency vs. energy are shown in fig.1.

Samples of bulk materials were counted, air dried, in standard 50 ml polyethylene flasks, which were filled up to a height of 35 mm. The weight of the samples was 30 to 80 grams. The dependence of the full photopeak efficiency for lead-210 (46.52 keV) and for Th-234 (63.29 keV) from the sample weight is given in the insert of fig.1. Counting times were usually 1200 minutes and sometimes as high as 4000 minutes, due to the low activity of many samples.

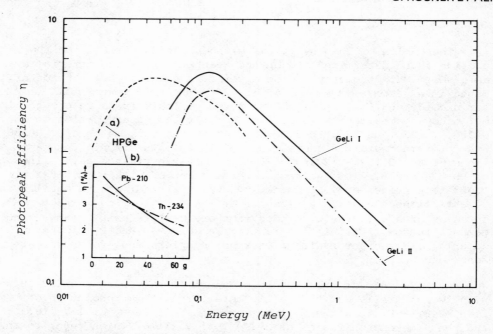

Fig.1 Response functions of the detectors used. I, II : Ge(Li) detectors. HPGe: High purity Germanium detector, a) η as function of gamma energy, b) η for Pb-210 (46.52 keV) and for Th-234 (63.29 keV) as function of the loose weight of spiked fly ash samples.

Lower limits of detection for the gamma spectrometric measurements are summarized in table 1. Detection limits[12] are given with a 95% confidence level (based on 4.66 times the standard deviation of the background) for a counting time of 1200 min. Sensitivities are based on a sample weight of 35 grams.

Statistical counting errors at the 95% confidence level range from about 50% at sample activities below 1 pCi/g to the order of 10% in the 10 pCi/g range and to 2-3% in the 100 pCi/g range. Systematic errors are mainly due to deviations from the efficiency curve and are estimated to be approximately ±5%. Below 150 keV, matrix effects may lead to errors of ±20% in the worst case. Another possible source of systematic errors is due to deviation from the radioactive equilibrium in the determination of U-238, Th-232 and Ra-226. An equilibration time of at least two weeks was allowed between filling the sample vial and counting.

Table 1   Detection Limits Obtained in Direct Gamma Spectrometric Analysis

Nuclide	Gamma Line Evaluated (keV)	Background Rate (cpm)	Lower Limit of Detection (pCi)[a]	Sensitivity (pCi/g)[b]	
Pb-210	46.52	1.0	41	1.2	
U-238	63.29	1.0	45	1.3	ca.4 ppm
Ra-226	295.22	2.0	28 ⎫		
	351.99	2.5	16 ⎬ ca.20	ca. 0.6	
	609.32	1.7	15 ⎭		
Th-232	238.63	6.0	15 ⎫		
	583.14	1.0	20 ⎬ ca.20	ca. 0.6	
	911.07	0.8	30 ⎭	ca. 5 ppm	
K-40	1460.75	1.5	77	2.2	

[a] Counting time, 1200 min
[b] Sample weight, 35 gram

PROCEDURES BASED ON RADIOCHEMICAL SEPARATION AND ALPHA SPECTROMETRY

It seems that radiochemical separations, followed by alpha spectrometry, have in very few cases been applied to analysis of samples related to coal-fired power plants[3,8]. Yet, such techniques are necessary in analysis of alpha emitters such as Po-210 and Th-230, which have no gamma rays suitable for direct gamma spectrometry under low level conditions. Further, current determination methods for Uranium and Thorium in samples from coal fired power plants, such as the most frequently used direct gamma spectrometry, or neutron activation or conventional inorganic methods, give no simultaneous information on the U-238 and U-234 content, or on the Th-232, Th-230 and Th-228 content, as does alpha spectrometry. Combined radiochemical-alphaspectrometrical techniques were therefore applied for determination of Po-210 in bulk materials and in escaping fly ash, as well as for the determination of U-238, U-234, Th-232, Th-230 and Th-228 in the escaping fly ash samples. In addition, chemical separation followed by low level beta counting was used for Pb-210 determination in escaping fly ash. Representative samples of escaping fly ash are available only in small amounts from several milligrams to, at maximum, one gram, collected on several grams of quartz wool, and therefore are not suitable for direct gamma spectrometry[10].

Decomposition of samples

Bulk materials were decomposed by a wet procedure comprising boiling of 1 g sample under reflux with 50 ml of concentrated nitric acid, treatment of the residues with hydrofluoric and perchloric acid, and finally taking up the entire sample in 50 ml of 6 N hydrochloric acid. Use of a small quantity (ca. 30 mg) of ammonium vanadate as an oxidation catalyst was helpful in many cases. Samples of escaping fly ash, which were collected on 2-4 grams of quartz wool, were allowed to stand with dilute (1+1) hydrofluoric acid until dissolution of the quartz wool. After evaporation, the residue was treated with perchloric acid and taken up as above. When the escaping fly ash samples were rich in carbon, they had to be treated, after dissolution of the quartz wool, like the bulk material samples. Whenever expected activity was high enough, solutions from escaping fly ash samples were divided into aliquots in order to allow the determination of several radionuclides.

Po-210

Polonium-210 was coprecipitated with bismuth sulfide on 20 mg of Bi^{3+}. The bismuth sulfide was centrifuged, dissolved in 9-N hydrochloric acid, and the Polonium was deposited on silver foil. After alpha spectrometric measurement, the Polonium-210 activity was calculated from the peak ratios of Po-210 and Po-208. The latter had been added as a tracer before starting the decomposition procedure.

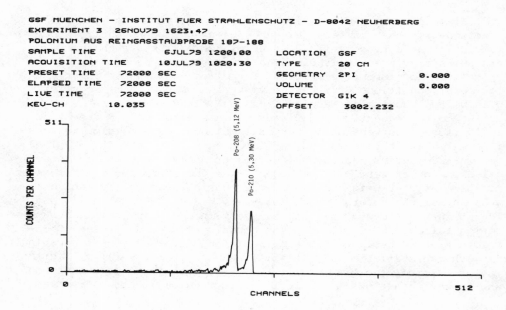

Fig.2 Alpha spectrum of the Polonium fraction isolated from an escaping fly ash sample (Sample weight 113.6 mg; aliquot used, 50 percent).

ANALYSIS OF RADIONUCLIDES

Corrections were applied for Po-210 build-up or decay, based on the known Pb-210 content of the samples. An example of the spectra obtained is shown in fig.2.

U-238, U-234

For determination of U-238 and U-234, an aliquot of the solution in 50 ml of 9-N hydrochloric acid was passed over an anion exchange column, 300 mm in length x 9 mm diameter, filled with Bio Rad AG 1 x 4, 100 - 200 mesh, in the chloride form. After washing with 50 ml of 9 N HCl, Iron was eluted with as little as possible 7-N nitric acid, and Uranium was finally eluted with 50 ml of 1.2 N HCl. An alpha spectrometry counting source was prepared by electrolysis of the Uranium from 10 ml of a slightly acidic ammonium sulphate solution onto a stainless steel disc with a diameter of the active deposit of 1.2 cm. U-238 and U-234 activities were calculated from the ratios of the areas under their respective alpha lines relative to that of U-232, which had been added as a tracer before separation. An example of the spectra obtained is shown in fig.3.

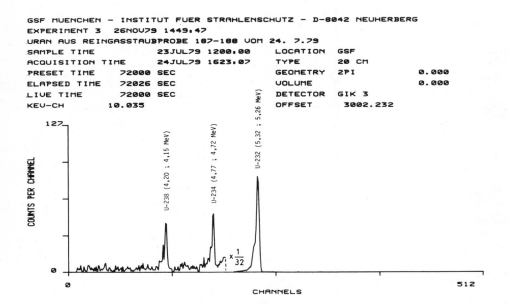

Fig.3 Alpha spectrum of the Uranium fraction isolated from an escaping fly ash sample (Sample weight 113.6 mg; aliquot used, 10 per cent).

Th-232, Th-230, Th-228

For determination of Th-232, Th-230 and Th-228, an aliquot of the sample solution in 50 ml of 7-$\underline{N}$ nitric acid was passed over a column which was identical to the one used in Uranium separation, with the resin in the nitrate form. The column was washed with 200 ml of 7-$\underline{N}$ HNO_3, the Thorium was eluted with 50 ml of 9 N HCl and electrolytically deposited as described for Uranium. The tracer used was Th-234 separated from a Uranium sample by anion exchange. For determination of the chemical yield, a separate counting of the tracer was performed either in a high purity Germanium detector or in a flow proportional counter after alpha spectrometric measurement of the Thorium isotopes.

Pb-210

After addition of 20 mg of Pb^{++} carrier together with 2 mg of Bi^{3+} and 2 mg of La^{3+} as a hold-back-carrier to another aliquot of the sample solution, Lead-210 was precipitated as the sulfide. It was then purified via the sulfate, precipitated and counted as the chromate in a low level beta counter through a filter paper after waiting at least 2 weeks for ingrowth of Bi-210.

Alpha counting

Alpha spectrometric measurements of the samples were carried out in Frisch grid ionization chambers, the details of which are described elsewhere[14]. The advantage of these chambers, compared to semiconductor detectors, which are more commonly used in alpha counting of chemically separated samples, is a higher detection efficiency. With standard point Po-210 sources[15], full widths at half maximum of 20.6 ± 0.4 keV, detection efficiencies of $40 \pm 1\%$ and instrumental backgrounds of 0.0083 counts/min, the two latter in an energy interval from 5.284 to 5.326 MeV, were obtained.

Statistical counting errors (including those from peak area ratio determination or from chemical yield determination) at the 95% confidence level are between 20 and 50 percent at activity levels below 1 pCi/g, and between 3 and 10 percent at the higher activity levels, in the analysis of bulk materials for Po-210. For samples of escaping fly ash, statistical errors in Po-210 analysis, as well as in the determination of Uranium and Thorium isotopes and of Lead-210, were between 5 and 20 percent in most cases, the small sample size being the limiting factor.

Lower limits of detection[12] (LLD's) for the radiochemical methods are summarized in table 2. The background given in column 3 is the sum of the instrumental background and of the reagents blanks, which were prepared with the same amounts of reagent as the original samples themselves. Detection limits are given for a 95% confidence

Table 2 Detection Limits Obtained in Combined Radiochemical-
Alphaspectrometric Analysis

Nuclide	Procedure	Background Rate (cpm)	Lower Limit of Detection (pCi)[a]	Sensitivity for Escaping Fly Ash (pCi/g)[b]
Po-210	Bi_2S_3 precipit. Ag foil depos. Tracer Po-208	0.054	0.16	1.6
U-238	Anion exchange	0.065	0.034	1.7
U-234	from 9 N HCl Tracer U-232	0.12	0.048	2.4
Th-232	Anion exchange	0.035	0.027	1.4
Th-230	from 7 N HNO	0.048	0.037	1.9
Th-228	Tracer Th-234 (β or γ counting)	0.020	0.024	1.2

[a]Counting time, 1200 min.
[b]Sample weight, 200 mg. For aliquoting see text. Sensitivity in bulk material analysis (sample weight 1 g) is equal to the lower limit of detection.

level, calculated from 4.66 standard deviations of the background at 1200 minutes counting time. Sensitivities are based on such aliquots as were usually taken for analysis (50% for Polonium-210, 20% for Lead-210, 10% each for Uranium and Thorium isotopes determination) from 200 mg samples of escaping fly ash, and on 1 gram samples of bulk materials.

DIRECT ALPHA SPECTROMETRY

An interesting application of the large area Frisch grid ionization chamber is that some materials, spread as thinly and uniformly as possible over a large surface, can be subjected directly to alpha spectrometric measurement without previous chemical separation[17]. In the present study, filters from the Andersen impactor used to determine the particle size distribution in the exhaust stream proved to be suitable for Polonium-210 determination by this technique. Therefore it was possible to determine the distribution of Po-210 activity on the various particle size fractions in the exhaust stream.

Combined samples were prepared from filters bearing the same particle size fractions collected on the respective sampling days. Further combined samples were prepared from the dust collected in the preimpactor, and from the dust collected on the final filter on the respective sampling days. Each combined sample was counted in a 20 cm diameter sampling dish in the ionization chamber. An example

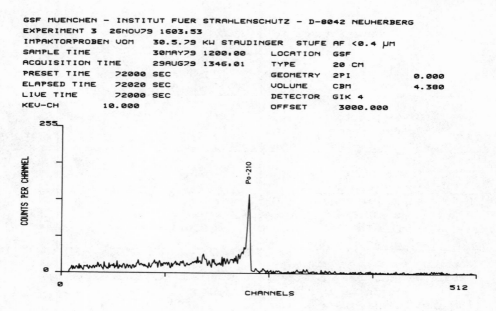

Fig. 4 Alpha spectrum of the dust collected on the final filter of the Andersen impactor (Samples combined from 3 operating days). Particle diameter $d_{50} < 0.4 \mu m$.

of the alpha spectra obtained is shown in fig.4. With an empirical detection efficiency of 0.20 ± 0.05 for Po-210 counted in the energy interval from 4.93 to 5.33 MeV, and an instrumental background of 0.083 cpm, a lower limit of detection for Po-210 of 0.087 pCi is calculated at 1200 minutes counting time.

DISCUSSION OF THE ANALYSIS METHODS

Briefly, advantages and disadvantages of the methods used may be summarized as follows:
Direct gamma spectrometry requires comparatively little manpower for sample preparation, as well as for counting and evaluation procedures, which are partly automatic. The sensitivity is sufficient even for materials of low specific activity, provided that enough material (ca. 35 g in the present study) is available. The method is nondestructive and allows multi-nuclide determinations from one sample. The nuclide to be determined, or one of its daughter nuclides, must emit a suitable gamma radiation. Indirect measurement via a daughter nuclide requires assumptions or experimental precautions concerning radioactive equilibrium.

Radiochemical separations, followed by alpha spectrometry, are necessary in analysis for alpha emitters, such as PO-210 and Th-230 which have no gamma rays suitable for direct gamma spectroscopic measurements under low level conditions. Moreover, radiochemical methods are most appropriate for analysis of the escaping fly ash samples. These are available only in amounts of up to 1 g, collected on several grams of quartz wool, and therefore not suitable for gamma spectrometric measurements. However, the low detection limits obtained with the radiochemical procedures allowed even very small samples and sample aliquots down to ca. 10 mg to be analyzed. Further, isotopes of Uranium and Thorium are determined directly, without assuming radioactive equilibrium. Disadvantages of the radiochemical procedures are the time-consuming dissolution and separation and the preparation of practically massless sources required for subsequent alpha spectrometry.

Several attempts were made to check the accuracy of the methods used: To obtain a picture of the accuracy of the direct gamma spectrometric method, an interlaboratory comparison was carried out involving analyses of 8 fly ash samples by 5 laboratories[18]. Results agreed within the statistical counting errors, which was considered to be satisfactory for the purposes of the present study. Further, for comparison of gamma spectrometric with combined radiochemical-alpha-spectrometric methods, the same 8 fly ash samples were analyzed chemically in our laboratory as described above. Again, satisfactory agreement within the statistical counting errors between gamma spectrometric and alpha spectrometric results was obtained. Furthermore, radiochemical analysis of the NBS Standard Reference Materials (1632a Coal) and (1633a Fly Ash) gave values of U-238 and Th-232 within 1 to 5 percent from the certified Uranium and Thorium mass concentration values[19].

RESULTS AND DISCUSSION

The numerical results of all the individual samples measured in the present study are reported elsewhere[13], as well as details on the individual techniques. Values for the radionuclide concentrations of K-40, Th-232 and Ra-226 are in agreement with values hitherto reported for coal, bottom ash and collected fly ash in the German literature for German coal [6,7,20]. Only few measurements have been reported for Pb-210[7], and no values for P.-210 in German coals. As far as values are published for German brown coal[5], they agree with ours. Generally, samples from the coal-fired power plant had a higher activity than those from the brown coal-fired plant.

Various tendencies to enrich in the finest sieve fractions of collected fly ash were found for the various radionuclides in the coal- and the brown coal-fired plant. Relative to the radionuclide content of original, air dried, coal, the highest enrichment was found for Pb-210 and Po-210 in the sieve fractions (from >10μm to <4μm) of fly ash collected from the last electrostatic filter stage, namely from 38 to 130 and from 36 to 240 in the coal-fired, and from 6 to 18 and from 8 to 30 in the brown-coal fired power plant. For other radionuclides, enrichment factors varied from 5 to 20 for the coal-, and from 4 to 9, for the brown coal-fired plant in the same sieve fractions. For all nuclides, the degree of enrichment was smaller in fly ash collected from the preceding filter stages, and in all sieve fractions thereof. Water content of the air dried material was ca. 7 percent for coal, and ca. 50 percent for brown coal, respectively[10].

The main aim of the present investigation was to obtain information on the radionuclide emission rates of the two power plants under study. The most direct way, towards this aim, is from the analysis of stack samples. However, since obtaining representative samles of escaping fly ash in sufficient quantity is difficult, in addition to analysis of escaping fly ash samples, the measurement of the radionuclide concentrations in collected fly ash as a function of particle size was made. As the particle size distribution is known in collected as well as in escaping fly ash[10], the radionuclide content of escaping fly ash can be calculated. In fact, measured specific activities of escaping fly ash were found to be in good agreement

Table 3 Annual Airborne Radioactive Emissions of a Coal-fired and of a Brown-coal Fired Power Plant (1000 MW$_e$, continuous operation)

Nuclide	Coal-fired Plant			Brown-coal Fired Plant		
	Spec. Activity Range (pCi/g)	Mean (pCi/g)	Emission Rate (mCi/GW·a)	Spec. Activity Range (pCi/g)	Mean (pCi/g)	Emission Rate (mCi/GW·a)
U-238, Th-234[a], U-234[b] each	8.1-12.9	9.8	16	1.8-3.0	2.3	3
Th-230	5.5-5.7	5.6	9	(0.5-0.9)[c]	(0.7)[c]	(1)[c]
Ra-226	6.2-8.4	7.3	12	1.4-2.0	1.7	2
Pb-210	52 - 74	66	108	3.3-7.7	5.5	7
Po-210	81 -159	125	205	6.1- 11	7.4	9
Th-232, Ra-228[a], Th-228[b], Ra-224[a] each	2.1-2.8	2.5	4	0.4-0.9	0.6	1
K-40	21 - 25	23	38	7 - 11	9	11

[a]Radioactive equilibrium assumed, [b]Radioactive equilibrium found
[c]Based on 3 values only.

with calculated specific activities, derived from the measured specific activity of fly ash sieve fractions from the last electrostatic precipitator stage.

From the radionuclide concentration in escaping fly ash the radionuclide emission rate can be calculated based on the known[22] dust content and volume stream of exhaust air. A calculation was carried out with dust contents of 60 mg/m^3 and 30 mg/m^3 for the 320 MW coal-fired and the 600 MW brown coal-fired power plant, with air throughputs of $1 \cdot 10^6$ and $2.7 \cdot 10^6$ m^3/h (1013 mb, 273 K)[13]. Table 3 summarizes the daily mean values of activity concentration in escaping fly ash, together with the resulting annual emission rates, normalized to an electrical output of 1GW/year.

Annual emission rates provide now a base for an estimation of radiation exposure of the population living in the vicinity of coal-fired power stations.

REFERENCES

1. M. Eisenbud and H.G. Petrow, Radioactivity in the Atmospheric Effluents of Power Plants that use Fossil Fuels, Science 144:288 (1964).
2. United Nations Scientific Committee on the Effects of Atomic Radiation, "Sources and Effects of Ionizing Radiation", United Nations, 1977, p. 86-89.
3. C. E. Styron, Preliminary Assessment of the Impact of Radionuclides in Western Coal on Health and Environment, 2nd National Conference on Energy Conservation, Rockville Pike, 1978.
4. H. L. Beck, C. V. Gogolak, K. M. Miller, and W. M. Lowder, Perturbations on the Natural Radiation Environment Due to the Utilization of Coal as an Energy Source, Proceedings of the Symposium on the Natural Radiation Environment III, Houston, Texas, April 23-28, 1978.
5. H. Kirchner, E. Merz, and A. Schiffers, Radioaktive Emissionen aus mit rheinischer Braunkohle befeuerten Kraftwerksanlagen, Braunkohle 11:340 (1974).
6. H. Schmier, Zur Frage einer Strahlenexposition aus natürlichen Quellen, insbesondere durch die Verbrennung und Verbreitung von Kohle, Kolloquium on Problems of Environmental Radioactivity and Radiation Protection, Federal Health Office, Berlin, Oct.7, 1976.
7. W. Kolb, Radioaktive Stoffe in Flugaschen aus Kohlekraftwerken, PTB-Mitt. 89:77 (1979).
8. J. W. Kaakinen, R. M. Jorden, M. H. Lawasani, and R. E. West, Trace Element Behaviour in Coal-Fired Power Plants, Environ. Sci. Technology 9:862 (1975).

9. VDI-Guideline 2066, Messen von Partikeln, Staubmessungen in Gasen, gravimetrische Bestimmung der Staubbeladung, Sheet 1 (October 1975), Sheet 2 (Draft- August 1979).
10. W. Jockel, Abschlußbericht des Forschungsvorhabens St.Sch. 695, Technischer Überwachungsverein Rheinland e.V., Köln (March 1980) here: p.15.
11. Ref. 10, p. 18.
12. J. H. Harley, (Ed.) HASL Procedures Manual, HASL-300, New York 1972, Section D-08-01.
13. B. Chatterjee, H. Hötzl, G. Rosner, and R. Winkler, Untersuchungen über die Emission von Radionukliden aus Kohlekraftwerken - Analysenverfahren und Meßergebnisse für ein Steinkohle- und ein Braunkohlekraftwerk -, German Report GSF S-617, Neuherberg (February 1980).
14. H. Hötzl and R. Winkler, Eine Großflächen-Gitterionisationskammer mit hohem Auflösungsvermögen zur Messung von Alphastrahlern in Proben mit niedriger spezifischer Aktivität, German report GSF S-474, Neuherberg (June 1978).
15. Laboratoire de Metrologie des Rayonnements Ionisants, Standards of Radioactivity, Catalogue 1977, C.E.A., Gif-sur-Yuette (1977).
16. Physikalisch-Technische Bundesanstalt, Radioaktive Standardsubstanzen, PTB 6.31, Braunschweig (1980).
17. H. Hötzl and R. Winkler, Large Area Gridded Ionization Chamber and Electrostatic Precipitator.Application to Low-level Alphaspectrometry of Environmental Air samples, Nucl. Instr. Meth. 150:177 (1978).
18. Ref. 10, Appendix Volume, p. 21
19. G. Rosner and M. C. Lapointe, to be published.
20. G. Keller, Die Konzentration natürlicher radioaktiver Stoffe in Saarkohle und Saarkohlenasche, Proceedings of the 12th Annual Meeting of the Fachverband für Strahlenschutz, Norderney, Oct 2-6, 1978, p. 1016.
21. D. H. Klein, A.W. Andren, J. A. Carter, J. F. Emery, C. Feldman, W. Fulkerson, W. S. Lyon, J. C. Ogle, Y. Talmi, R. J. van Hook, N. Bolton, Pathways of Thirty-Seven Trace Elements Through Coal-Fired Power Plant, Env. Science. Technol. 9:973, (1975).
22. Ref. 10, p. 29 and p. 40.

SECULAR EQUILIBRIUM OF RADIUM IN WESTERN COAL

V. R. Casella, J. G. Fleissner, and C. E. Styron

Monsanto Research Corporation
Mound Facility*
Miamisburg, Ohio 45342

ABSTRACT

Concentrations of radium-226, radium-228, and thorium-228 in coal from six Western states have been measured by gamma spectroscopy. The existence of secular equilibrium was verified for radium-226 and previously measured uranium-238 and also for radium-228 and thorium-228. The measured radionuclide concentrations for Western coal averaged about 0.3 pCi/g for radium-226 and 0.2 pCi/g for radium-228 and thorium-228. These average values are not greatly different from those in coal from other provinces of the United States.

INTRODUCTION

In order to reduce the dependency of the United States on foreign oil and meet the growing energy demand, it will be necessary to turn to the nation's abundant reserves of coal. Coal deposits in the Western states are of particular importance because Western coal reserves (∼200 billion tons) represent 72% of the identified U. S. coal resources and Western coal has a low sulfur concentration.[1] Furthermore, most of the coal minable at current prices is in the West.[2] However, there is an increasing concern over the potential problems associated with large additions of trace elements to the environment from burning coal.[3,4] In addition to nonradioactive trace elements, coal contains uranium and thorium and their decay products, and some Western coal reserves have been reported to contain much higher concentrations of uranium than most Eastern coal.[2]

---

*Mound Facility is operated by Monsanto Research Corporation for the U. S. Department of Energy under Contract No. DE-AC04-76-DP00053.

A project has been initiated at Mound Facility to evaluate the potential radiological impact of coal utilization. In the first phase of this project, preliminary studies were carried out on Western coal.[5] Coal samples from 19 mines in six Western states, representing 65% of the total coal mined in those states, were collected and analyzed for uranium. Only large Western mines (>0.4 million metric tons per year) were sampled and mines producing both lignite and subbituminous coal were included in this study. The locations and names of the mines are shown in Figure 1.

The radium-226 concentrations in these coal samples are of particular importance. Radium-226 is one of the more toxic radionuclides and is an element of concern identified in federal and state standards. Also, radium-226 concentrations can be related to activities of its relatively short-lived daughters, such as radon-222. The purpose of the present work was to measure the radium-226 concentrations in these Western coal samples and compare the results with the previously measured concentrations of the parent radionuclide of the naturally occurring uranium series, uranium-238. Since gamma spectroscopy was the method chosen for the radium-226 measurements, radionuclides from the naturally occurring thorium series, in particular radium-228 and thorium-228, could also be conveniently measured. A comparison of the concentrations of the radionuclides measured in these Western coal samples with concentrations reported for other coal provinces in the U. S. would be useful in determining whether the projected increase in Western coal usage will result in an increased radiological impact.

EXPERIMENTAL

The procedures for selection of the Western coal mines and for sample collection are described in a previous report.[5] After arriving at Mound Facility, the samples were ground to 60 mesh, dried at 110°C overnight, and riffled to assure homogeneity. Each coal sample was packed into an aluminum can (volume $\sim$93 cm^3) and hermetically sealed to preclude radon emanation. The samples were then aged for at least three weeks to ensure that the daughter activities of radium-226, radium-228, and thorium-228, which emit the measured gamma radiation, were in secular equilibrium. Gamma-ray spectra were obtained from each sample using a 70 cm^3 Ge(Li) spectrometer, and spectral analysis was performed with a PDP 11/34 computer using the computer code (GRPNL2) described by Fleissner.[6]

A spectrum of a Wyoming coal sample which was counted for 200,000 seconds is shown in Figure 2. Gamma-ray peaks at 242(^{214}Pb), 295(^{214}Pb), 352(^{214}Pb), and 609(^{214}Bi) keV were used for radium-226 measurements, peaks at 338(^{228}Ac) and 911(^{228}Ac) keV were used for radium-228 measurements, and peaks at 238(^{212}Pb) and 583(^{208}Tl) keV were used for thorium-228 measurements. The 186 keV peak from radium-226 was used as a monitor to verify that

RADIUM IN WESTERN COAL

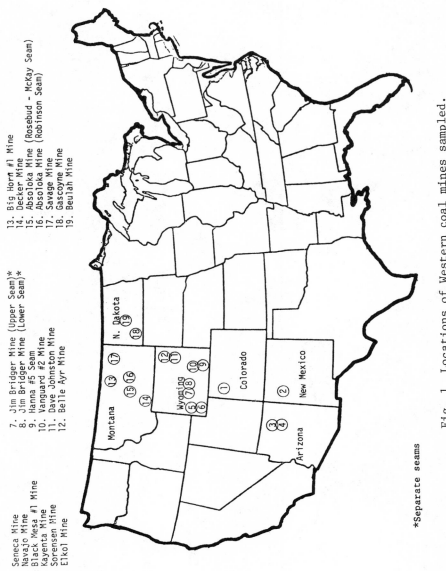

1. Seneca Mine
2. Navajo Mine
3. Black Mesa #1 Mine
4. Kayenta Mine
5. Sorensen Mine
6. Elkol Mine
7. Jim Bridger Mine (Upper Seam)*
8. Jim Bridger Mine (Lower Seam)*
9. Hanna #5 Seam
10. Vanguard #2 Mine
11. Dave Johnston Mine
12. Belle Ayr Mine
13. Big Horn #1 Mine
14. Decker Mine
15. Absoloka Mine (Rosebud - McKay Seam)
16. Absoloka Mine (Robinson Seam)
17. Savage Mine
18. Gascoyne Mine
19. Beulah Mine

*Separate seams

Fig. 1 Locations of Western coal mines sampled.

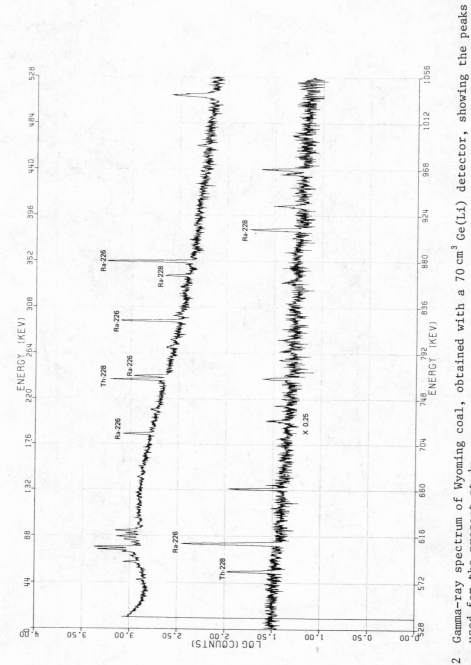

Fig. 2 Gamma-ray spectrum of Wyoming coal, obtained with a 70 cm³ Ge(Li) detector, showing the peaks used for the present study.

equilibrium was achieved. Since this peak and the 185 keV peak of uranium-235 are not resolved and since a correction for the 185 keV would introduce additional errors, the 186 keV peak was not used in the radium-226 measurements.

Reference coal samples were used to calibrate the gamma spectroscopy system. These coals had been previously analyzed by radiochemical methods for uranium-238, uranium-234, thorium-234, and lead-210 of the uranium series, and thorium-232 and thorium-228 of the thorium series. Secular equilibrium of these naturally occurring series was observed within experimental error.

Corrections were applied to the observed count rates of the Western coal samples analyzed in this study for the self-absorptions between the different samples. Most of these coal samples were counted for 200,000 seconds, but a few samples were only counted for 50,000 seconds, resulting in poorer counting statistic uncertainties.

RESULTS AND DISCUSSION

Concentrations of uranium-238 and uranium-234 measured previously,[5] and of radium-226, radium-228, and thorium-228 determined in the present work are given in Table 1. Gamma-ray spectra could not be obtained on 2 of the 19 Western coal samples listed in Table 1 because nearly all of these samples were used for destructive analyses. Radionuclide concentrations in the coal samples ranged from 0.05 to 1.20 pCi/g for uranium-238 and uranium-234, from 0.09 to 1.45 pCi/g for radium-226, and from 0.06 to 0.66 pCi/g for radium-228 and thorium-228. The radium-226 concentrations agreed with the uranium-238 and uranium-234 concentrations within errors (2σ) as did the radium-228 and thorium-228 concentrations.

These results showed that radium-226 and uranium-238 and -234 from the uranium series, and radium-228 and thorium-228 from the thorium series are in secular equilibrium in representative samples of coal from six Western states. Our results are in agreement with other studies that have been reported. Coles, et al.[7] analyzed coals from two Western coal-fired plants and Beck et al.[8] analyzed coal samples from several coal provinces (one Western coal) and also found these radionuclides to be in secular equilibrium.

The radionuclide concentrations given in Table 1 for coals from Western coal mines are roughly comparable to values for Eastern coal as reported by Bedrosian,[9] Martin et al.,[10] and an EPA study.[11] Average concentrations for the uranium series and thorium series radionuclides in the present study were 0.3 pCi/g and 0.2 pCi/g, respectively. These concentrations are below the national average of 0.6 pCi/g for uranium-238 and of 0.5 pCi/g for thorium-232 in coal reported by Swanson et al.[12] Some rare Western

Table 1. Concentrations of U-238, U-234, Ra-226, Ra-228, and Th-228 in Coal from Mines in the Western United States

Site Name and State	U-238[a] (pCi/g)	U-234[a] (pCi/g)	Ra-226 (pCi/g)	Ra-228 (pCi/g)	Th-228 (pCi/g)
1. Seneca, Colorado	0.29±0.02	0.29±0.02	0.27±0.05	0.31±0.10	0.38±0.05
2. Navajo, New Mexico	0.58±0.02	0.61±0.02	0.63±0.07	0.66±0.12	0.58±0.06
3. Black Mesa, Arizona	0.16±0.02	0.17±0.02	0.15±0.06	0.16±0.10	0.18±0.05
4. Kayenta, Arizona	0.16±0.01	0.19±0.01	0.20±0.06	0.20±0.10	0.17±0.06
5. Sorensen, Wyoming	0.06±0.02	0.07±0.02	0.11±0.04	0.09±0.08	0.09±0.03
6. Elkol, Wyoming	0.05±0.01	0.05±0.01	0.09±0.04	0.06±0.08	0.06±0.03
7. Jim Bridges Mine (Upper Seam) Wyoming	0.16±0.01	0.17±0.01	0.21±0.03	0.14±0.09	0.12±0.03
8. Jim Bridges Mine (Lower Seam) Wyoming	0.21±0.02	0.23±0.01	0.30±0.04	0.19±0.09	0.24±0.03
9. Hanna, Wyoming	1.17±0.06	1.19±0.02	1.45±0.16	0.49±0.08	0.48±0.03
10. Vanguard, Wyoming	1.13±0.06	1.12±0.02	b	b	b
11. Dave Johnson Mine, Wyoming	0.45±0.02	0.49±0.02	0.64±0.07	0.42±0.08	0.44±0.03
12. Belle Ayr, Wyoming	0.12±0.01	0.13±0.01	0.16±0.04	0.11±0.10	0.16±0.04
13. Big Horn, Wyoming	0.12±0.01	0.12±0.01	0.16±0.08	0.15±0.08	0.15±0.04
14. Decker, Montana	0.13±0.01	0.13±0.01	0.16±0.04	0.09±0.10	0.06±0.04
15. Absoloka (Rosebud), Montana	0.22±0.03	0.23±0.02	0.29±0.05	0.22±0.08	0.21±0.04
16. Absoloka (Robinson), Montana	0.20±0.02	0.20±0.02	0.24±0.04	0.14±0.10	0.12±0.04
17. Savage, Montana	0.16±0.02	0.17±0.01	b	b	b
18. Gascoyne, North Dakota	0.23±0.01	0.24±0.02	0.24±0.04	0.14±0.10	0.12±0.04
19. Beulah, North Dakota	0.10±0.01	0.11±0.01	0.13±0.04	0.07±0.09	0.10±0.03

[a] Data for U-238 and U-234 are taken from reference 5.
[b] Insufficient amount of sample for gamma analysis.

coal reserves have been reported to contain uranium at 10 to 100 times greater concentrations than Eastern coal,[2] but as shown by the present study, they are generally not representative of most coal found in the West. Therefore, the projected increased usage of most Western coal, rather than coal from other coal provinces of the U. S., should not result in an increased radiological impact on the environment.

REFERENCES

1. D. E. Barber and H. R. Giorgio, "Gamma-Ray Activity in Bituminous and Lignite Coals," Health Phys. 32(2):83 (1977).
2. J. D. Vine, "Geology of Uranium in Coaly Carbonaceous Rocks," U. S. Geological Survey, Professional Paper 356-D, 1962.
3. D. P. Rall, "Report of the Committee on Health and Ecological Effects of Increased Coal Utilization," Federal Register, 43(10):2229 (1978).
4. M. Eisenbud and H. H. Petrow, "Radioactivity in the Atmospheric Effluents of Power Plants that Use Fossil Fuels," Science, 144:288 (1964).
5. C. E. Styron, V. R. Casella, B. M. Farmer, L. C. Hopkins, P. H. Jenkins, C. A. Phillips, and B. Robinson, "Assessment of the Radiological Impact of Coal Utilization I. Preliminary Studies on Western Coal," MLM-2514 (February 12, 1979).
6. J. Fleissner and R. Gunnick, "GRPNL2: An Automated Program for Fitting Gamma and X-Ray Peak Multiplets," to be published.
7. D. G. Coles, R. C. Ragaini, and J. M. Ondov, "Behavior of Radionuclides in Western Coal-Fired Plants," Environ. Sci. Technol. 12:442 (1978).
8. H. L. Beck, C. V. Gogolak, K. M. Miller, and W. M. Lowder, "Perturbations on the National Radiation Environment Due to the Utilization of Coal as an Energy Source," presented at the DOE/UT Symposium on the National Radiation Environment III, Houston, Texas, April 1978.
9. P. H. Bedrosian, D. G. Easterly, and S. L. Cummings, "Radiological Survey Around Power Plants Using Fossil Fuel," Report EERL 71-3, U. S. Environmental Protection Agency, Washington, D. C. (1970).
10. J. E. Martin, E. D. Harward, and D. T. Oakley, Radiation Doses from Fossil Fuel and Nuclear Power Plants, in: "Power Generation and Environmental Change," D. A. Berkowitz and A. M. Squares, ed., American Association for the Advancement of Science (1971).
11. H. Krieger and B. Jacobs, "Analysis of Radioactive Contaminants in By-Products from Coal-Fired Power Plant Operations," EPA-600/4-78-039 (1978).
12. V. E. Swanson, et al., "Collection Chemical Analysis, and Evaluation of Coal Samples in 1975," U. S. Geol. Surv. Open-File Rep. 76-468 (1976) draft.

ELECTRON BEAM IONIZATION FOR COAL FLY ASH PRECIPITATORS*

R. H. Davis, W. C. Finney and L. C. Thanh

Department of Physics
Florida State University
Tallahassee, Florida 32306

INTRODUCTION

Sulfur is the element most frequently cited in discussions of emissions from coal combustion or conversion. In pulverized coal combustion, the coal utilization technology for the electric power industry today, the emission SO_2 is directly related to the sulfur content of the coal. Many utilities place a premium on low sulfur coal for the purpose of meeting SO_2 emission standards. But is low sulfur coal a panacea for pulverized coal combustion? In practice, no. In fact, a few utilities have added sulfur oxides to the feed stocks for the purpose of overall compliance with both sulfur and particulate emission standards.

The paradox lies in the role which sulfur plays in the control of particulate emissions. While low sulfur content insures correspondingly low production of SO_2, sulfur also plays a crucial role in the conductivity of the fly ash produced by pulverized coal combustion. Low sulfur coal produces a fly ash of correspondingly low conductivity or high resistivity. The performance of conventional electrostatic precipitators is successively degraded as resistivity values increase above $\approx 10^{10}$ ohm cm.[1]

The origin of the "high resistivity problem" in the electrostatic precipitators and the motivation for electron beam ionization development are evident in the operating principles of corona driven precipitators.[2] A schematic of the conventional precipitator is shown in Fig. 1 in which the plates are normally at ground and the

---

*Work supported in part by D.O.E. Contract #DE-AC21-81MC16229.

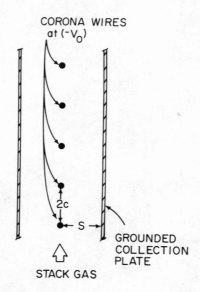

Fig. 1. Typical wire-plate geometry for electrostatic precipitator.

wires are maintained at a negative voltage sufficient to produce corona. As shown in Fig. 2, electrons produced by the corona are typically captured by electronegative oxygen molecules to form O_2^- ions. Particle charging takes place by means of two mechanisms: field charging which is dominant for particles >0.5 μm and diffusion charging which is the more probable mechanism for particles <0.2 μm. The two mechanisms compete in the size range 0.2 to 0.5 μm. Both charging mechanisms strongly depend on the ion current density and, in the case of field charging, explicitly on the local electric field.

The charged particles are driven into the collector plates by the electric field and adhere until a layer is built up which is sufficiently thick for mechanical removal by vibration or "rapping". High resistivity of the fly ash becomes a problem when the potential drop across a layer of adhered fly ash becomes sufficiently large for an electrical breakdown in the fly ash called "back corona". Such discharges reentrain some trapped fly ash but more important introduce positive ions which initiate breakdown between the plate and the corona wires. Such breakdown forces a reduction in the operating voltage of the precipitator, loss of efficiency and possibly the derating of the entire steam plant system.

Three attacks on the "high resistivity" problem are underway

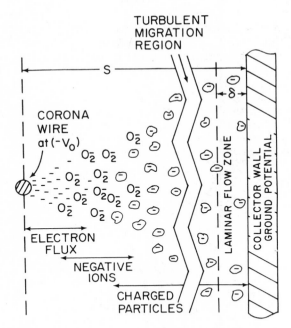

Fig. 2. Schematic details of charge carriers in the vicinity of the corona wire and in the laminar flow zone at the collector plate. The turbulent migration region occupies most of the separation (S) between the wire and plate.

currently: Two-stage operation,[3,4,5] pulsed excitation[6,7] and the use of additives which modify the resistivity including SO_3.[1] The motivation for electron beam ionization lies in the first option, the separation of the control function into two stages of particle charging and subsequent collection in a separate unit.[8,9] The advantage is that the special problems of particle charging are handled in the precharger without concern for the different problems of collection. In fact it may be advantageous to suppress collection in the precharger. On the other hand, the collector is required to provide nothing more than a migration electric field with little or no ion current to produce back corona. Because of the very high ion current densities, electron beam ionization is a promising technology for particle prechargers.

ELECTRON BEAM IONIZATION

Energetic electron beams (energies <3 MeV) are produced by

small commercially available units which are currently used in the
number of industrial applications such as paint curing, industrial
x-ray work and non-vacuum electron beam welding.[8] Reliability and
performance are essential in these applications. A 3 MeV research
accelerator was used to obtain the results reported here.

The equipment schematic for ion current density measurements
is shown in Fig. 3. Current density measurements are made in the
test volume between plates C and C' and the results of one set of
measurements are shown in Fig. 4. The measured values in excess
of 100 mA/m^2 or 500 times larger than typical operating values for
corona driven precipitators.[9]

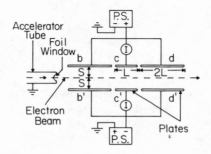

Fig. 3. Schematic diagram of the experimental apparatus. The
electron beam is delivered between the upper and lower
sets of plates. S = 5 cm, L = 10 cm.

The current versus voltage curves of Fig. 4 are curious in
two respects: the I-V relationship is more or less Ohmic with no
evidence for saturation i.e. a current limit due to complete
removal of ions, and the measured current values are far in excess
of those expected from primary ionizing events alone. While this
bonus of current density bodes well for particle charging applications, it is also clear that secondary processes which can lead to
breakdown are operating and may require suppression when aerosols
are used. An investigation of the conditions for producing ion
currents by primary ionization alone was initiated.

ELECTRON BEAM IONIZATION

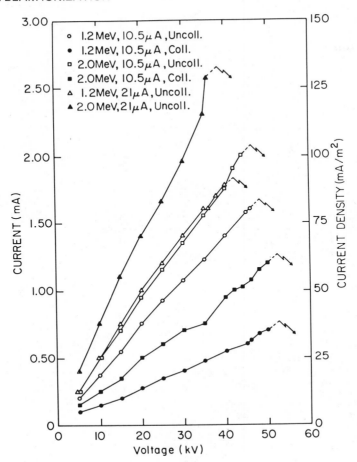

Fig. 4. The effects of beam energy, beam current, and collimation on the ion current characteristics. Plots are for the current read from the anode (plate c). Uncoll. - Uncollimated beam; Coll. - Collimated beam.

SATURATION CURRENT MEASUREMENTS

The first saturation current measurements[10] were made with the equipment shown in Fig. 5. The objective of the wedge collimator and honeycomb collimators is to limit the ionization zone in the test volume between plates C and C'. The observation of saturation is evident in the relatively flat segments of the I-V curves in Fig. 6 and 7.

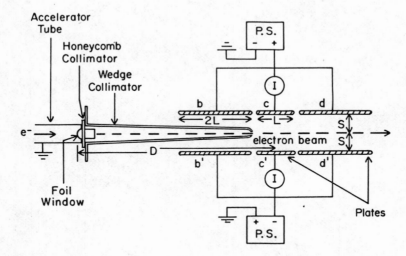

Fig. 5. Schematic diagram of the experimental apparatus. The electron beam passes through either the collimators or baffles with apertures and into the interelectrode region. L = 10 cm, foil window to center of plates c, c', D = 55 cm.

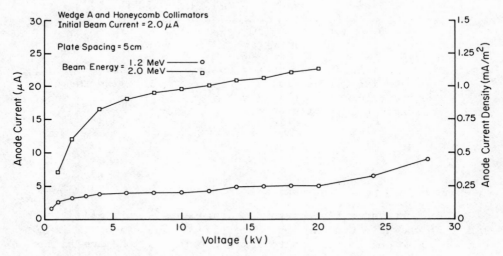

Fig. 6. The effects of beam energy on ion current saturation, using Wedge A and honeycomb collimators.

ELECTRON BEAM IONIZATION

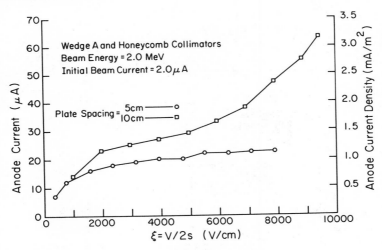

Fig. 7. The effects of plate spacing on the ion current vs. electric field relation, using Wedge A and honeycomb collimators.

Saturation was achieved equally well by replacing the collimator wedge and honeycomb wedge by a sufficiently small aperture in a thick plate so that the beam current admitted through the aperture was small. See Fig. 8. The dominant requirement for the observation of saturation is the restriction of the energetic electron flux to a sufficiently small value.

Before comparing the value of the saturation currents with that expected from primary ionization by the energetic electron beam, it is necessary to estimate the recombination rate. This was done by using the relation[11,12]

$$I = I_S(1 - \frac{1}{6} m^2 \frac{IS^3}{V^2}) \tag{1}$$

the quantity m in an empirical value which depends on the high mobilities and the recombination coefficient for ions in air at standard temperature and pressure and S is the electrode spacing. An example of the analysis is plotted in Fig. 9. Extrapolation of the curve backwards yields a saturation value of ~25 µA. The measured value of the current on the saturation plateau is 22 µA. Consequently 90% of the primary ionization is collected.

The measured value of the ionizing electron current delivered to the interelectrode volume in Fig. 9 is I_b=30nA. Consequently

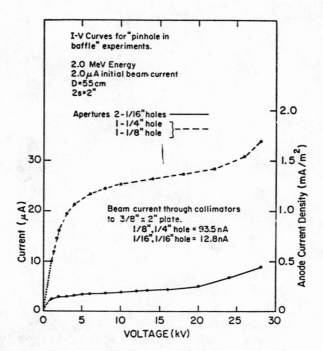

Fig. 8. Ion current vs. plate voltage curves comparing two aperture in baffle configurations.

the amplification factor is

$$I_S/I_b = 833 \tag{2}$$

The path length of the electron beam in the test volume is $\Delta L = 10$ cm. From the range-energy relations for energetic electrons,[13] the increment in range dR (in mg/cm^2) per a decrement of energy dE:

$$\frac{dR}{dE} = 530 \text{ mg/MeV cm}^2 \text{ @ 2 MeV beam energy.} \tag{3}$$

ELECTRON BEAM IONIZATION

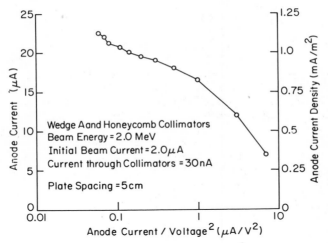

Fig. 9. Ion current vs. ion current/voltage^2 curve for Wedge A and honeycomb collimators.

The energy reduction ΔE for the range interval ΔR is given by:

$$\Delta E = \Delta R / 530 \text{ MeV} \tag{4}$$

where:

$$R = \Delta L \text{ (air density)} = 12.9 \text{ mg/cm}^2 \tag{5}$$

which yields $\Delta E = 12.9/530$ MeV for a beam energy of 2 MeV.

Taking the specific ionization for air as 32 eV per ion pair, the amplification factor calculated from the energy absorbed is given by:

$$\frac{\Delta E}{32 \text{ eV}} = \frac{I_s}{I_b} = 760 \tag{6}$$

which compares with the measured value of 833.

Similarly, for a beam energy of 1.2 MeV the following values are obtained:

$$(I_s/I_b) \text{ calculated} = 786 \tag{7}$$

and

(I_s/I_b) measured = 870 \hfill (8)

The agreement between the calculated and measured values is good.

Additional investigations of the conditions for saturation are under way, but several conclusions are clear from the preliminary results. First, the initial energetic electron flux must be restricted to very small values for the observation of saturation. Second, rather substantial ion current dentisties are achieved with primary ionization alone. Third, the recombination rate is small enough to be neglected in these experiments. Fourth, secondary ionization processes make significant contributions in the open geometry measurements previously discussed.

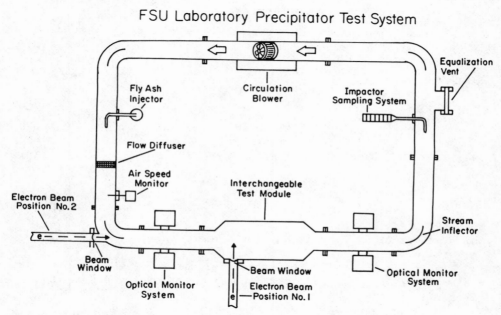

Fig. 10. Electron Beam Precipitator Test System.

TEST SYSTEM DESIGN

The schematic for the precipitator test system now under construction is shown in Fig. 10. This system will provide a controlled aeorsol with prompt loading monitors before and after the precharger-collector assembly. The precharger geometry to be tried first (Fig. 11) is one in which the function of particle charging and collection are spacially well separated. The electric field established between the anode and the cathode will produce a negative monopolar boundary of the ionization zone in the vicinity of the anode. Here advantage is taken of the very different mobilities for electrons, ions and charge particles. A requirement is that the stack gas velocity be sufficient to carry forward the bulk of the charged particles while unused ions are collected at the anode. The design result is the production of charged particles with very few "waste" negative ions. The charged particles are then collected in a wire precipitator geometry which is operated at voltages below that for corona emission.

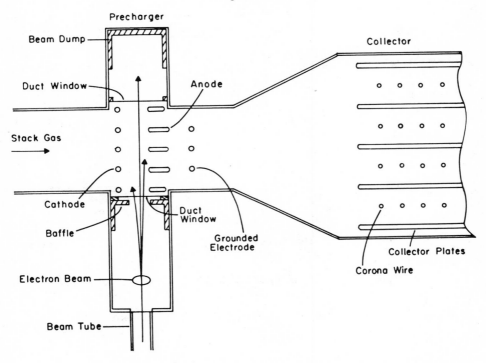

Fig. 11. Schematic of separated Precharger Test Module

Space charge may be a problem. Also, in the consideration of retrofit applications, it is important to make the electron beam system as compact as possible. The schematic of an immediate precharger is shown in Fig. 12. The ionization zone immediately precedes the entrance to the collector plates. A considerably larger ionization zone is required and in this case the gas velocity is that for typical precipitator operation, 1m/s or less.

One problem which is anticipated is the build up of dust deposits on the anodes or in the case of the immediate precharger on the plates themselves. Guard electrodes which will function in somewhat the same fashion as the grid in the SRI precharger are shown in Fig. 12. Another possible solution to the problem is radiation induced conductivity which will be taken up first in the section on related problems.

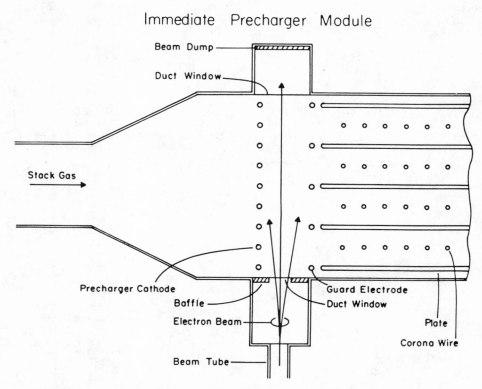

Fig. 12. Schematic of Immediate Precharger Test Geometry.

RELATED PROBLEMS

Energetic electrons ionize the atoms of a solid or the atoms of a liquid just as well as they do the atoms of a gas. The temporary ionization of a solid insulator is the basis for modern particle detectors. When subjected to a flux of energetic penetrating electrons, even the resistivity of diamond is drastically reduced. Consequently, one suggestion for solving the anode leading edge problem is to maintain a sufficient flux of energetic electrons that the induced conductivity of the deposited dust is kept at a useful value.

A more provocative question raised by preliminary experiments on radiation induced conductivity is that of persistence. A marked reduction in the resistivity, more than an order of magnitude was observed for an insulating dust sample under irradiation. What was unexpected was the persistence of this reduction for tens of hours. The persistence is surprising enough. It does raise a question as to whether electron beam treatment of the fly ash while still entrained will have a similar persistent effect on the resistivity. It should be emphasized that any estimates based on conventional solid state relaxation mechanisms are orders of magnitude too short for such an expectation. At the moment the persistence observed for three millimeter thick samples of fly ash and alumina dust remains unexplained.

Another interesting problem is the question of direct electron charging or, in the nomenclature of solid state physics, electron implantation. For a given size of particle there is an upper limit on electron energy such that the electron will be trapped in the particle rather than pass through. Consequently, as the ionizing flux produces a swarm of electrons of increasing number and decreasing energy, a certain fraction of the electrons will have sufficient energy to penetrate a particle but not to escape. There is already some evidence that direct electron charging takes place at elevated temperatures.[14,15] One advantage of such a mechanism in the case of electron beam ionization is the ability to drive the charge on the particle up to field emission limit or any other limit since electrons are available with sufficient energy to overcome any achievable Coulomb barrier.

An intriguing area for further investigation is the filtration of charged particles.[16] Fabric filters are currently arranged in bag houses for the purpose of removing particulate pollutants from stack gas. The major drawback in the use of bag houses is the fact that the pressure drop increases as the fabric becomes loaded with a dust cake. The increase in pressure drop as fabric loads with dust is substantially reduced when the particles are charged with little or no loss in collection efficiency.[17,18,19,20] The physics of this filtration improvement is not well understood. Some combination of particle electrification and filtration may constitute a new technology for particulate emission control

REFERENCES

1. S. Oglesby, and G. B. Nichols, <u>Electrostatic Precipitation</u>, Marcel Dekker, Inc., New York, Pub. 1978.
2. H. J. White, <u>Industrial Electrostatic Precipitation</u>, Addison-Wesley, Reading, Massachusetts, 1963.
3. D. H. Pontius and L. E. Sparks, APCA Journal <u>28</u>, 699 (1978).
4. S. Masuda, M. Washizu, A. Mizuno, and K. Akutsu, Conference Record of IAS-IEEE Annual Meeting, Toronto, Canada, October 1978, p. 16.
5. H. W. Spencer, J. J. Schwab, and J. Tassicker, Test Program for an Ionizer-Precipitator Fine Particle Dust Collection System,

Electric Power Research Institute, Palo Alto, California, Paper No. 77-2.1.
6. K. S. Kumar, P. L. Feldman, H. I. Milde, and C. Schubert, Conference Record of IAS-IEEE Annual Meeting, Cleveland, Ohio, October 1979, p. 1333.
7. P. Lausen and H. Henriksen, Conference Record of IAS-IEEE Annual Meeting, Cleveland, Ohio, October 1979, p. 163.
8. R. H. Davis and W. C. Finney, Energy Research $\underline{2}$, 19 (1978).
9. W. C. Finney, L. C. Thanh, and R. H. Davis, Proc. of the Second Symposium on the Transfer and Utilization of Particulate Control Technology, Denver Research Institute, Denver, Colorado, July 1979.
10. L. C. Thanh, W. C. Finney and R. H. Davis, Conference Record of IAS-IEEE Annual Meeting, Cincinnati, Ohio September 1980.
11. J. W. Boag, "Ionization Chambers", Radiation Dosimetry, Vol. II, Ed. F. H. Attix and W. C. Roesch, Academic Press, p. 17, 1966.
12. S. Masuda, K. Akutsu and N. Ishiga, "Control of gaseous pollutants by the use of a high energy electron beam", Special Research Project, Faculty of Engineering, University of Tokyo, 1975.
13. L. Katz and A. S. Penfold, Rev. Mod. Phys., Vol. 24, 1, p. 28, Jan. 1952.
14. C. C. Shale, "New concept of electron detachment for air in negative corona at high temperature", Circular 8353, U. S. Bureau of Mines, 1967.
15. J. R. McDonald, M. H. Anderson, R. B. Mosley and L. E. Sparks, Change Measurements on Individual Particles Exiting Laboratory Precipitators, Report of Southern Research Institute, Birmingham, Alabama 1979.
16. Control of Emissions from Coal Fired Boilers, Vol. I of Proc. Second Symposium on the Transfer and Utilization of Particulate Control Technology, EPA Report 600/9-80-039a September 1980.
17. G. W. Penney, "Using Electrostatic Forces to Reduce Pressure Drop in Fabric Filters," Powder Separation, 18, 1977, pp. 111-116.
18. T. Ariman, "Progress in Fine Particulate Control Technology", Joint Symposium on Fibers, Electrostatics and Filtration, Princeton, N. J., November 1979.
19. D. J. Helfritch, "A Summary of Performance Testing of the Apitron Electrostatically Augmented Fabric Filter," Joint Symposium on Fibers, Electrostatics and Filtration, Princeton, N. J., November 1979.
20. K. Innoya and Y. Mori, "Experimental Advances in Fabric Filtration Technology in Japan," 2nd Symposium on the Transfer and Utilization of Particulate Control Technology, Denver, July 1979.

INDEX

Activation analysis, see Neutron activation analysis
Aliphatic carbon
  and shale oil yield, 356
Alpha spectrometry
  determination of radionuclides in coal ash, 463-468
Aluminum salts
  effect on kaolinite properties, 248-250
Anions
  determination by ion chromatography, 177, 178
Aromaticity, 356
Arsenic
  in coal, 90, 119, 127, 129, 143, 169-172, 326
  in coal conversion products, 86, 90
  in fly ash, 203
  in oil shale products, 18, 20
  in oil sands, 54, 55, 58
  in petroleum, 10, 58
  in shale oil, 10
Ashing
  low temperature (LTA), 443-447, 455
  wet, 36, 37
Asphaltenes
  coal
    composition, 301, 302
    molecular weights, 315
    structure, 295
  definition, 296
  elemental composition, 298-302, 412, 413

Asphaltenes (continued)
  heteroatom contents, 308, 309, 311-314, 412, 413
  oil sand
    composition, 412, 413
    content in bitumen, 412, 413
  petroleum
    composition, 301, 302
    heteroatom contents, 311-314
    model structures, 302, 304-306, 310, 311
    separation, 297
    trace element contents, 29, 30
  spectroscopic studies, 295-321
Atomic absorption spectroscopy
  flame
    trace elements in oil shale products, 2, 11-13
  graphite furnace
    determination of lead in oil, 263
    trace elements in oil shale products, 3, 13, 14
  for trace elements in coal
    comparison with other methods, 115-131
Autoclave, 210

Bacteria
  and coal mineral formation, 324, 325, 329
Barite, 142, 148
Biodegradation of petroleum, 50

Boron
  in coal, 118, 119, 129, 143
  in coal conversion products, 84, 119
  in coal gasification waters, 186, 190
  in oil shale products, 18, 20
Bromine
  in coal, 90, 118, 119, 127, 143
  in coal conversion products, 90, 119
  in oil sands, 55, 57
  in oil shale products, 10, 18, 20
  in recycled oil, 266
Brown coal
  radionuclide contents, 470

Cadmium
  in coal, 118, 119, 143
  in coal conversion products, 119
  in oil shale products, 18, 20, 23, 24
  in petroleum, 31, 38, 39
Californium-252
  neutron source for activation analysis, 39, 152
Capture gamma-ray analysis, see activation analysis, prompt
Carbenes, 297
Carboids, 297
Carbon
  in coal, 119
  in coal conversion products, 119
  in oil sand bitumen, 51
  in oil shale products, 22
  types in oil shale, 356
Carbon-13 NMR, see nuclear magnetic resonance
Carbon-14
  reaction of compounds with coal, 230-235
Carboxylic acid
  functional groups in coal reaction with labelled compounds, 230-235

Catalytic hydrogenation
  of coal, 369
Catalysts
  petroleum refining, 271-283
Chalcopyrite, 142
Chlorine
  in coal, 90, 118, 119, 138, 143, 153
  in coal conversion products, 90, 119
  determination by prompt activation analysis, 153
  in oil sands, 51, 57, 58
  in oil shale materials, 10
  in petroleum, 10, 31, 58
  in recycled oil, 265
  species
    determination by ion chromatography, 266
Clay Minerals
  in coal, 142-144, 146
  determination by X-ray methods, 242
Cluster analysis, 168
Coal
  asphaltenes, see asphaltenes
  blending, 160
  combustion
    analysis of feed coal, 161
  conversion processes
    fate of trace elements, 84
  conversion products
    determination of nitrogen, 69-82
    determination of oxygen, 69-82
    determination of trace elements, 88, 89, 92, 94, 117
    determination of uranium, 88, 89, 94
    nature of titanium species, 89, 92, 93
  demineralization, 128, 129
  density separation, 166
  determination of sulfur, 152, 153, 156-162
  determination of trace elements, 83-96, 115-131, 133-140, 163-174

INDEX

distribution of inorganic species, 116
drying methods, 77-79
elemental analysis, 156-162
factor analysis of trace element data, 163
gasification
  in-situ, 175-188
  trace element fate, 84
gasification waters
  composition model, 180-187
  determination of trace elements, 177, 178
  distribution of boron species, 186
gasification wastes
  distribution of iron species, 185-187
hydrogenation, see also coal liquefaction
  analysis of products by GC/MS, 211
  catalytic, 369
  deuterium isotope studies, 207, 212-215
  hydrogen sulfide formation, 216
  kinetics of tetralin reactions, 368
  mechanism, 208
  nuclear magnetic resonance studies, 211, 212, 365-380
liquefaction, see also coal hydrogenation
  application of nuclear magnetic resonance, 365-380
  fate of trace elements, 84, 85, 89-92, 126
  mineral transformations, 86
liquids
  elemental analysis, 69-82
macerals, 325, 326
minerals
  authigenic, 145, 148
  correlations with macerals, 325, 326
  determination by scanning electron microscopy, 142

Coal (continued)
  minerals (continued)
    detrital, 145, 148
    factor analysis, 163-174
    geologic controls in origin, 323-335
    origin, 144, 145, 323-335
    role of bacteria, 324, 325, 329
    trace element contents, 323-325
  reactions of functional groups, 230-235
  radionuclide contents, 459-471, 473-479
  radium content, 473-479
  reduction with sodium-potassium, 342
  structure, 337-348
  trace element methods
    comparison of methods, 115-131
  wastes
    effect of seawater leaching, 193, 195
    measurement of trace elements, 192
  water content, 80
Coalification, 331
Cobalt
  determination in catalysts, 271-283
Convolution difference spectroscopy, 338
Coquimbite, 451, 455
Cross polarization NMR, 337-348

De-asphalted oils, 41
Demineralization of coal, 128, 129
Desulfurization catalysts
  determination of composition, 282
Detection limits
  for inductively coupled plasma, 12, 178
  for uranium by fission track, 67
Deuterium
  distribution in coal liquids, 376

Deuterium (continued)
  exchange with hydrogen in coal, 207-208
  nuclear magnetic resonance, see nuclear magnetic resonance
  transfer reactions in coal hydrogenation, 372
Diazomethane-^{14}C
  reaction with coal, 229-235
Dichlorocarbene-^{14}C
  reaction with coal, 229-235

Electron
  accelerator
    use in fly ash precipitator, 484
  beam ionization
    in fly ash precipitator, 481-494
  microprobe analysis
    matrix corrections, 288, 289
    of particulates, 285-294
Electrostatic precipitators, 482, 483
Emission spectroscopy
  direct current plasma arc
    detection limits, 12
    determination of trace elements, 2, 9-11
  inductively coupled plasma, 177, 178
Errors
  in fast neutron activation analysis, 71-72
  in X-ray fluorescence, 273
Exchange index, 218

Fabric filters, 493
Factor analysis
  use in coal trace element data, 163-174
Fast neutron activation analysis, see activation analysis
Field ionization mass spectroscopy, 413-416
Fission track
  analysis
    detection limits for uranium, 67

Fission track (continued)
  analysis (continued)
    determination of uranium in coal, 88, 92, 94, 143
    determination of uranium in petroleum, 61-68
    fast neutron, 64, 65
  counting
    automated methods, 65
  detectors
    etching conditions, 65, 88
    macrofol, 63, 64
Fly Ash
  particles
    analysis by electron microprobe, 285-294
  precipitators, 481-494
  radionuclide contents, 460, 464-471
Formation permeability
  effect of fluid composition, 244
Freeze drying
  for oxygen in coal, 71, 76-79
Frisch ionization chamber, 467
Functional groups
  in coal
    reactions with labelled compounds, 231

GAMANAL computer code
  comparison with other codes, 97, 106-110
Gamma-ray spectroscopy
  comparison of data reduction codes, 99-114
  for radionuclides in coal ashes, 460-463
  for radium-226, 228 in coals, 474-477
  for thorium-228 in coals, 474-477
Gas chromatography-mass spectroscopy
  analysis of coal hydrogenation products, 211
  analysis of oil sand bitumen, 409-441
  analysis of oil shale waters, 383-397

INDEX

Graphite furnace atomic absorption, see atomic absorption spectroscopy
Groundwater
  trace elements from in-situ gasification, 175–188

High performance liquid chromatography
  with graphite furnace AAS for lead in oil, 263
Hydrocarbons
  in oil sand bitumen, 392, 393
  from oil shale kerogen pyrolysis, 401, 402
  in recycled oils, 267
Hydrodesulfurization catalyses
  analysis by X-ray sluorescence, 271–283
Hydrogen scrambling, 378
Hydrogen sulfide
  formation in coal hydrogenation, 216
Hydrogenation, coal, see coal hydrogenation
Hydrogenation index, 218

Illinois basin coals, 115–131
Inductively coupled plasma
  determination of trace elements in coal gasification waters, 177, 178
Inorganic
  affinity of trace elements in coals, 126
  constituents in coal gasification waters, 175–188
Instrumental neutron activation analysis, see activation analysis
Interferences
  in fast neutron activation analysis, 71–73
  in neutron activation analysis, 4, 5
  in X-ray analysis, 273, 286
Ion chromatography
  for anions in coal gasification waters, 177, 178

Ion exchange
  role in trace element origin in coal, 328
Ion microprobe
  determination of trace elements in coal, 142
Iron
  in coal, 90, 118, 119, 121, 138
  in coal gasification waters, 185
  in coal conversion products, 86, 90, 119
  in oil sands, 55, 57, 58
  in oil shale products, 10, 18, 20, 23, 24
  minerals
    formation by low temperature ashing, 455
  in petroleum, 10, 31, 58
  species in coal
    determination by Mossbauer spectroscopy, 447–458
  sulfates in coal
    identification by Mossbauer spectroscopy, 448–454
Isotopic Exchange
  in coal hydrogenation, 218
Isotopic composition
  of deuterated solvents, 224
  of hydrogenation products, 222
Isotopic tracers
  in coal hydrogenation studies, 207–228

Jarosite, 447, 451, 455

Kaolinite
  effect on sediment permeability, 241–256
  effects of aluminum salts on colloidal properties, 248–250
Kerogen
  analysis by mass spectroscopy, 399–407
  concentrates from oil shale, 352, 400

Kerogen (continued)
  decomposition, 359
  extraction from oil shale, 400
  generation of hydrocarbons from, 401, 402
  types in oil shale, 352
  vacuum pyrolysis, 400
Kinetics of hydrogenation, 368
Kornelite, 449, 451, 455

Lead
  in coal, 119, 144
  in coal conversion products, 84
  in oil shale products, 10, 19, 21
  in petroleum, 10, 31
  in recycled oil
    determination by atomic absorption, 260, 261
    determination by HPLC-AAS, 263
Liquefaction, coal, see coal liquefaction
Liquid chromatography, see also high performance liquid chromatography
  of organics in oil shale retort water, 386
Low temperature ashing
  of coals, 443-447
  effects on iron minerals, 455

Magic angle spinning NMR, 354
Mass balances
  for trace elements in oil shale processing, 26
Mass spectroscopy
  analysis of oil shale kerogen, 399-407
Matrix corrections
  in electron microprobe analysis, 288
Mercury
  in coal, 90, 119, 144
  in coal conversion products, 16, 90
  determination by fast activation analysis, 14
  in oil sands, 55, 57, 58

Mercury (continued)
  in oil shale materials, 10, 16, 24, 25
  in petroleum, 9, 10
  separation scheme, 13, 15
  in shale oil, 18, 20
  in solvent refined coal, 16, 90
  in standard reference materials (NBS), 16
Metal complexes
  in oil sands, 56
  in petroleum, 32, 56
  in solvent refined coal, 89, 92, 93
Metalloporphyrins
  in oil sands, 52, 56
  in petroleum, 32
Minerals
  in coal, see coal minerals
  in oil sands, 51
  reactions in coal liquefaction, 86
Minicomputers
  use in gamma-ray spectroscopy, 97-114
Molecular weight
  of asphaltenes, 315
Molybdenum
  in catalysts, 275-283
  in coal, 144
  in oil shale processes, 18, 20,
  in petroleum, 10, 31
  in shale oil, 10
Monte Carlo methods, 285, 287
Mossbauer spectroscopy
  determination of pyritic and sulfate iron, 451
  determination of sulfur species in coal, 443-458

Naphthol-^{14}C
  reaction with coal, 232
National Bureau of Standards SRM's, see standard reference materials
Neutron activation analysis
  automated, 133-140
  coincidence - non-coincidence counting, 5-7

INDEX

Neutron activation analysis (continued)
  comparison of data reduction codes, 97–114
  comparison with other methods, 115–131
  delayed neutron determination of uranium, 134
  determination of trace elements
    in coal, 115–131, 167
    in coal gasification waters, 177, 178
    in oil sands, 52–54
    in oil shale materials, 1, 2–7, 10
    in recycled oil, 259, 265–267
    in standard (USGS) rocks, 99–114
  fast
    interferences, 71–73
    nitrogen determination, 69–82
    oxygen determination, 69–82
    sample preparation problems, 71, 76–79
  interfering reactions, 4, 5
  prompt gamma-ray analysis
    determination of trace elements in coal, 151–154
    on-line, 152–162
    sequential, 134–136
Nickel
  in catalysts, 271–281
  in coal, 90, 119, 123, 127, 129, 144
  in coal conversion products, 90, 119
  in oil sands, 52, 54–58
  in oil shale products, 19, 21, 24, 25
  in petroleum, 10, 31, 58
  porphyrins
    in oil sand bitumen, 52, 56
    in petroleum, 30, 52, 56
  in shale oil, 10
Nitrogen
  bases
    in oil shale waters, 388, 389

Nitrogen (continued)
  in coal, 70, 119
  compounds
    from kerogen pyrolysis, 404
    in fossil fuels, 69–82
    in oil sands bitumen, 51
    in oil shale products, 22
    in petroleum asphaltenes, 299, 311–314
Nuclear magnetic resonance (NMR)
  and asphaltene structure, 295–321
  carbon-13
    of coal hydrogenation products, 211, 212, 365–380
    and coal structure, 337–348
    and oil shale, 349–363
  coal liquefaction mechanisms, 365–380
  cross polarization, 349
  deuterium
    and coal hydrogenation products, 211, 212, 374
  magic angle spinning, 349–363
  and oil shale retorting waters, 360

Oil, see petroleum
Oil sands
  asphaltenes, see asphaltenes
  bitumen
    analysis by GC/Ms, 409–441
    asphaltene content, 51, 413
    composition by GC/MS, 432–440
    compound class contents, 51, 412, 413
    content in, 51
    elemental composition, 53, 410, 412
    hydrocarbon contents, 51, 413
    phytane content, 438
    resins content, 413
    trace element content, 50, 52, 54–58
  chemical properties, 51
  deposits, 49–52, 54–58, 409, 410

Oil sands (continued)
  origin, 50
  physical properties, 30, 51
  reserves, 50, 409, 410
  trace element contents, 50, 54–58
Oil Shale
  elemental composition, 22
  evaluation
    and aliphatic carbon, 356
    by nuclear magnetic resonance, 349–363
    carbon aromaticity, 356
    transformation ratio, 358
  kerogen, 352, 399–407
  potential oil yield, 358
  reserves, 1
  retort waters
    chromatography, 386
    hydrocarbons by GC/MS, 392, 393
    nitrogen bases, 388, 389
    organic acids, 390, 391
    trace element contents, 9
  retorting
    mass balances, 23, 26
    NMR study of products, 359, 360
    trace element distribution, 1–27
  trace element determination
    by flame atomic absorption, 2, 11–13
    by neutron activation analysis, 2–7
    by plasma arc emission, 2, 7–9
    by X-ray fluorescence, 2, 7–9
Oil spills
  origin by trace element contents, 43
On-line analyzer, 157–162
Organic acids
  from oil shale retorting, 390, 391
Organic association
  of trace elements in coal, 126, 143
  of trace elements in solvent refined coal, 92, 93

Organometallic compounds
  as trace element standards, 35
Oxygen
  in coal, 119
    determination by fast neutron activation analysis, 70
    drying methods, 77–79
    sampling problems, 77–79
    standards for, 73
  in coal liquids, 69–82
  in fossil fuels, 69–82
  in oil sand bitumen, 51, 410, 412
  in oil shale products, 22
  in petroleum asphaltenes, 299, 311–314

Partial polarization spectroscopy (NMR), 340
Particle analysis
  by electron microprobe, 283–294
Peat formation
  on coal mineral origin, 324
Permeability of clay media, 241–256
Petroleum, see also oil sand, oil shale
  analysis
    of aromatic fraction by GC/MS, 417–422
    of saturates by GC/MS, 422–430
  asphaltenes, see asphaltenes
  chromatographic separation, 416
  determination of
    cadmium by neutron activation analysis, 38, 39
    trace elements by atomic absorption, 2, 11–13
    trace elements in de-asphalted oil, 40, 41
    uranium by fission track analysis, 61–68
    vanadium by neutron activation, 38, 39, 41
  metalloporphyrin contents, 32, 45, 52, 56
  origin
    role of bacterial degradation, 50
  phytane content, 426

INDEX

Petroleum (continued)
  pristane content, 426
  trace element content, 9, 10,
     30-32, 39-42, 44
  wet ashing, 36, 37
pH
  and coal mineral formation,
    324
Phenolic groups in coal
  reaction with labelled com-
    pounds, 230-235
Phytane
  in petroleum, 426
Proton induced X-ray emission
    (PIXE)
  elemental analysis of coal,
    189-204
Polonium-210
  in coal ashes, 464, 465
  radiochemical separation, 460,
    464-471
Porphyrins, see metalloporphy-
    rins; nickel porphyrin;
    vanadium porphyrin
Potassium-40
  in coal ashes, 460
Preasphaltenes, 369
Pristane
  in petroleum, 426
Prompt gamma-ray analysis, see
    neutron activation
    analysis
Proton microprobe
  determination of elements in
    coal, 189-204
Proton NMR, see nuclear mag-
    netic resonance
Pyrite
  in coal, 142, 148, 328, 448-
    454
  in coal conversion processes,
    86
  determination by Mossbauer
    spectroscopy, 448-454
  role of bacteria in formation,
    328

Radionuclides
  in coal, 473-479
    determination by alpha spec-
      troscopy, 463-468

Radionuclides (continued)
  emission from power plants,
    459-472
Radiotracer studies, 230-235
Radwin-226, 228
  in coals, 473-478
Rare earth elements (REE)
  in coal, 90, 119, 127, 138,
    144, 147, 148
  in coal conversion products,
    90, 119
  in oil sands, 55, 57
  in oil shale products, 18-21,
    31
Recycled oil, 255-270
Redox potential
  and coal mineral formation,
    324
Reduction of coal, 342
Resins
  in oil sand bitumen, 413
Rozenite, 449, 451
Rutherford back scattering, 189-
    204

Scanning electron microscopy-
    X-ray fluorescence
  determination of trace ele-
    ments in coals, 142, 146
Selenium
  in coal, 90, 119, 122, 127, 144
  in coal conversion products,
    90, 119
  in oil sands, 55, 57, 58
  in oil shale products, 10, 19,
    21, 24
  in petroleum, 10, 31, 58
Shale oil, see also oil shale;
    petroleum
  trace element contents, 9, 10
Size exclusion chromatography
    (SEC)
  of coal conversion products,
    88, 92
Solvent extraction
  of organics from retort waters,
    385
Solvent Refined Coal Processes,
    84, 88
Sphalerite, 142

Standard
  fly ash
    analysis by microprobe, 292
  glasses, 291
  reference materials (NBS)
    comparison trace element data, 128
    nitrogen contents, 70, 75, 77, 80
    oxygen contents, 70-80
    trace element concentrations, 16, 42, 52, 127, 131-139
  rocks (USGS)
    trace element contents, 106-110
Standards
  for nitrogen determination, 73, 74, 76
  for oxygen determination, 72, 73, 76
  for prompt neutron activation analysis of coal, 151-153
  for trace element analysis, 35, 52
Sulfates
  in coal
    determination by Mossbauer spectroscopy, 450-454
Sulfur
  in asphaltenes, 298-300, 311-314
  in coal
    blends, 160
    comparison of methods, 450
    determination by prompt neutron activation, 152, 153
    determination of species by Mossbauer, 443-458
    effect on fly ash precipitators, 481
    origin, 329, 330
    role of bacteria in origin, 324, 325, 329
  in Illinois Basin coals, 115
  in oil sand bitumen, 412
  in petroleum asphaltenes, 311-314

Sulfur (continued)
  species
    effect of low temperature ashing, 444, 445
    from kerogen pyrolysis, 406
Szomolnokite, 451, 455

Tar sands, see oil sands
Target transformation, see factor analysis
Tetralin
  deuterated, 209, 210
Thiourea
  adduct, 422-430
  non-adduct, 430-436
Thorium-228
  in coals, 473-479
Thorium decay products
  in coal combustion, 460
Thorium isotopes, 230, 232, 228
  radiochemical separation, 466
Titanium
  in coal, 89, 90, 92, 119, 138, 144
  in coal conversion products, 90, 91, 119
  complexes
    in coal conversion products, 92, 93
  in oil sands, 55-58
  in oil shale products, 19, 21, 24, 25
  in petroleum, 10, 31, 58
  in solvent refined coal, 89, 92, 93
Toxic elements
  in coal, 115
Trace element analysis
  comparison of techniques, 24, 25, 120-125
  detection limits, 12, 67, 178
  sampling problems, 33, 35, 36
Trace elements
  analysis of coals
    by automated activation analysis, 133-140
    comparison of methods, 120-125
    by ion microprobe analysis, 142

INDEX

Trace elements (continued)
 analysis of coals (continued)
  by neutron activation analysis, 151-154, 167
  by PIXE, 189-204
  by proton microbeam analysis, 189-204
  by Rutherford backscattering, 189-204
  by SEM-EDX, 142, 146
  by X-ray fluorescence, 189-204
 in asphaltenes, 29, 30
 in coals
  use of factor analysis, 163
  geochemical controls, 325, 326
  modes of occurrence, 141-149
  organic-inorganic affinities, 126, 143
  role of coalification, 331
  role of ion exchange, 328
  washability studies, 126
 in coal conversion products, 88-94, 117
 in coal gasification waters, 181-187
 in demineralized coals, 129
 in oil sands, 49, 52, 53
 mass balance in oil shale processing, 26
 in oil shale, 9, 10
 in petroleum, 9, 10, 30, 32, 33, 39, 41, 42
 in shale oil, 9, 10
Tracers
 radioactive
  in coal hydrogenation studies, 229-240
 stable
  in coal hydrogenation studies, 207-208
Transformation ratio, 358

Uranium
 in coal, 88, 92, 92, 94, 119, 127, 138, 144
 in coal conversion products, 88, 92, 94, 119

Uranium (continued)
 determination by
  delayed neutron omission, 134, 137-139
  fission track analysis, 88, 92, 94, 143
 in oil shale products, 19, 21, 24, 25
 in petroleum, 31, 61-68
Uranium-234
 in coal, 477, 478
 radiochemical separation, 465
Uranium-238
 in coal, 477, 478
 decay chain series
  in coal combustion products, 460
 radiochemical separation, 465

Vacuum pyrolysis
 of kerogen, 400
Vanadium, see also vanadyl
 in coal, 90, 119, 138, 144
 in coal conversion products, 90, 91, 119
 complexes, see vanadyl porphyrins
 in oil sands, 52, 54-58
 in oil shale products, 10, 19, 21, 24, 25
 in petroleum, 10, 31, 38, 39, 58
Vanadium-nickel ratio
 and age of petroleum, 30, 32
Vanadyl porphyrins
 in oil sands, 52, 56
 in petroleum, 30
Vitrinite, 342

Washability
 of coal, 126
Water
 determination in coal, 80
Wet ashing
 of petroleum, 36, 37

Xenotime, 142
X-ray
 absorption
  in particles, 386, 287

X-ray (continued)
  diffraction
    determination of asphaltene structure, 305
    mineral species in rocks, 241
  fluorescence, see also proton induced X-ray emission; scanning electron microscopy
    analysis of clay minerals, 242
    determination of trace elements
      in catalysts, 271-283
      in coal, 115-131, 189-204
      in oil shale materials, 2, 7-9
      in petroleum, 44
    errors, 273
    multiple source excitation, 7
    radioisotope sources, 272

Zircon, 142, 148